Indian Geography in the 21st Century

Indian Geography in the 21st Century: The Young Geographers Agenda

Edited by

Ravi S. Singh
Banaras Hindu University
Varanasi, India

CAMBRIDGE SCHOLARS PUBLISHING

Indian Geography in the 21st Century: The Young Geographers Agenda,
Edited by Ravi S. Singh

This book first published 2009

Cambridge Scholars Publishing

12 Back Chapman Street, Newcastle upon Tyne, NE6 2XX, UK

British Library Cataloguing in Publication Data
A catalogue record for this book is available from the British Library

ISBN (10): 1-4438-1365-6, ISBN (13): 978-1-4438-1365-5

In the memory of Bernhard Varenius (1622-1650 CE),
a genius geographer who died young.

❋❋❋❋

Dedicated to the spirit of
the deans and doyens of Indian Geography.

❋❋❋❋

For the future generation of Indian geographers.

VIEWS ON THE BOOK

"Past intellectual trajectories and knowledge productions require both, a sense of historical embeddedness of the discipline and a constant questioning and reassessing the constructs therein. Agendas and visions have to be continuously questioned and changed, reconfigured and renegotiated and the onus rests primarily on the younger generation of the scholars. The present book is an important contribution towards this responsibility. The critical questioning of some of the established concepts and practices in terms of theory and methods is the highlight of the book. In doing so, there is a desire to be creative and bring in what can be termed as 'insiders' view'. The book is a refreshing break from some of the the past writings in the discipline".

—**Saraswati Raju**,
Professor, Jawaharlal Nehru University, India

"It is refreshing to see the awakening, concerns and marching of young geographers in India those involved in making and creating geography as beautiful lifeways in the contemporary era of global humanism and cross-cultural understanding. This book presents a radiant path for realization and action, illustrated with various ways and forms of current practices in contemporary geography".

—**Rana P.B. Singh**,
Professor of Cultural Geography, Banaras Hindu University, India

"No other discipline represents the project of modernity better than geography in India. It flourished with spatial and regional planning, and continues to excel in positivism and applications of remotely sensed data with the aid of Geographical Information System. At the same time, Indian geography reflects age-old and deep-rooted hierarchies that ail Indian social milieu. Age hierarchy – or geriatric rule - supersedes metropolitan dominance and gender disparities and disempowers innovative approaches in the name of respect for experience. By encouraging a set of younger geographers to think about their discipline, and by bringing their work together within its covers, Ravi S. Singh has done a service to Indian geography. I hope that this book will begin a reflexive engagement with geography, without which we cannot set the agenda for the future of the discipline."

—Kuntala Lahiri-Dutt,
Fellow, The Australian National University

"....While editing essays, Ravi S. Singh, one of the dynamic young Indian geographers and also a keen observer of the transformation of Indian geography, has not only proceeded with conceptual rigour but also with deep understanding of diverse and complex Indian geographies. Indian Geography has to be considered as multi-dimensional and multi-sectoral. The editor has brilliantly integrated different fields of Indian geography. This volume provides a timely resource on present geographical practices and potential solutions for developing and sustaining high quality geographical education systems in India. It includes the main element of a roadmap to improve India's geography education in response to the demand of growing economic liberalization and rapidly changing Indian societies. These papers are essential readings for both students and teachers of geography in India. The volume offers sustainable educational capabilities to various cross-sections of Indian Geography."

—R.B. Singh,
Vice-Chair: IGU Commission on Biogeography & Biodiversity,
Secretary General: NAGI, &
Member, IUGG-IGU Joint National Committee, INSA

"The volume attempts to bring to the fore ideas being grappled by young and mid-career geographers from key institutions in India. Such interdisciplinary approach helps to situate key arguments about the existing status of geography into perspective. The thematic exposition provides ample scope for a balanced approach towards issues that plague the future existence of geography in India. The eternal conundrum of 'space' versus 'region', the ever deepening divide between 'physical' and 'human' and the constant poaching of themes 'geographical' by all others to pass it off as 'environmental' demands critical debates. Defining the future agenda is possible only by going beyond the stranglehold of past 'personalities' and ideas and attempting a critical assessment of the state of Indian geography."

—M. Satish Kumar,
Coordinator India Initiative
School of Geography, Archaeology & Palaeoecology
Queen's University Belfast, United Kingdom

CONTENTS

List of Tables

LIST OF FIGURES

PREFACE

The 21st century is the beginning of a new world in which 'reality' is passé and 'realities' of multiple possibilities are accepted norm. In academe too it needs open-hearted welcome, acceptance and adoption without any sort of prejudiced reservation. Disciplinary water-tight compartmentalization similarly is passé. Under the sway of emerging approach to knowledge creation, disciplinary boundaries tend to collapse; at best disciplines provide simply the context and orientation. The overarching and all embracing stretch of the virtual world rejects the age related superiority/inferiority. It seems again that edge of age over ability is passé as the present era is of ageless concepts. Schwartzberg (1983/2009) and Mukerji (1992/2009) openly acknowledged how the (undue) dominance of old geographers by virtue of their sheer seniority is taking its toll in India. No wonder, then, that younger professionals do not get their due place (both in informal interactions and formal academic space) and encouragement to express their minds. There are end numbers of instances of even snubbing a new idea by age-old and tested technique of posing the rotten remarks like, "Is it geographical? What is geographical in it?" Above all, the ultimate important issue is the professionally correct stand and creative line of action. No matter, whether that is of a seniormost or a professional beginner. If a discipline has to exist with full-life and academic vigour, youngsters will have to be given their space, opportunity and required support. It is the time to acknowledge many turns that were perhaps not even imagined before they actually happened.

If the youth is being given chance in other walks of life, e.g. sports and politics, what is wrong in adopting the same in academics? The same is needed in academics too where youngsters' ideas are rubbished not because they lack substance but they are said by someone who is not "enough old". Does it not lead to ask a natural question, "Is wisdom a function of age?" The academe, especially Indian Geography, will have to do away with the *Yayati* principle (of the ancient Indian tradition) that gives edge to experience over ability and adult age over youth. If we take journalists' voice representing the people, Mrinal Pande, the Editor of *Hindustan* (a Hindi daily), titled one of her articles as *Kripaya anen den unhen aage*! ("Please let them come forward"). Her advocacy of the youth therein, indicates required attitudinal change towards them, and speaks scores about the change in the offing (cf. Pande 2007). That also reflects faith in youngsters' ability in making things happen and also an assurance that the future is in worthy hands. It is already happening in the West where younger academics continue to contribute the most in the recent debates. Such examples are in geography as much as other disciplines. Noble ideas have universal adoption and there is no reason why such a thing can not happen in Indian Geography.

But, so far, it has not happened. On top of that there are a variety of valid limitations which together disallow space to young professionals to freely/liberally express their ideas on the past, the present and the future. Whereas the matter of fact is they are the ones who have to face the near future. Hence, it is crucially relevant to know their views. Ron Johnston (1985, 25) observation, "Future of geography depends on its future practitioners", the source of inspiration for the present work, is perfect statement clarifying the need and the value of thoughts of younger professionals. That explains the worthiness of this book. A total number of sixteen contributors belonging to different institutions located in various corners of India have written this book jointly. It need to be mentioned explicitly that the purpose of the book is neither to present a review of the works nor to present a progress report as such, all contributors try to raise issues aiming at the future scenario. This book could have been authored singularly but was not done so purposefully. Seeking the future of an academic discipline is not an individual enterprise and therefore impossible singularly. After all how a discipline should be or ought to be is matter of collective vision of those who will take the same into the future. Positive criticism is incomplete as long as not complemented with a future agenda. And, the future agenda again should not be high-pedestaled and merely prescriptive. Rather it needs to be practice-able in a sustainable manner to ameliorate Indian geography from its blamed, rightly to a great extent, status for being copy of "westernised" to clear identity of its own. Such statement(s) of action can come from younger generation of practising professional geographers who, as refereed earlier, have to the take the Indian geography into future.

Dealing with present and the past is common while looking into future is rare. Recently, I did an anthology *Indian Geography*: *Perspectives, concerns and Issues* (Rawat Publications, Jaipur/New Delhi, 2009) with the sole purpose to bring together scattered previous works on Indian Geography so that emerging ideas from past works are read as interconnected train of thoughts. In this book should be read as a sequel to that which looks into future in the backdrop of an assessment of the past and present feats of academic achievement. Respective contributors have not assessed the individual makers of modern Indian Geography instead their academic ideas pertaining to the works considered in individual essays. Thus, this volume is an attempt to establish dialogue between the past, present and future Indian geography. Honestly speaking there has been clear lacking of such dialogues.

This book begins with an introductory chapter subtitled "envisioning the future Indian geography". Rest of the essays are organised into five parts which are not equal either in terms of consisting chapters or their length. In away it is a reflection of prevalent lopsidedness within the discipline as in practice in India. Part I, "The Contextual Orientation", has three essays, by Ravi S. Singh, A.P. Mishra and R.K. Maiti, which in a way provide a background to the following discussions. In part II, "Reshaping Geography Education", the included three essays by Sarfaraz Alam, Yemuna Sunny and P.K. Ryngnga, respectively, look at the state of geography education and also delve into the ways of "reshaping". Part III "Resurrecting Physical Geography" too comprises of three essays. These essays, authored by Sudip Dey, L.N. Satpati and N.C. Jana, though examine issues with reference to different sub-fields, advocate "resurrecting" of physical geography by bringing it closer to human and society. The following part IV "Retrieving Human Geography" is the largest part of the book with eight essays by Taruna Bansal, Sohel Firdos, Ravi S. Singh, Aparajita Chattopadhyay, Archana K. Roy, Vijaya P. Khairkar, Arun K. Singh and Bikramaditya Kumar Choudhary, respectively. They all seek to "retrieve" respective sub-fields. Emerging ideas in different essays in an edited volume often are left unattended and they are thus lost for sure! We did not want it happen with the present one and therefore a separate part V: "The *Summum Bonum*" is included. It has the concluding essay, titled "Designing a working agenda for the 21st Century", that attempts an interlinking of thoughts emerging from the preceding essays. At the end, an appendix is provided which contains the select events and publications. It is perhaps the first attempt of its kind and part of an ongoing work. Hence, it is quite likely that many deserving entries have not found place. I would be glad to receive and acknowledge any such information, by learned readers, which improves this list.

I can not claim the book in your hand to be all-inclusive, neither in terms of themes and sub-themes covered nor the issues on agenda outlined and discussed, as no attempt is complete and conclusive. Yet the success lies in the relevance and strength of the central objective, how honestly are they pursued and how do they contribute to the advancement of the discipline concerned; and, that way to the greater body of knowledge. That would be decided by the readers and reviewers. The earnest wish of all contributing authors is to see this book contributing to the dialogue, particularly among the young geographers, for which Indian Geography has shown a very low-appetite.

References

Johnston, R.J. 1985. Introduction: Exploring the future of geography. In *The Future of Geography*, ed. R.J. Johnston, 3-26. London and New York: Methuen.

Mukerji, A.B. 1992/2009. What ails Indian Geography? Some statements. *National Geographical Journal of India*, 38 (1-4): 205-218. Reprinted in *Indian Geography: Perspectives, Concerns and Issues*, ed. Ravi S. Singh, 223-241. Jaipur/New Delhi: Rawat Publications.

Pande, Mrinal 2007. Kripaya anen den unhen aage! ("Please let them come forward"), in Hindi. *Hindustan*, September 30, Manthan section, Varanasi edition, p. 12.

Schwartzberg, Joseph E. 1983/2009. The state of South Asian geography. *Progress in Human Geography* 7 (2): 232-253. Reprinted in, *Indian Geography: Perspectives, Concerns and Issues*, ed. Ravi S. Singh, 305-330. Jaipur/New Delhi: Rawat Publications.

Singh, Ravi S. (ed.) 2009. Indian Geography: *Perspectives, Concerns, and Issues*. Rawat Publications, Jaipur/New Delhi.

ACKNOWLEDGEMENTS

Honestly speaking, the idea of doing a volume exclusively by young geographers, who have recently made their entry in academia as professional geographers and shall play pivotal role in unmaking/remaking of the Indian Geography in coming decades, came to my mind while reading the *Future of Geography* edited by Prof. R.J. Johnston—one of the leading prolific international geographers. As mentioned in the preface, his comment that "future of geography depends on its future practitioners" became the guiding idea to imagine and work on this book. Hence, my first indebtness is towards him though we have never met in person. It would have simply not been possible to do this work if the ready support from contributing authors were not received. Their thoughts and concerns have ultimately shaped the book as the readers find. If this book is able to get good response, the credit goes simply to all of them. It is my privilege to thankfully acknowledge all the contributors to the present book. I also take this opportunity to appreciate the cooperation extended by many colleagues who could finally not contribute for one reason or the other.

In my school days, my basic interest lied in (Hindi) literature and history owing especially to my father—an avid literature-lover—who had studied literature and history at the university. The credit of turning me to, geography goes to my first geography teacher (Late) Dr. K.N. Upadhyay, and my interest towards geographical thought to my teachers and academic mentors (Late) Prof. Aijazuddin Ahmad and Prof. Rana P.B. Singh. I bow my head to them in reverence as words can not convey my obligation to them. In my professional life, interactions with seniors like Prof. K.R. Dikshit (Pune), Prof. Kashi N. Singh (Varanasi), Prof. M.H. Qureshi (Delhi), Prof. H.N. Sharma (Guwahati), Prof. Saraswati Raju (New Delhi), Dr. Kuntala Lahiri-Dutt (Canberra), Dr. R.B. Singh (Secretary General: National Association of Geographers India, NAGI, Delhi), and Dr. M. Satish Kumar (Belfast); contemporaries like Dr. Anu Kapur (Delhi), my colleagues at the Banaras Hindu University Prof. Rana P.B. Singh, Prof. P.R. Sharma, Deepayanji, Arunji, Sarfarazji, Bikram, and Aparajita (formerly at our department and now with IIPS, Mumbai) have been gratifying learning experience. I would like to give additional and special thanks to Dr. Arun K. Singh and Dr. Sarfaraz Alam who helped me in the editing of two essays of this volume.

Books are springs of ideas which play direct role in the thought formation. My *adda* for relishing the multi-disciplinary pabulum is "Harmony"—the bookshop at the famous Asi Ghat of Varanasi—always accessible courtesy the ever-smiling proprietor friend Mr. Rakesh Singh. I acknowledge my indebtness to him. My thanks are due to our departmental library in-charge Mr. Abhishek Kumar (SPA), and Mr. Chhabi Nath Ram and Mr. Vinod Kumar Yadav of the National Geographical Society of India for generously providing ready assistance. I need to acknowledge the help from Mr. Ram Kumar Tiwari, research scholar at our department, for redrawing a few illustrations used in this book. I express my gratefulness to my parents Shri Birendra Prasad Singh and Smt. Shanti Singh who have always been the source of my moral strength. It is my sincere duty to express my gratitude to Cambridge Scholars Publishing, CSP for showing interest in this project. I am personally thankful to Carol Koulikourdi, Amanda Millar, and Soucin Yip-Sou, at the CSP, who provided every help and extended ready support at every stage to get this book published in the present form in a very short period of time. My wife Dr. Pratibha Singh and son Abhisht Adityam deserve special thanks for bearing with me while the editorial work of this book was on. Success of the book is a joint achievement of we all contributors; whilst the moral responsibility of all commissions and omissions, if any, rests with me.

Ravi S. Singh
Banaras Hindu University
Varanasi

"Future of geography depends on its future practitioners".
—Ron Johnston in *The Future of Geography* (1985, 25)

CHAPTER ONE

INTRODUCTION: ENVISIONING THE FUTURE INDIAN GEOGRAPHY

RAVI S. SINGH,
BANARAS HINDU UNIVERSITY

You can see things; and, say 'why?' But I dream things that never were; and I say 'why not?'
—George Bernard Shaw (1856-1950) in *Back to Methuselah* (1921)

Introduction

A group of scholars believe that "promoting innovation and Exciting Geography is essential component for the future success of the subject" (http://econgeog.misc.lit_u.ac.jp/icgg/statement_icgg.html). The promotion of innovativeness could not be thought of in the absence of "criticality" and thus "questioning" becomes imperative. In other words, "new" takes shape in the womb of "old"; however, questioning the old is a necessary process for the birth of new, and acceptance and promotion of the new nurtures that. Simultaneously, accepting the non-conventional or new is equally necessary (of course, a well considered demarcation of challenging the conventional would be needed in certain specific instances, e.g. the conventional signs and symbols used in the making of maps?). That is how innovation is inculcated and imbued. Both, discarding of the part of "old" that becomes redundant and accepting the "new" are acts of courage and matter of vision. Most of the works we know harp on the present taking the past as the resource. Musings on future have been overlooked by and large. Gradually, a new trend is shaping in Indian Geography—basic and critical issues which used to be the matter of informal backdoor gossips are being brought to formal discussions in the class rooms and corridors and also in academic space(s). Previously, discussing such matters was generally the privilege of the honourable "seniors" during their conversations and occasionally in their addresses. While a few appraisals by overseas' "Indianist" geographers were creative in orientation, A.B. Mukerji (1991/2009) showed the mirror to an identity called Indian Geography which one may like to call blasphemous in terms of its content, tone and pitch. This unfolding of major lacunas "ailing" Indian Geography is worthily followed by a few other attempts (cf. Ravi Singh, ed. 2009). Criticism without alternative ideas create vacuum. Following this idiom, the present book, done by young geographers in India, is a humble step forward in this direction. The purpose of this chapter is to present an interlinking introduction to the seventeen main essays, in self-introspectory mode of discussion, organised into five parts and the central idea of 21st century agenda of Indian Geography running across. The present chapter discusses the future vision in five main sections to follow: do way with the *Yayati* principle, the contextual orientation, education matters, envisioning future physical geography, and rebuilding human geography.

Do Away With the *Yayati* Principle

All that is achieved by a society do not form the tradition. A tradition is conscious construction by the concerned society. Naturally that can not be fixed for all the times to come. Accordingly, we find changing traditions over time as much as by space. Academic traditions too undergo the similar process. Emergence of criticality in the field of geography, during the last leg of the 20th century in the Western scholarship, proved to be a conformed change and also attested greater role of young scholars in turning the *Kālacakra* (literally the "time-wheel") on. Interpretation of the past and envisioning the future is a prerogative of the present generation. Hence, questioning past practices should not be taken as malicious and ill-intended; though, in practice both criticism and praise are taken personally. Criticism (and praise too) in any intellectual or academic dialogue is of the ideas not individuals per se. We all know science is the craft of making impossible imagination come true. That is how science never believes in the "eternity of ideas" rather treats knowledge as partial and tentative, and keeps on improving what we know. These beliefs in the scientific world have ensured vibrancy in all sciences and made them highly spirited.

If we take pride in calling our discipline "scientific", we need not claim for a "spurious science" tag, merely for a "high image" in the society or ensuring (better) grants for our departments or funding of our research proposals, as already indicated by some (cf. Lahiri-Dutt 2005/2009).

Instead we should look forward to inculcate, encourage, practice and promote the scientific temper which is objective, rationale and progressive in its basic nature. While reviewing an edited volume, K.R. Dikshit did not appear happy with the "uncritical" lot among young geographers which engages in admiring the seniors and reminded them to follow the spirit of science, "Let us remember, a critical assessment is far more rewarding than an uncritical acceptance of every thing that the elders do (Dikshit 2002, 112). In this spirit, the elders are expected to take younger geographers and their ideas in a constructive spirit and treat their imagination as creative instead of rubbishing away as immature, impossible, and utopian. Students of geography know at least about Bernhard Varenius (1622-150 CE) who exemplifies the generation of young geographers. He is credited with "distinguishing between general and special geography; and, the mutual dependence of these approaches". He was a futurist par excellence with quite a clear conception of the nature and purpose of future geography (cf. Dickinson 1969; James 1972, 124-126). Faiths in the young voices have been expressed in many contexts even before the onset of the present century. Indian Geography too awaits expression of such a faith in young geographers. After all, how long we can continue with the Yayati syndrome and deny the youth its due. Indian society, badly suffering from the Yayati syndrome, appears emerging out of this regressive problem. And, the upcoming results have also been encouraging. It is not only the young (cricket) team India which have made the nation proud during last few years with its historic achievements; rather it is happening all around—the information technology sector, business, and also the Indian politics which had least scope for representation by youth at the upper level. In addition to the re-elected young parliamentarians, the entry of a good number of youth as the first-timers in Lok Sabha is an expression of faith by the political parties and also the people in the youth. As far as qualification and competency is concerned, they are surely better than many of their seniors when they were youth.

So, hoping the same happen in academic disciplines in India, particularly in geography in the present case, is in every sense genuine and justified. It is not to say that young geographers have not been getting opportunity to work and publish research results. Often than not their work is published with their research supervisor, in case of a student, or seniors, in case they are employed. After all we all know the credibility of a tenured, called permanent in Indian parlance; faculty member gets a smoother passage of article/paper in the publication pipeline. As, generally, the quality of paper and significance of ideas become of secondary importance and the person's status becomes primary. It is one of the issues concerning the professional ethics and morality. Such a situation is indication of lacking transparency and non-adherence to standard professional practices which Schwartzberg (1983/2009) referred to. Honestly speaking, the problem still persists. The West has been able to come out of it to a large extent; it seems the developing countries like ours however will take time.

The Contextual Orientation

> You can not argue with someone who denies the first principles.
> —Anonymous

No endeavour is without a context. Hence, interpreting the context is as much essential as the endeavours. The following portion of the book begins with the introductory part I entitled "Contextual Orientation". As it is self-explanatory, its motive is to provide a comprehensive background for the discussions to follow in other sections consisting of essays on different themes. Though, the tradition of critical appraisal of the progress taking place in Indian Geography has been rather feeble, yet one can now refer to over a hundred of writings of different genre available. They contain statements, generally of the seniors, on the content and purpose of Indian Geography. Chapter 2 is an effort to elaborate Indian Geography's trajectory, in somewhat different fashion. Instead of presenting a general narrative of the evolutionary history, emerging critical points through the battery of musings are projected. General impression is we have "progressed" considerably during a span of over last eight decades forgetting the fact that this word means not merely "growth"—of institutions, students' enrolment, employment and teachers, books and journals published, professional academic bodies, doctorates in the subject, funding, computers and other equipments, etc.—it carries more important value-loaded meaning involving *quality*. After all, in spite of these numerical achievements, there is a sense of loss we inherit from the immediate past century. Critics and the concerned feel aggrieved. The loss inherited is immense and varied too attest the works known till date. Worrying truth is, in general, we do not seem interested in the recovery of the loss incurred and identified well. Not knowing the ailment is forgivable but what to say about not taking care even after being informed of that clearly! The second part of this chapter looks for the emerging directions in the present 21st century.

The following Chapter 3 could be read as continuation of the debate introduced in the preceding chapter. Mishra sees the contemporary geography as "a discipline which deals with the emerging regional spatial dynamics". He well impresses that such a regional dynamics is thrusting up new from of inequalities and injustices which can not be satisfactorily addressed through development strategies in the conventional mode of investigation and research. It enfolds new issues, forces and processes which open up various possibilities for the subject and its practitioners. The need is to radicalise geography, believes the author and argues for "a search of politics… (to)…support and develop in tandem with progressive and emancipatory political commitment". He too expresses faith in "creativity" for the future geography as shown explicitly by Kapur (2004). The geographers are urged to reorient own research work towards ensuring justice to the people and their welfare. That kind of commitment to people's agenda holds the key to raise the status of geography in the developing societies in the near future.

Hindrance in the growth of geography in its basic unified form camouflages its true nature and makes its identity enigmatic (cf. Singh 2008) which is responsible for a confused public image and failure in attracting and retaining talents in higher studies and research. Maiti offers a different orientation in Chapter 4. He argues for resolving and integrating, what he calls "conflicting ideas", essential to upgrade geographic research and enhance geographers' academic status and intellectual esteem and recognition. Based on the his personal experience in working on different themes, this chapter presents discussion on evolution of the methodological philosophy and the present trends, nature of present researches, potential future research in the postmodern context, and the possibility of coherence through cognitive approach is assessed with a few examples from empirical researches.

Reshaping Geography Education

…geography is first of all knowledge gained…
—Carl Ortwin Sauer (1889-1975) in 'The Education of a Geographer' (1956)

The underlying importance of education primarily lies in the fact that it has direct and strong implications for individual advancement. And, in turn, it has immense prospect for the realization of a society's amazing potential and aspirations for economic and technological development. Geography for its direct utilitarian value has been one of the integral elements in the education of a person and also to all human inquiry (cf. McDougall 2000). At many occasions the issue of geography education in India has been addressed. Most of them address the problems at tertiary education level. School education automatically gets neglected. One will have to admit that majority of us have not taken the pain of taking a dip deeper in the water. The sickness reported and solutions prescribed hence prove to be quite general and shallow. Cosmetic treatments do not cleanse the system to make that healthy. And, that is what has happened to geography education in India.

In this backdrop, the part II seeks to bring together thoughts of a few concerned young geographers who feel the need of revamping geography education. The modern school geography as taught contemporarily owes to the British educational system through which it got implanted in India on the European rather the British pattern. Thus, "flourished" the subject since the British period but for a reported decline in the early 1950s. The onset of 21st century has not shown good signs, thinks Alam, in the following Chapter 6 which addresses the state of school geography in India. This paper concentrates on the geography praxis in the upper primary and lower secondary school curricula. It is organised into four main sections. The first presents a brief account of evolution of geography as a school subject. In the next section the recent trends in school geography in India are analysed followed by a scrutiny of the present status and concerns. These discussions lead to identification of the future agenda for young geographers "towards the strengthening the place of geography in school curricula". The author rightly underlines the need of reflecting the rapidly changing world in school geography curricula and urges young geographers to develop an understanding of the "politics behind the curriculum construction".

Text books are very crucial when it is the matter of imparting education especially at school level. While what a text book should contain and what not could be a contested issue on several grounds, role of reading in learning is acknowledged by all. Hence it is essential to make them appealing so that children read them with interest. The issue of text books so far has been treated at shallow level. Sunny in Chapter 5 with a non-conventional approach goes deeper into the problem. She examines text books, especially looking into how geography texts deal with the relationship of human-nature, through her experiences in the course of some workshops and reviews of geography school textbooks. Her essay is organised into three sections. The first of them is a "reflection on the rich enhancement of literature and how text book traditions depart from those". The second examines some nuances of human-nature relations as portrayed in school geography text books. And, the last section is devoted to scan text books of Eklavya—a voluntary organisation working in the field of school education—created on the basis of pedagogic and social science critiques of existing practices in schools.

Over the years, many positive changes have been noticed in Indian Geography but they tend to be higher education oriented, precisely speaking university-centric. Since the (undergraduate and postgraduate) degree colleges, in terms of the structure of Indian education, are part of higher education; they have benefited from the changes implemented through the University Grants Commission. Otherwise, in practice, to a great extent their position continues to be peripheral and inferior too compared to the universities. The next chapter in this part continues with the spirit of preceding two chapters and also addresses new issues. Central concern with school geography in Chapter 7 marks the continuity and the inserted context of higher education makes the addition to that. The actual focus is on the continuing gaps between geography education in schools and institutions of higher learning. Ryngnga tries to analyse, the prime position of geography in primeval days of its emergence, hallmarks of Indian Geography, quality in school education, "rejuvenation" of the subject at higher level, job market for geography graduates, awareness regarding geography's relevance and potential, and strategies to strengthen school geography, in brief. The last line of her conclusion is a positive note, "...there is hope not just that geography will grow in prominence, but that the students and faculties at Indian colleges and universities will be increasingly enriched by what the discipline can contribute to their institutions' intellectual life and pedagogic mission". Well, this remark could be used to begin an altogether new debate!

Resurrecting Physical Geography

If we do not find anything pleasant, at least we shall find some thing new.
—Voltaire (1694-1778) in *Candide* (1759)

Physical geography—constituted basically by geomorphology, climatology, and biogeography—is fizzling out in Indian Geography should be taken now not merely as caution rather as a hazard hanging around to become disaster! Except in the specialized journals like Indian Journal of Geomorphology, feeble and sporadic frequency of research works is testimony to further slumping down of the state and status of physical geography. One should therefore be not surprised to see a regressive pull in this broader and significant field of geography. Pushing physical geography towards the dark corner is complained and bringing it back into curriculum prominently as it used to be is suggested by many. But for

A.B. Mukerji perhaps no body else has taken the pain of looking into the problem beyond indicating and suggesting some generalities. Negation by senior and senior-most non-physical geographers is capable enough of relegating it to oblivion and is truly termed as "catastrophic". His prophecy "negligence of physical geography will lead to the final collapse of geography in India" (cf. Mukerji 1992/2009) may prove to be right. However, it can not be allowed to happen. Right at this point, will it not be a worthy question to ask that what has been the role of Indian physical geographers in this relegation process? Did the kind of physical geographies we have doing connect to the society and social purpose, etc.? Jog's (2008, 28) remark summarizes the current craze for a technique and unfound genuine dedication for the problem investigated, "The techniques and tools of research gain importance …at times one wonders if the original topic of research is getting camouflaged in the description of these tools and techniques". To protect our propriety over the territory of physical environment-society interface, we will have to resurrect physical geography. Validity of Mukerji's points is unquestionable and awaits response from physical geographers working in India. The three chapters (8, 9, and 10) constituting the part three offer some ways to achieve this higher goal.

Chapter 8 is very spirited essay dealing with micro-geomorphology. Dey has used innovative personal experiential mode of explaining. He analyses and discusses current geomorphic issues and learning geomorphology, search for new dimension in landform science, geomorphic studies in different micro-regional environments, technical applications: some recent works, and planning for the journey towards the dream destination. In the last section, he urges to reach the current international level target despite a few negativities acting as hurdles. Our aim should be finding out the way(s) to rise above the barriers. That will enable us to produce better works in geomorphology and make us worthy claimant of prestige, higher status and laurels at international level in the future. "Let us dream and work for the brighter future of landform science in India".

Significance of water in a civilisation needs no elaboration. Studies on water from geographical perspective need to be done considering complimentary dimensions: physical environment and the social context. Geology and climatology can benefit our understanding of the nature of water availability; while the social context will provide with demand-supply issues dependant on prevailing water related practices—use/misuse, storage/wastage, and also conservation. In a modern society, the role of state too is seen important as it is the strongest player in policy formulation and implementation. Studies advocate an integrated approach toward water management in view of ever growing demand and declining per capita availability of the same. Satpati takes up the case of hydro-geography in Chapter 9. Taking a very brief note of development in Indian Geography, he goes ahead to justify the need for a new specialisation termed "hydro-geography". The discussion to follow in this chapter is focused on water resources–importance and current status, anthropocentric use of water, an overview of water resources of the world, global water use and problems of concern, water of India–issues and concerns, hydro-geography in India–planning for future agenda, and a suggested curriculum framework for introducing a course on hydro-geography.

It is true that human civilisations have progressed with certain natural menaces. Disaster studies itself is late beginner in India and the geographers are among the band of academics who have started working in this field recently. Jana has taken up the issue of hazard and disaster management in the last Chapter 10 of this part. Starting with a comprehensive introduction, he analyses the contribution of Indian geographers in the study of a variety of issues related with hazards and disasters. The other following sections have been devoted to a brief appraisal of geography's role in disaster management and place of geographers in policy making in this regard; and, the role of young geographers in the frame of future agenda. He concludes with a final call to include disaster studies and management in the advance geography curricula as our discipline is appropriately equipped to handle disaster management which needs cross-disciplinary expertise.

Retrieving Human Geographies

The urge for destruction is also a creative urge!
—Michael Bakunin (1814-1876) in *Jahrbuch für Wissenschaft und Kunst* (1842).

Proliferation in the Indian Geography literature so far has been in "favour" of human geographies. Then, the title of part four "Retrieving Human Geographies" may make a reader uncomfortable. In view of repeatedly impressed dissatisfaction with directionless empirical studies mushroomed during last century, reclamation of human geography too becomes imperative. Lopsidedness towards human geography is manifested in the present book as well with around half of the total main essays dealing a human geographic sub-theme. The arguments forwarded in these eight chapters here explain the rationale of the title of this part.

Nations across the globe tend to evolve gender-sensitive than they were at any previous point of time. This statement however can be well debated as some ancient societies are known to be "women-friendly". To a layman, women studies and gender studies may appear no different; despite the fact that both are not treated synonymous or coterminous in academic scholarship. There has been spurt of such studies at least since the last quarter of the 20th century. In Chapter 11, Bansal looks into the issue of women's presence in Indian Geography. Her arguments sequentially fall into four main sections on feminist approaches in geography, place of women in geography, recovering women in geography, and the study of women as a method in geography.

In the following Chapter 12, Firdos devotes his ideas to "redesigning social geography". This chapter begins with a quote from Ahmad (1999), a mixture of despair and hope, which conveys poor grounding of social geography in India. The author has attempted to develop an understanding of the problem through his reflections on the fragmented geography, some peculiar characteristics of geography, social geographical perspectives on "difference", coming to know difference through social geography, geographic responses to difference: nature-culture and the body, social geography of India: challenges before it, and social geographical perspectives on public policy. At the end, he concludes that the geographers have a strong armoury of skill and expertise through which they have been contributing to the public policy.

The issues of social and cultural enquiries are quite overlapping and as one would expect, so is the field of social and cultural geography. This overlapping nature is clear in the case of India. For example, both caste and tribe are studied by students of social and cultural geography. Similarly, language is an issue of enquiry in both of these sub-fields. Actually, what separates them is their respective orientation and academic perspective. In the broader context, cultural geography appears as a larger canopy. Chapter 13 looks into developments in cultural geography in India vis-à-vis the global trend. This essay is organised into five major sections: the nature of historical roots, recent development (since late 20th century), and the Indian scene, what we continue missing, and tasks ahead. Increasing engagement with critical issues at the international level is not found attended by Indian geographers who are still continuing with descriptive approach. Though dominated by settlement and their diffusion studies, a few recent themes of research like geography of religions, cultural symbolism, cultural astronomy, etc. prove a little headway made for the fact that they are interpretative and not simply descriptive. A few potential areas of future research are also identified and evaluated by the author. The discussions in this chapter sufficiently suggest immense possibilities for a prosperous cultural geography in India—after all it is a country of thousands of years old civilisation in which a great number of culture groups have interacted for centuries.

People matter not in terms of their cultural traits alone. Their number in a territory and change(s) therein and consequences are equally important. Probably people have mattered since ancient times when their ultimate economic "worth" got recognition and accordingly they were valued for the society. With the passage of time, human beings' recognition as "resource" has dramatically altered academic perspective and the state policy on population. It no more requires to be proved that population matters as it has direct relationship with development as it can potentially affect the later in both ways—positive and negative. So, the population size and associated issues further justify why there should be a branch of human geography devoted to the exclusive study of population. Chattopadhyay in her essay (Chapter 14) notes that the growth of population geography in India is in its formative years. Perhaps, the practitioners in general are not aware of the great variety of aspects which could be profitably explored for a richer population geography. As blamed at several places in different contexts, geographers in India continue with their only concern for spatial description. That is how they get stuck in the preliminaries of the disciplinary approach forgetting the larger goal of understanding and explaining human behaviour over space through rigorous and innovative methodologies—the demand of "complex" subject like geography with it multi-dimensionality. She has tried to reflect on population geography with reference to three fundamental issues: teaching content, published research works, and the future in precise manner.

Migration is one among the three principal concepts around which population geographic studies are organised at the highest level. In one hand it directly affects population size and indirectly by affecting fertility and mortality characteristics as it changes the general composition of the population which have bearing on them. Migration is also directly related with regional development as it involves movement of people belonging to the working age-group. In the traditional literature, migration is treated very broadly overlooking the deeper issues and underrating their importance. Roy's concern in Chapter 15 is to highlight the common features of prevalent migration studies and illustrate that what is done so far is highly insufficient. Globalisation which is growing stronger as a process adds yet another forceful dimension to be covered by scholars working on migration. She has first attempted a comprehensive and critical survey of the status of migration Studies in India; and, then discussed two vital issues—geographical perspective of migration and a critical review of migration study in syllabi of India universities.

In the general observation, migration has immensely supported rather accelerated urbanisation world over. Economic compulsions and attractions simultaneously act to make people land in the urban centres, especially in the larger ones which are believed to have better opportunities. In the Asian experience, including India, overgrowth of mammoth size metropolitan centres may be attributed to it. When India marches towards a greater urban future, urban issues will naturally assume significance. Though a recent survey (cf. Misra 2008) shows satisfaction with the general progress taken place in urban geography in India, majority of the studies have not gone beyond traditional concerns. Liberalisation and globalisation together have wrought a new urban India that continues to evolve and change the economic and social space. Urban landscape is being continuously modified as new infrastructure development continues rapidly. Deteriorating living conditions, poverty, environment, etc. like problems remain as frequently covered ones in geography literature. One may note that the social and cultural elements fail to attract Indian geographers' attention whilst new phase of urbanism could be found already set in India. By the mid-21st century, when almost half of the Indians will live in the cities; geographers will have to engage with emerging issues like urban governance and management; and also, the patterns of urban culture and social problems. Two essays in this book deal with urban geography.

In Chapter 16, Khairkar's ideas are organised into six sections. This essay could be read in two parts. She first discusses the rationale for studying urbanization and then follows to analyse different techniques of urban analysis: the role of qualitative techniques and their use, use of remote sensing and geographical information system, and the role of focus group. The last part covers future directions where a few major issues are examined.

Arun Singh looks for the shifting paradigms in urban geographic studies in the next Chapter 17. Taking a different approach from the previous essay, he begins with an analysis of pattern and impact of urbanisation in India. The remaining part of this chapter is devoted to the enquiry into India's urban vision and the agenda of urban management, urban management in India: paradigm shift, 74th CAA: a watershed in the history of decentralised urban management, geography, geographer and urban management, and the way forward. The author finds "great employment opportunity and huge job market for geographers in the changed scenario" after the enactment of the 74th CAA. And, thinks that it will promote geography among the youth. The role of geographer can be seen as Enabler, Developer and Facilitator beside the conventional ones.

India is a large country and urban centres are islands of different size and varying characteristics. Regions as a matter of fact differ from each other irrespective of the assigned context. To address the issue of regional equality in terms of development, the Indian government has been making efforts since the attainment of Independence in 1947. Still, a lot need to be done. Choudhary's attempt is to "deconstruct some of the myths and analyse the conception of regional development" in Chapter 18. He begins with analysing the theories of regional development with special reference to India. Further, his concern is to argue for the role of regional dimension in development process and expose how in realistic terms regional development agenda lurched while the governmental version of the story always presents a story of concerned and concerted efforts in this direction. In the rest two sections, space is devoted to discuss the required focus in regional development in India and what could be the role of geographers in this context.

The *Summum Bonum*

Two roads diverged in a wood, and I –
I took the one less travelled by,
And that has made all the difference.
—Robert Frost (1874-1963) in "The Road Not Taken" (1916)

Conventionally, an edited volume does not have a concluding chapter. The thought of having it in the last part the *Summum Bonum* (literally "the highest good") is to bring out the *highest good* flowing out of all included essays. The chapter title "Designing a working agenda for 21st century" further makes the content and intent clear. It is no longer a secret that many prescriptive suggestions have been made with reference to very many vital issues concerning Indian Geography. But they never got implemented and translated into action. A serious explorative exercise might provide a few probable accountable reasons. This essay kick offs with the issue of forgotten agendas. It needs to be admitted that this section is quite selective in analysis as a complete exercise is out of its scope. And, to do that exhaustively, a multi-volume group work would be needed. Later in this essay, a humble effort to find out answers, is made drawn mainly from the essays in the preceding sections, to three questions: What to do? Why to do? How to do?

Prioritising thrust areas and accordingly shifting research focus with changing times need to be reinforced. It is openly admitted that though all issues on the agenda outlined are not new; but, the covered finer ideas are inevitably fresh and perhaps offer better insight into the problems. The tone and pitch of arguments do not show sign(s) of being "too ambitious". Most notable is the embedded fact that this agenda is not for "others" to follow. Rather it is a personal date, for all contributors and their generation of Indian geographers, with respective research area to be pursued tomorrow. The 21st century is full of potentialities, prospects and opportunities for the Indian Geography's identity free from external biases. "To attain this goal a lot of rigour is required" (Singh 2009, 20). Hope the dreams come true!

References

Ahmad, Aijazuddin. 1999. *Social Geography*. Jaipur: Rawat Publications.

Dickinson, Robert E. 1969. *Makers of Modern Geography*. London: Routledge and Kegan Paul.

Dikshit, K.R. 2002. Book Review of Indian Geography: Voice of Concern (by Anu Kapur). *Transactions, Institute of Indian Geographers* 24 (1-2): 108-112.

James, Preston. 1972. *All Possible Worlds*: *A History of Geographical Ideas*. Indianapolis, NY: The Odyssey Press.

Jog, S. 208. Geomorphology. In *Progress in Indian Geography 2004-2008*: *A Country Report*, ed. D.K. Nayak, 28-29. New Delhi: Indian National Science Academy.

Kapur, Anu. 2004. Geography in India: A languishing social science. *Economic and Political Weekly* 39 (37): 187-4195.

Lahiri-Dutt, Kuntala. 2005/2009. Geography as a marginal science. *Economic and Political Weekly*, 40 (7): 689-691. Reprinted in *Indian Geography*: *Perspectives, Concerns and Issues*, ed. Ravi S. Singh, 242-250. Jaipur/New Delhi: Rawat Publications.

McDougall, Walter A. 2000. You can't argue with geography, Footnotes: The Newsletter of FPRI Marvin Wachman Fund for International Relations, 6 (5).
http://www.fpri.org/footnotes/065.200009.mcdougall.cantarguegeography.html (Accessed on 24 August 2008).

Misra, H.N. 2008. Urbanisation. In *Progress in Indian Geography 2004-2008*: *A Country Report*, ed. D.K. Nayak, 46-50. New Delhi: Indian National Science Academy.

Mukerji, A.B. 1992/2009. What ails Indian Geography? Some statements. *National Geographical Journal of India*, 38 (1-4): 205-218. Reprinted in *Indian Geography: Perspectives, Concerns and Issues*, ed. Ravi S. Singh, 223-241. Jaipur/New Delhi: Rawat Publications.

Schwartzberg, Joseph E. 1983/2009. The state of south Asian geography. *Progress in Human Geography* 7 (2): 232-253. Reprinted in *Indian Geography: Perspectives, Concerns and Issues*, ed. Ravi S. Singh, 305-330. Jaipur/New Delhi: Rawat Publications.

Singh, Ravi S. (ed.) 2009. *Indian Geography: Perspectives, Concerns, and Issues*. Jaipur/New Delhi: Rawat Publications.

Singh, Ravi S. 2009. Introduction: Concerns and debates in Indian Geography. In *Indian Geography: Perspectives, Concerns and Issues*, ed. Ravi S. Singh, 1-22. Jaipur/New Delhi: Rawat Publications.

—. 2008. Challenges to Indian Geography: A contextual analysis. *Geographical Review of India* (Platinum Jubilee Volume) 70(1): 57-70.

Web Source

http://econgeog.misc.lit_u.ac.jp/icgg/statement_icgg.html

PART I:

THE CONTEXTUAL ORIENTATION

You can not argue with someone who denies the first principles.
—Anonymous

CHAPTER TWO

INDIAN GEOGRAPHY: SOME RECENT VIEWS ON INHERITANCE AND THE SENSE OF LOSS

RAVI S. SINGH,
BANARAS HINDU UNIVERSITY

Introduction

The colonial past of a country has an overarching impact in different spheres of national life. Impacts are visible not only in terms of what all gets created in the process of imperialistic rule—laws, governance, infrastructure, etc.—but also the mind and thought of the people for generations to come. Influences of colonisation on the landscape could be easily deconstructed to pave the path for a totally new design of what a national society thinks desirable for it; but, the same is not possible if the minds get colonised. The education system in India, like country with a colonial past, in general continues to carry nuances of the British imperialism and geography is no exception. Neglect of geography is perhaps rightly alleged to be due to some kind of risk felt, rooted in the imperial political interest, in imparting knowledge of geography, by the British to us despite being aware of the importance of geography (Singh, Ravi 2009c) in (higher) education. Hence, geography in primary education said to have enjoyed good position entered higher education late (cf. Ravi Singh 2009c). Even then, the 21st century marks the completion of over three quarter centuries since the institutionalisation of geography in India. In these years, many developments have taken shape to form the 'identity of Indian Geography'. Around more than a hundred attempts are known addressing the nature or needs, weaknesses or strengths, past achievements or the future tasks. In majority of them, the sense of loss in what we have inherited and weakness in present endeavours is not difficult to trace. The present essay is a brief effort to analyse the nature of inheritance and sense of loss together with the directions we look for in the 21st century—entirely different in terms of intellectual as well as technological contexts.

Nature of Inheritance and Sense of Loss

The fields of knowledge are plagued by public image and naturally geography is no exception (Carter 1968, 2). The public image is the outcome of various processes and circumstances (Ravi Singh 2009c). Geography, for larger part of its evolutionary history, faced a typical situation—justifying its position amidst the systematic branches of knowledge. In India, this problem still haunts us making us lose time, energy and creativity ultimately leading to lose the ground quite often. Therefore, we need to keep in mind how geography and geographic space is subjugated in academic mainstream (see Soja 1989, Singh 1998). In this section, we try to know through critical survey that what we have inherited from the past whether that is gainful or sheer loss! Though the history of quest for geographic knowledge, geographical awareness and concern in India are as old as the civilisation and undoubtedly rich as attested by the ancient wisdom we have inherited and possess in our traditions; the history of professional geography is comparatively recent in India.

India, being one of the living and continuing civilisations of the world, has always attracted the Westerners and it attests India's prominence, beyond South Asian sub-regional context. Though the British link had been quite strong, and continued well beyond India's Independence, the French one had been relatively feeble during the colonial period (cf. Thakur 1994a). After the Independence, the credit goes to O.H.K. Spate (1911-2000) for perhaps being the first overseas' geographer to discuss the nature of modern geography with reference to strides made by Indian Geography by the mid-1950s. Spate basically gave an address, after accepting the prestigious Jawaharlal Nehru Medal of the National Geographical Society of India (NGSI), published same year in the NGSI's mouthpiece—the *National Geographical Journal of India* (Spate 1956) with obvious limitation in scope. Its relevance in the present essay's context is justified by a substantial portion devoted to what the community of Indian geographers need to do. His initial emphasis was on looking at unity within geography, as illustrated by great geographers in all times that provides "complete understanding". Discussing "relevance of geographical studies to Indian problems", he underlined the significance of geographical studies apparent utility for a common man. He stressed "demand-mapping" rather than overworked "problem-mapping", which is more relevant in the present sense when contemporary world moves quicker to globalise further. Calling upon Indian geographers, he suggested to look into "highly desirable and indeed essential" fields of research—industrial decentralisation, planning of towns and rehabilitation of old ones, forestry, soil etc.—immense and potentially most fruitful.

The American interest gradually grew especially after the mid-1950s. Perhaps the slowly growing interest owes to the changed world reality—while the US emerged as the world-power, it started becoming the first choice for potential immigrants including the Indians, with new interest in the world affairs. Both of these facts played role in promoting studies on India by the American scholars including geographers comprising of the expatriate Indians and 'Indianists'. Sopher (1923-1984) was one of the 'Indianist' American geographers with a deeper concern for Indian Geography illustrated by the list of his major research works. Contributing to the *Geographers Abroad*, edited by Marvin W. Mikesell, focused on the "the problems and prospects of research in foreign areas", he expressed deeper concerns (Sopher 1973), particularly the lack of 'Indianness' in Indian Geography which should have been voiced by Indian geographers. In his characteristic academic style, Sopher's remarked, '....whether there can be or ought to be indigenous Indian geography' (Sopher 1973, 110). We will have to admit even now this question hardly attracts serious audience. Even before this challenging piece of writing, we know about scores of concerns shown about Indian Geography (cf. Kapur 2004a) but majority of them chose non-critical issues to address—not hitting the bird's eye! If the Europeans invaded the Indian sub-continent as imperialist colonial powers, pursuance of a colonial agenda and design and known prejudices should not surprise any one. But, as 'dweller' geographers, we should have devoted ourselves to unravel 'the popular landscape symbols, mental maps, and "sense of place" of the dwellers...." (Sopher 1973, 112) and make up for the loss incurred during the British rule. Quite telling is the fact that sense of pace in Indian has been quite strong but that could not get expressed in we Indian geographers work. Sopher (1973, 115) rightly observes, "An Indian geographer on home ground may prefer to play stranger in his writing, while the stranger from abroad seeks to stimulate the dweller's perceptions". He highlights where Indian Geography went wrong and indicates to what could be done to recover.

In the first half of 1980s, two writings were published one each by Joe Schwartzberg (1983/2009) and R.P. Misra (1983). Incidentally both devoted to the analysis of the state of Indian Geography but they did not deal with same issues. Schwartzberg dwelt largely upon the geography as practiced in India. Prior to India's Independence, in pursuit of colonial interest, a good deal of spade work led to data and maps generation and also institutional system of procuring and producing them, educating and training, and so on. But the attainment of Independence meant altogether a different perspective—burgeoning hopes and dreams—varied ways and means to realize them! After all it was the time to build a post-colonial India. End numbers of possibilities were to be turned into reality. In the process many good things in the form of numerical increase and expansion happened—the number of institutions, jobs, teachers and students, etc. These could however not help the issue of qualitative lacking, discussed later in this essay, overlooked by majority of us. The larger issue of constructing a post-colonial Indian Geography to address the greater goal of making an Indian identity and "Indianness" never got due recognition. It has now become a major ailment, as we will see later in this chapter. Schwartzberg indicated why the quality of geographic research is pathetic and has demonstrated the process of its worsening despite number of geographical societies and all of them having a journal of their own in south Asia (Schwartzberg 1983, 235). Though he tried to analyse the condition of Indian geography objectively not done previously by the Indian geographers, some instances of lop-sidedness could be found as well.

Misra's (1983) was an editorial "Introduction", to the first volume of *Contributions to Indian Geography,* of which he was the series editor, wherein he tried "to point out the main weaknesses…and to suggest a few initial steps…to make further strides in the years to come" (Misra 1983, 1). He did point out crucial issues, but the fact remains that he could not hit the roots of the problem, and instead chose to address "safer" areas, which after all involves the risk of "hurting" and antagonising the concerned not considered desirable healthy practice in the Indian situations.

Bharadwaj (1987) in his article observed, "….geographers writing on India, whether indigenous, expatriates, or non-Indians, seem to be generally aware of the fundamental problems of Indian geography, and some have already proposed guidelines for the future". He further identified "several working categories of themes" like seasonality (and its socio-economic implications), *jati* hierarchy (and its various implications with reference to resource utilisation, productivity, etc.), family orientation (with reference to migration especially), linguistic complexity (with reference to inter-regional communication and also language/dialect induced regionalism), pretty long historico-cultural dimensions, and special focus on the large middle class amidst the poor common mass. But then, should we discard everything that is coming from the Occident. No.

The author quite rightly observes at the end of his essay, "Western ideas ….need not be rejected, uncritically accepted or contorted or watered down to fit the Indian reality. They need to be accepted as *a*, not *the* way of doing geography." It appears that exceptions apart, the 'dweller' geographers are either satisfied with whatsoever they are practising and do not feel the purpose of calls given by senior or fellow workers or we can simply attribute it to the lack of capacity for one reason or the other. In general, we lack a proper grounding in the philosophical, conceptual, and theoretical underpinnings of our discipline (Dikshit 1994, vii-viii) despite the introduction of a compulsory course/paper on geographical thought at advance level, thanks to a few concerned colleagues. Truly speaking, a course on geographical thought which makes us understand the development of geographical ideas and geographic methods, are seen as ritualistic by most of we serving geographers. In this way, the purpose gets defeated whereas it has sufficiently been impressed and elaborated that the grounding in geographical thought can help educator, researcher and students of geography to contextualise his/her academic endeavour.

The famous essay, by A. B. Mukerji (1992/2009) one of the senior most geographers, 'What Ails Indian Geography' published (and reprinted) at several places, made candid confessions in the process of diagnosing what has been "ailing" Indian Geography. The sectional titles well reflect his concerns: gerontocratic grinding, missing methodology, fizzling physical geography, sterile stereotypes, funders' blunders, geography – a science of choroplething?, feeble fieldwork, pax britannica and pax Americana, the bogey of relevance, no place under the sun, and throwing the baby with the bath tub. His essay was surely a brave and truthful admission by a senior professional geographer who did not hesitate in putting the issues straightway. It is difficult for anybody, including his critics, to disagree with him and the genuine issues he shares with us. Many critical efforts after him are known but none can match Mukerji's sharpness till date.

In the first half of 1990s a worthy effort in the form of *The Roots of Indian Geography* (Singh and Singh 1992) appeared. A collection of 29 essays, it covered progress in many subfields and contribution of a few major and active Indian geographical societies. Though basically consisting of progress report type essays, it is noteworthy for the fact that many new themes like pilgrimage and cultural animal geography got covered (cf. Singh 1995b). Two essays, by Mukerji (discussed above) and Karan, respectively, in it covered the progress of Indian Geography and persisting problems. The essay by Karan began with outlining the history of geographical thought in India and in the remaining part focused on the post-1947 developments divided into three phases: pre-colonial, colonial (up to 1947) and the period of 1947-1990 through which he tried to analyse individual "quality" contributions "towards advancement of science of geography in India". The process of institutionalisation of geography in higher education began during the colonial rule in the 1920s and spread out successively. The pioneers had a clear cut missionary purpose on their minds which they carried out through establishing the university level centres and production of a train of students who worked as the emissaries of geography throughout the country. They had the opportunity to prove themselves as inspiring teachers and pioneer researchers too. As the real expansion and growth of geography could take place only after the attainment of Independence, Karan rightly elaborated the post-1947 developments in detail through the tabular diagram presenting a synoptic view of the 'principal research clusters and major scholars' in modern Indian Geography. One will have to agree to the fact that in a short span of time, Indian geographers produced works in almost all sub-fields—from geomorphology to geography of religious belief systems. In one sense, it is commendable achievement indeed! The challenges, like the lack of infrastructural facilities or rigour in post-graduate training, nevertheless need serious audience.

Singh's (1995) attempt was futuristic not merely by the title but also the content-wise. A brief critic of a few earlier works followed the search for identity which the author found lying in sensitivity towards the "Spirit pf Place" present in the sacredscape. Advocating the "ecopsychological approach", he suggested, "If we have to evolve our geographical identity, most likely we have to perceive of life and action as the moral principles of nature…Geographical identity be taken as synonymous to ecological identity" (Singh 1995, 277). His contention advocated the return to cultural roots for "intellectual exploration and experiential learning in which lies the prospects of the 21st century Indian Geography. These ideas created a milder reaction in Ahmad's presidential address to the Indian Geography Congress who contended the concept of Indian Geography itself and disagreement to Rana Singh's arguments (Ahmad 1997).

The work of Kapur (1998), perhaps the last of the 20th century, too tried to look at the future of Indian Geography from the perspective of alliance with public policy. We know about several presidential addresses delivered in the annual meetings of different active geographical bodies in India. Of course, they are important in the sense that they reflect changing concerns and emerging issues in the eyes of senior professional geographers, e.g. one can see Kapur (2002) to compare the presidential addresses of the National Association of Geographers, India (NAGI); they often end up prescribing the *dos* and the *do nots* (cf. Kapur 2004a) instead of invoking new thoughts and instilling inspiration in the young practitioners who look for the same. They tend to reflect the aura of the mightiest personalities and their academic specialisation overshadowing a critical and objective approach. Additionally, the huge number and relative inaccessibility (if not published) of such addresses also makes it impossible to consider them here.

Apart from the tradition of country level Progress Reports and Surveys of Research, we know about a few arduous attempts of stock-taking (of Indian Geography) e.g. Karan (1966), Sinha (1986), Singh and Singh (1992); Thakur (1994a &b), and Singh (1996). It is not the time and place to elaborate upon them; yet one can summarise them together as documentation of who did what where and when without a critical vision. One may find several instances of repetition of general issues already raised by their predecessors. But, the point that how far they have been able to do justice to the job of true introspection is open to questioning in academic spirit.

21st Century Directions

With changing times, circumstances and challenges facing a society change, and, so do the response(s) to them determined by the contemporary perception, state of knowledge and cutting-edge technology available for application. Writing in the last quarter of the 20th century, Haggett (1990, 163-164) remarked, "....geographical literature is also strewn with the spent cartridges of old projections and any survey of past forecasts of future spatial patterns shows a similar chastening mixture of delusions and insights". He further comments that "we seek to blend events to avoid future we dislike or to attain a future we seek". Going by the general human psychology, this statement is obviously true. But, it is possible to do so in view of all what happens around us within our discipline. He identified seven research areas of future: revolutionised computer aided cartography, new paradigm for climatic evolution, new regional geography supplemented by integrated quantitative models, end of the gap between micro-and macro-studies, man-land ethic as the philosophical basis for geography, global land-use inventory (read "GIS"), and establishment of a space-proofed set of statistical tests.

How relevant is this envisioned future for Indian Geography is a curious question. We can seek answers from the works attempted in the 21st century by the Indian geographers. Perhaps, the first one was by R.D. Dikshit (2001) with the purpose "...to review and diagnose the status of geography in India both as a branch of learning and as a profession" (Dikshit 2001, 1). He tried to raise many noteworthy issues a few of which need serious consideration. In the first part he addressed the changing philosophy of geography since mid-20th century and focused the later one towards 'rebuilding Indian geography' with proposal on how the restructuring should be done. In the following year, an edited volume of the presidential addresses of NAGI was published (Kapur 2002) organised into two parts: a lengthy first introductory part of ninety pages and the second part containing twenty-two presidential addresses delivered, preceded by an introductory curriculum vitae of the respective presidents, at the Indian Geography Congresses (1980-2000). Reviewing the same, K.R. Dikshit's remark is a caution for the young geographers, "…younger geographers in India look at the senior geographers, venerably and even uncritically. This may be an extremely polite gesture of good will, but, certainly, it doesn't represent the spirit of science or scientific enquiry. Let us remember, a critical assessment is far more rewarding than an uncritical acceptance of every thing that the elders do" (Dikshit 2002, 112).

M.S. Gill threw light on geographic research in India during the later half of the last century. He noted that now despite several re-doings, fermentation of ideas, wide focus (on the production of varied spaces), precision and sharpness in research problem formulation, etc. have severely lacked in the practice of geography in India. What appears very important and strong point, raised perhaps for the first time by him, is self-plagiarism (Gill 2003) taken heavy toll in Indian Geography. Plagiarism is one of the most denounced illegal academic acts. Therefore we generally refrain from that. But, another dimension of it called "self-plagiarism" is a bigger issue involving question of professional and personal ethics. Further, it has implication for research too wherein old wine is repackaged in new bottles. It is one of the major problems, in humanities and social sciences, in a society like ours which lack effective and full-proof tracking system for such a rampant malpractice.

The year 2004 in a way proved to be rich. Writing in the *Progress in Indian Geography 2000-2004*, a country report presented every four year to the International Geographical Union, Kapur (2004a) explored the possibility of seeking the future for Indian Geography through creativity. Running into valuable thirteen pages, the author impressing the importance of creativity in knowledge formation discussed its scope in (India's) geographical enterprise and called it the right time to inject the ailing Indian Geography with creativity. And, also identified the future shifts. In the same year 2004, Saraswati Raju, one of the leading Indian geographers of her generation, published an article in *Geoforum* devoted to 'emerging research and praxis' in India (Raju 2004). Mostly based on her personal experience, she highlighted the mainstream tendency, still stuck to the conservative colonial conventions, discouraging works on a non-conventional research problem related to emergent realities. The conservative Indian Geography thus continues to prevent the unfolding of alternative geographies and critical academic enterprises. Another article by Kapur (Kapur 2004b) was published in the September 11, 2004 issue of *Economic and Political Weekly, EPW.* The explicit thesis of this article was why geography is not given the recognition as a social science and what makes it a "languishing" social science in India. In the course of forwarding her logic, Kapur tried to highlight negligible presence of the discipline in the common (social science) space. This work reminded us of the persistent poor position of geography. But, the (over-)stressing on the social science tag for geography is termed to be 'obsession' by colleagues (Dikshit 2006, 142-143). Kapur's observations in her article (2004) could evoke three responses only, quite low against expectations. So far, only three direct responses are known Prasad (2005), Lahiri-Dutt (2005/2009), and Landy (2005). The last two were from abroad–Australia and France, respectively. Interdisciplinarity, an accepted important reality in contemporary academics, the inner strength of our discipline should be marketed professionally argued Prasad (2005, 79). The responses naturally were mixed type with additional issues emanating from personal experiences. Lahiri-Dutt (2005/2009) raised many valid questions intertwined with her own engagement with geography a student-teacher-researcher in India where she spent two decades' long term as a faculty at Burdwan University's geography department before migrating to Australia. She provides peep into the sufferings of geographers, particularly those who are doomed to be working in non-elite institutions and located in peripheral areas along with several marginalities. Other issues raised pin-pointedly by her include the exclusion of geography in the academic business of social sciences, arbitrary branding of a work as 'geographic'/'non-geographic', lack of theoretical perspective in Indian works, over-emphasis on GIS like techniques taking its own toll, ridiculed recommendations, the false 'science' tag used to extract extra grants and claim boastful status, over-reliability on statistical techniques, out-dated school-curriculum, questions quality 'in teaching, of teachers, and students, the quality of the curriculum….', and the great gender divide. Dikshit (2006, 142-143) too have responded briefly on the issue raised in a sub-section.

K.R. Dikshit published two papers (Dikshit 2005, 2006) recently. The former of them treats the matter of evolution of geography in India from the ancient to the modern (colonial) period and tries to reason out our lagging behind and presents the accomplishments comprehensively. No work of this standard is known prior to it. In the latter one, the author looked at the progress taking place in the western and Indian geography. First he describes the Indian heritage (of geography), then discusses modern, postmodern and contemporary in the western perspectives. To put in brief, Dikshit (2006) has tried to explain changing paradigms in geography in the first part and the focus is in the second part is on Indian geography. The sub-title of the second major section "the discomfiture of geographers" presents the author's reading in precise manner. The treatment of the issues attended well reflects deeper thinking by a seasoned and concerned geographer, indeed a treat to all interested in the debates taken/taking place in Indian geography. His remark (Dikshit 2006, 150) regarding the virtually useless "geography–non-geography" debate asks for serious consideration: "The geography–non-geography debate would lead us nowhere. There should be more debate on the spirit, purpose, content, approach and methodology of the subject, relevant to the Indian context. So long we have a purpose in doing the kind of geography we are doing, it is academically acceptable. Dedication is important. Let it be clearly understood that geography is neither mathematics nor physics, the subject requires description and interpretation and the findings of a geographer cannot be put in some mathematical equation that could revolutionise our thinking."

The issues concerning geography education, especially in schools, have been raised by geographers in indicative manner. No body perhaps had tried to look into the intricacies of pedagogic problems before Sunny (2006/2009). She tried to analyse why geography languishes as a social study subject in at present. Her analysis exposed the persistent internal dualism in both geographical framework and our approaches. It is no longer a secret that contemporary geographies have radically moved away from the traditional ones simply in response to the contemporary world that continues to evolve. That requires change in the educational focus and practices. Accordingly, Sunny attempted an appraisal of the nature of geography education with special reference to the schoolbooks of State Council of Education and Research and Training (Madhya Pradesh). The conclusions drawn, however, are relevant for the country's school discourses on geography, ask for overhauling our overall approach to geography teaching in schools. Logically speaking, geography in tertiary education will have to take a new form in response.

In a way, 2007 proved to be a quiet period as we do not know about any published thought-oriented work concerning the Indian Geography.

An article has appeared in the Platinum Jubilee issue of the *Geographical Review of India* (Singh 2008) released recently. The author has tried to present his own assessment of Indian Geography with reference to the recent intellectual developments taken place at the global level. Some issues like discomfort with numerical increase quality of teaching and initiative fatigue, reluctance to 'new' and centre wise specialisation and consultancy addressed may not be the new ones but the points made are not explicitly indicated earlier. Other issues—hesitance in introspection, review or eulogy–deitification and glorification, lack of space for internal debate(s), over emphasis on computer based techniques, the new mantra—however were not formerly discussed.

In 2009, an edited volume, consisting mainly of previously published papers, tried to bring the main concerns and debates together (Ravi Singh 2009a) which have been refereed in the above discussion. Rana Singh (2009a) has taken a critical look at the evolution of modern Indian geography with special reference to "Identity of 'India' & 'Indianness' in Geography in India". Beginning with the phase-wise evolution of modern Indian Geography, he laments the pathetic situation of geographical works in India which prevented Indian Geography forming own respectable identity! He pushes the case, of practicing a tradition of geography embedded in our soil and culture, quoting "….Indian geographers…., don't have to be anxious to tread the path followed by their distinguished colleagues in the USA, Europe, Africa, Australia or China. Each region has its own socio-economic or political problems and they can not be uniform for all regions endowed with varying natural and human resources and varying forms of political set up. Indian geography could establish its own paradigm and priority in research and teaching of geography" (cf. Dikshit 2006, 153). We will have to admit that the question of indigenousness in Indian Geography has still not been given sufficient attention (cf. Singh 2009a).

A recent book by Rana Singh (2009b) is indeed a welcome addition to the existing stock. The whole book should be read in the single spirit of making a case for Indian Geography with clear and strong identity of itself. The first two chapters directly address Indian geography. The opening chapter is an appraisal of critical issues related to place of "India" and identity of Indian geography; whereas, the second chapter captures the evolution and growth of Indian geography. Though issues raised by Sopher (1973/2009), Schwartzberg (1983/2009) and Mukerji (1992/2009) serve as the foundation on which new interpretations are offered; at occasions the author has gone on to question the academic practices of even these scholars reflected in their major work(s) as well and exposes crippling weaknesses. In a way, they are critics' critic. Rest of the book is devoted to scientific cultural analyses of the modern ecological crisis, ethical issues, and also spirituality with an urge for environmental ethics and exploration into ancient Indian consciousness expressed in symbols and traditions. Exclusive discussions on ecospirituality as represented in the ancient Indian (Hindu) tradition, perspectives emerging from Hindu pilgrimage, Mahatma Gandhi's vision, and the East-West are quite rich and expository. A reader can understand a connected idea flowing through and the central message of building the identity of Indian thought and Indianness in the context of cultural roots and traditions running across the book in the reconciliatory approach which aims at developing 'the frame of harmonious synthesis' between the East and the West—serving the cause of humanity. In another soon to be released book, *Geographical Thoughts in India: Snapshots and Vision for the 21st Century*, Rana Singh (2009c) continues with his mission to search the (ancient) cultural roots, link them with scientific knowledge and develop a vision the present century's Indian Geography. The whole book is an endeavour to explain the cultural roots of the Indian identity and the gaps requiring "re-consideration" in the present century. Out Academic exercises are good and done essentially but we hardly try to know what people think of our discipline and we practitioners. Many works on public image of geography are known save for the Indian efforts. Ravi Singh (2009c) has attempted a study recently to see the people's perception of geography in India. It is apparently concluded that, though based on their school experience, people's responses confirm most of the ailments reportedly associated with Indian Geography.

Conclusion

The above presentation can rightly lead us to ask where Indian Geography is today. We are aware of geography's struggle for existence and survival and legitimate recognition in the academe. As revealed by the above discussions, the Indian situation is apparently worst. In one hand Indian Geography, shares some of the universal predicaments, and almost simultaneously continue with additional sufferings unique to it on the other. Indian geography is still grappling with basic questions debated around the mid-20th century by our western counterparts. The ailments are noticeable and responsible causative factors too are marked.

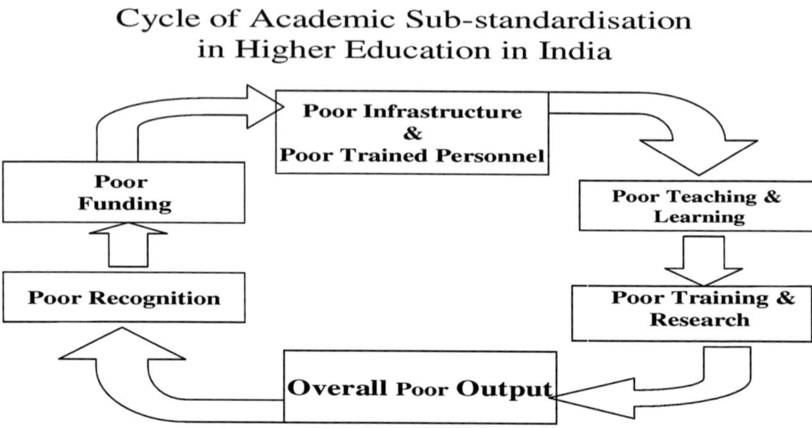

Fig. 2-1

If considered in a broader context, they together ultimately develop into a viscous circle of academic sub-standardization explained by Fig. 2-1 in the wider context of Indian higher education. The main emerging point concerns our response. Despite all sorts of academic politics towards marginalising geography, its value, stressed by the academics and common people as well, can not be undermined (Ravi Singh 2009c). We require responding both the global challenge and our Indian issues and problems. In this process adherence to digression will not be desirable. Hence "campaigning from the core geography but pushing in new directions" (Carter 2002, 156) would be essential. Creating scope and space for debate-dialogue—the fertile ground for breeding of (new) ideas—is imperative to save intelligent geographical minds from getting virtually blunt and hone them further. Highlighting and fortifying the 'geographic point', and resolving the inherent contradictions too are important in this enterprise. "The new awakening" through "self-realisation" holds the key to restoration of the lost glory, respect and recognition of Indian Geography and reclamation from the low image (cf. Rana Singh 2009b, 9). Of course, the 21st century has new challenges to offer, it also has opportunities for the "making of geography, in India too" (Rana Singh 2009b, 5) in the offing.

References

Ahmad, Aijazuddin. 1997. Fifty years of geography in India and abroad. Presidential address delivered at the 20th Indian Geography Congress, December 28-31, in Thiruvananthapuram (Kerala).

Bhardwaj, Surinder M. 1987. Toward an Indian Geography agenda. *National Geographical Journal of India* 33 (3): 327-334.

Carter, George F. 1968. *Man and the Land: A Cultural Geography*. 2nd ed., NY: Holt, Rinehart and Winston Inc.

Carter, Roger. 2002. Subject specialisms. *Geography* 87 (2): 156.

Deshpande, C.D. 1983. Discovering Indian geography. In, *Contributions to Indian Geography*, vol. I. *Concepts and Approaches*, ed. R.P. Misra, 11-14. New Delhi: Heritage Publishers.

—. 1991. Indian Geography in 2001 A.D. In *Emerging Trends in Indian Geography*, ed. Jayamala Deddee, 156-165. Jaipur: Rawat Publications.

Dikshit, K.R. 2005/2009. The roots of Indian Geography. *Transactions, IIG,* 27(2): 45-80. Reprinted in *Indian Geography: Perspectives, Concerns and Issues*, ed. Ravi S. Singh, 65-108. Jaipur/New Delhi: Rawat Publications.

—. 2002. Book Review of Indian Geography: Voice of Concern (by Anu Kapur). *Transactions, Institute of Indian Geographers* 24 (1-2): 108-112.

—. 2006. The changing western perspective on geography and the Indian context. T*ransactions, Institute of Indian Geographers* 28 (2): 123-155.

Dikshit, R.D. (ed.) 1994. *The Art and Science of Geography: Integrated Readings*. New Delhi: Prentice-Hall of India.

Dikshit, R.D. 2001/2009. Indian Geography: an encounter with reality. *Transactions, Institute of Indian Geographers* 23 (1-2): 1-18. Reprinted in *Indian Geography: Perspectives, Concerns and Issues*, ed. Ravi S. Singh, 172-197. Jaipur/New Delhi: Rawat Publications.

Gill, M.S. 2003/2009. Geographic research in India: its strength and weaknesses. *Geographical Review of India* 65 (2): 194-198. Reprinted in *Indian Geography: Perspectives, Concerns and Issues*, ed. Ravi S. Singh, 251-256. Jaipur/New Delhi: Rawat Publications

Haggett, P. 1990. *The Geographer's Art*. Oxford: Basil Blackwell.

Hartshorne, Richard. 1939. *The Nature of Geography: A Critical Survey of Current Thought in the light of the Past*. (Seventh imprint 1960; Indian reprint 2000), Jaipur/New Delhi: Rawat Publications.

Kapur, Anu. (ed.) 2002. *Indian Geography: Voice of Concern*. New Delhi: Concept Publishing Company.

Kapur, Anu. 1998. *Indian Geography: A Future with a Difference*. New Delhi: Allied Publishers Limited.

—. 2004a. Creativity: The future of Indian geography. In, *Progress in Indian Geography 2000-2004: A Country Report*, ed. H.N. Sharma, 5-17. New Delhi: Indian National Science Academy.

—. 2004b. Geography in India: A languishing social science. *Economic and Political Weekly*, 39 (37): 4187-4195.

Karan, P. P. 1992/2009. Development of geographic thought in India. *National Geographical Journal of India*, 38 (1-4): 179-193. Reprinted in *Indian Geography: Perspectives, Concerns and Issues*, ed. Ravi S. Singh, 372-391. Jaipur/New Delhi: Rawat Publications.

—. 1966. Recent contributions to the geography of south Asia. *Cahiers de Geographie de Ouebec,* 9: 317-32.

Krishan, Gopal, H.N. Sharma and Anu Kapur 2000. Indian Geography: Tasks ahead. *Annals of NAGI*, 20 (1): 131-154.

Lahiri-Dutt, Kuntala 2005/2009. Geography as a marginal science. *Economic and Political Weekly*, 40 (7), February: 689-691. Reprinted in *Indian Geography: Perspectives, Concerns and Issues*, ed. Ravi S. Singh, 242-250. Jaipur/New Delhi: Rawat Publications.

Mukerji, A.B. 1992/2009. What ails Indian Geography? Some statements. *National Geographical Journal of India* 38 (1-4): 205-218. Reprinted in *Indian Geography: Perspectives, Concerns and Issues*, ed. Ravi S. Singh, 223-241. Jaipur/New Delhi: Rawat Publications.

Raju, Saraswati 2004. Contextualizing critical geography in India: emerging research and praxis. *GeoForum*, 35: 539-544.

Singh, L.R. 1996. A hundred years of Indian Geography: Development and change. In *New Frontiers in Indian Geography*, ed. L.R. Singh, 1-57. Allahabad: R.N. Dubey Foundation. Reprinted in *Indian Geography: Perspectives, Concerns and Issues*, ed. Ravi S. Singh, 109-171. Jaipur/New Delhi: Rawat Publications.

Singh, Nina and Sangwan, Binu 1999. Challenges in geography teaching and research. *Transactions, Institute of Indian Geographers* 21 (2): 85-88.

Singh, R.L. and Rana P.B. Singh (ed.) 1992. *The Roots of Indian Geography: Search and Research*. Varanasi: National Geographical Society of India.

Singh, Rana P.B. 1995. Identity of geography in India beyond 2000: A search of sensitivity for future. *National Geographical Journal of India* 41 (3): 263-281.

—. 2009a. Identity of "India" and "Indianness", and geography in India: sensitive appraisal from the past and move to the 21st century. In

Indian Geography: Perspectives, Concerns and Issues, ed. Ravi S. Singh, 25 - 64. Jaipur/New Delhi: Rawat Publications.

—. 2009b. *Uprooting Geographic Thoughts in India: Toward Ecology and Culture in 21st Century*. Newcastle upon Tyne: Cambridge Scholars Publishing.

—. 2009c. *Geographical Thoughts in India: Snapshots and Vision for the 21st Century*. Newcastle upon Tyne: Cambridge Scholars Publishing.

Singh, Ravi S. 1998. Book Review of *Indian Geography* (by Anu Kapur) in *National Geographical Journal of India* 44 (1-4): 249-250; also in, *North Eastern Geographer* 29 (1-2): 79-81.

—. 2008. Challenges to Indian Geography: A contextual analysis. *Geographical Review of India* (Platinum Jubilee Volume) 70 (1): 57-70.

—. 2009a. Introduction: Concerns and debates in Indian Geography. In *Indian Geography: Perspectives, Concerns and Issues*, ed. Ravi S. Singh, 1-22. Jaipur/New Delhi: Rawat Publications:

—. 2009b. Image of geography in India; In *Indian Geography: Perspectives, Concerns and Issues*, ed. Ravi S. Singh, 198-220. Jaipur/New Delhi: Rawat Publications.

—. 2009c. Identity and image of Indian geography: The people's perspective. *Journal of Geography in Higher Education* DOI: 10.1080/03098260902734984.

Sinha, B.N. 1986. *Trends in Geographical Research in India: Published Papers in Indian Geographical Journals, 1928-1985*. Bhubaneshwar: Indian Council of Geographers (Utkal University).

Soja, E.W. 1989. *Postmodern Geographies: The Reassertion of Space in Critical Theory*. London: Verso.

Sopher, David E. 1973/2009. Towards a rediscovery of India: Thoughts on some neglected geography. In *Geographers Abroad: Essays on the Problems and Prospects of Research in Foreign Areas*, ed. Marvin W. Mikesell, 110-133. Chicago: Department of Geography, Research Paper No. 152, University of Chicago. Reprinted in *Indian Geography: Perspectives, Concerns and Issues*, ed. Ravi S. Singh, 305-330. Jaipur/New Delhi: Rawat Publications.

Spate, O.H.K. 1956. Modern geography and its meaning for India. *National Geographical Journal of India* 2 (4): 175-185.

Sunny, Yemuna. 2006/2009. Analysing current practices in geography education. *Economic and Political Weekly* 41 (3): 270-278. Reprinted in *Indian Geography: Perspectives, Concerns and Issues*, ed. Ravi S. Singh, 257-278. Jaipur/New Delhi: Rawat Publications.

Thakur, Baleshwar. 1994a. Indian Geography: Development, trends and prospects, Part I. *Transactions, Institute of Indian Geographers* 16 (1): 67-85.

—. 1994b. Indian Geography: Development, trends and prospects, Part II. *Transactions, Institute of Indian Geographers* 16 (2): 177-195.

CHAPTER THREE

GEOGRAPHICAL EXPLANATION OF CONTEMPORARY SPATIAL DYNAMICS

ANAND PRASAD MISHRA,
BANARAS HINDU UNIVERSITY

Orientation

In primitive stage of development process, the spatial identity of a region was characterized by simple mode of production, distribution pattern and its overall surroundings space with less interaction of human beings and their developmental activities have been confined with minimum contradictions. But on the other hand, the contemporary development processes are restricting our socio-economic sphere of life and landscape in a bigger way. In this context, urban-rural divide and other space related emerging conflicts now become the vital issue for geographers to relate these complexities of space in geographical paradigms. To initiate any discussion on the issue of place vs. geography, first of all we should be clear in our mind about the concept of our discipline. Geography is not just about place but also about space: about how the distinctive economic characteristics of places, and the conditions of economic possibility that they face, depend on their economically geographical interdependencies with other places (Barnes and Sheppara 2003).

Moreover, geography in the contemporary world may be described as a discipline which deals with the emerging spatial dynamics of regions. This regional dynamics has many new things for geographers and geography. It has many possibilities for and challenges to young and imaginative geographers. The complexities and the nature of spatial dynamics need to be looked into an investigative perspective. This new perspective in the field of geography is now becoming vital issue for discussion and searching for an alternative research methodology in the subject. The agenda of "spatial dynamics" is related, in terms of explaining injustice and inequality, and developing strategies, to social changes. Peet (1977, 20) explained, "…it was not just a matter of radicalizing geography, but first and foremost the search of a politics of transformation and the scripting of a "Social Science" that would support and develop in tandem with a progressive and emancipatory political commitment". The required research design can be fulfilled as many geographers favour historical-geographical materialism. Historical-geographical materialism starts from the premise that things (as objects and phenomenon) exist, but that these objects or phenomenon are the product of relationships, and this process has a dialectical mechanism in a geographical space. In this context, Ollman (1993, 11) pits his observation: "Dialectics restructures our thinking about reality by replacing the common sense notion of "thing" as something that has a history and has external connection to other things, with a notion of process, "which contains its history and possible futures and 'relation', which contains as part of what is its ties with other relations".

These philosophical dimensions have added many new chapters in the field of regional studies, especially in geography. In early periods, most of the geographical explanations were confined to physical aspect of natural phenomenon i.e. river, mountain, lake, ocean, etc. Eighteenth and nineteenth century geographical studies and geographers laid their emphasis on to investigate the various aspects of landscape, its process, structure and time related spatial dynamics. Physical geography became a prominent subject for geographers. Moreover, the increasing nature of the innovative capacities of human being expands economic activities, mode of production and level of technological interaction with society and nature. Between 1820 and 1982, the real value of output in the world's largest industrialized economies increased by a factor of severity. The population of these economies also grew, nevertheless the average worker produced fourteen times more real output at the end of this period than at the beginning (Maddison 1982). The great invention of nineteenth century was the evolution of knowledge and its production. This production of knowledge and its application have changed the period of the space considerably over time. The evolution of technology is closely linked to the economic and institutional character of a region. The emerging dynamics of a particular region/space has been produced by its own socio-political and cultural dynamics. The emerging territorial structure of development is contradictory (that is, full of tension and conflict), and the socio-spatial dynamics of production are inherently unstable. In this context, the geography of uneven development and its histogenesis are now the emerging agenda for contemporary geography and geographers.

The contemporary development process and its impacts on space/regions have many new dynamics. Skelton (1998) observed that culture, globalization and development have intersected in a number of complex ways. The modern development have obviously accelerated dramatically in the last half a century, spurred on by technological innovations, including the green revolutions and other developments during cold war (Perkins 1997, 12). Geographers have attempted in many ways to examine the forces and process of economic development-competition, the drive to accumulate profits, the evolution of technology and labour process, the tendency towards the concentration and centralization of capital-generate spatially differentiated outcomes, in terms of growth, prosperity and employment (Martin 2003). In this context, to examine the space, geographers need to understand the role, impact and evolution of modern development institutions. Geography is now becoming more important in its role of coordinating socio-economic development of nations, locating their productive forces and improving the economic structure. The consequences of modern development have generated the problems of protection, conservation and transformation of eco-systems. It has become a universal truth that the above emerging problem can be solved through inter-disciplinary research, which requires considerable contributions on the part of scientists and specialists in many fields of science and technology. In this regard, erstwhile Soviet geographer I.P. Gerasimov's (1983, 15) observation for geography and geographers roles is quite important. He said, "In our opinion, because of the entire course of its preceding development, geography is prepared better than any other science for such research activities conducted on the basis of joint participation of many branches of science. Geography has accumulated a huge volume of information on the natural conditions and researchers of the entire world and on the degree and forms of their development and economic use. It is no less important that in handling these new formidable tasks geography can rely on the complex system of its branches. And finally, geography possesses an integrated approach to natural or socio-economic phenomenon which is especially necessary for a successful scientific formulation of the problems posed" (Gerasimov 1983, 15). This approach of geography has proved the significant role of geographers in solving the present day problems.

Another aspect of contemporary development is lopsided development and inequality among the nations and people. This emerging phenomenon has severe consequence for the issue of sustainable development and also a new agenda for geographical inquiry and research. On this account Human Development Reports have made the complexity of global space (among different nation) clear and it should be taken into account by geographers in their research. The Human Development Report (1993) is an important fresh agenda for the discipline of geography. The Report said, "Policy makers are often mesmerized by the quantity of growth. They need to be more concerned with its structure and quality. Unless governments take timely corrective action, economic growth can become lopsided and flawed. Determined efforts are needed to avoid growth that is jobless, ruthless, voiceless, rootless and futureless" (UNDP Human Development Report 1993, 2). The above mentioned facts are the new emerging agenda for spatial analysis, and should be taken into account by young and future looking radical geographers. For this, new look for geographical inquiry is very much required.

Recent increasing role of market and minimization of state intervention in contemporary development strategy has originated another new dimension in spatial analysis. "State-centric political identity no longer monopolizes but shares the stage with a growing number of non-territorial claimants. Sub- and trans- national social movements transgress territorial boundaries in favour of identities 'grounded' on ecological anti-nuclear, ethnic, feminist, religious and other non-state based commitments. Uneven flows of information, culture, people, technology and capital involve human migrations and diasporic identities that complicate territorial frames of reference. The globalization of production and finance undercuts national economic planning eroding state sovereignty and the political identities it presupposes, even as supra-national forces alter state power and sub-national conflicts expose the illusion of homogeneity promoted in nationalist narrative (Peterson 2007, 112). The expanding space of global production network and its linkages towards micro geographical regions have created new demands for reorganization of space, society and economy. This micro level reality transforms the complexities of space and also poses problems for society and people. It can be seen at global as well as regional level. The contemporary development processes have made pressure on world spatial structure and its territory These emerging developmental dynamics have changed the geography of world from simple to complex. There are complex geographies operating in these matters, geographies that need to be thought thoroughly and more carefully too. The threat from global actions is seen by many as the source of numerous environmental dangers in specific places–where habitat disruption endangers species and modern economies encroach on ecosystems and aboriginal landscapes, simultaneously threatening cultures and their ecologies (Mander and Goldsmith 1996).

Considering the modern development and its impact on nature, there is the requirement for a scientific approach to development of geographic science. This emerging science of geography should also be concerned with transformation that took place in a region. It should also change its thrust area for research, which may be useful for studying transformation of region and society. We have such previous evidences in our discipline. von Thünen (1826) was first to study the agricultural land use in a scientific way, later applied to industrial location by Weber (1929) and ultimately used to explain the distribution of services and settlements by Christaller (1966). All these efforts by different scholars are significant for the development of geographical research. Presently, our world economy is changing very fast. These processes are predominated by the evolution of knowledge and technology. The increasing space of transport, electro-communication and quick and free movement of capital have created the demand of new world order. Such emerging reality and facts are other issues for geographers, setting a fresh agenda for future research and development of geography. In this vein, the increasing complexities of regions should be analyzed in the structuralist approach. The structuralist approach, explaining the location of economic activity and the way it changes, represents a more holistic approach than behaviouralism. The behaviour is shaped by wider processes in the social, political and economic spheres. Notions of culture and of class rather than individual ideas determine the spatial structure of economic activity. Firms work within a capitalist society in which production is essentially a social process structured by the relationship between capital and labour (Daniels 2005). The existing nature of relation between capital and labour in a space for production is another upcoming geographical reality, which requires fresh attention from geographers to look into the matter in a scientific way. This emerging space is not a simple geographical element; but due to free market competition, the prevailing equilibrium of space may get undermined by market forces. The evolutionary process of competitive space in different parts of a region and its history are now becoming an important agenda for geographical studies. This agenda should be taken into account with the following two objectives: to generate complete geographical information about emerging competitive regions, i.e. geographical information about virgin land and its comparison with market oriented space; and, to analyze industrial space, its production process and of market expansion, and also geographers should consider, in their study, finding out the correlation between production, space and market.

The concept of competitive space is controversial and it originated from new technological site of region during the period of modern industrial growth. It changed the geography of regions. Under these emerging situations, new geographical approach is the need of the hour to develop the discipline in the contemporary context. For this, a systematic change for geographical study should be considered in spatial study of regions. The new geographical themes therefore could be the study of geographical information about regions in terms of their geographic-historical materialism, analysis of the changing pattern of mode of production in a specific geographic region, study of the emerging spatial dynamics in terms of its spatial reorganization, and elaborate and investigate the emerging challenges of spatial conflicts and conceptualisation of dichotomy in capital-labour-land in the region concerned. On the basis of above facts, the emerging areas in geography may be identified as the following.

Competitive Space and Geography

The process of industrial development along with divergent paths of growth creates the problems of disruption of all established geographic elements. Under these circumstances, successive industrial development have built up the great cities, transport system, the landscape of production and market network that influenced nature and society. This emerging contemporary phenomenon liquidates the geography of the past initiating vibrant discussion in geography and also poses challenges in the area of filed inquiry. Moreover, the spatial structures of production have added many new dimensions in the area of industrial development. In this context Massey (1984) analysed the dynamics of industrial development which represent "the competitive space" in capitalist economy. He observed, "….development is a social and conflictual process; the geography of industry is an object of struggle …new spatial division of labour are more than just pattern, a kind of geographical re-shuffling of the same old pack of cards. They represent whole new sets of relations between activities in different places, new spatial patterns of social organization, new dimensions of inequality and new relations of dominance and dependence. Each spatial division of labour represents a real and thorough spatial restructuring. There is more than one kind of "regional problem" (Massey 1984). The changing pattern of spatial dynamics in a competitive capitalist economy has created many new thrust areas for geographers in contemporary development studies. These emerging areas are extremely challenging and interesting particularly for young students of geography. It requires a scientific inquiry of an industrial space and this effort should be taken into account to solve the emerging conflict between people and development. In this context, a restructuring in the discipline of geography and its methodology by the researchers and field workers in this discipline now becomes imperative. It should be based on the process of social transformation, its conflict and change that emerged in a competitive space through development. It is more or less class-based perspective that needs to take into account not just the role of geography, nor that of history in isolation; but of the simultaneous co-existence of place and time. This emerging concept in development and spatial studies has generated the demand for growing relationship between history and geography. Driver (1988, 14) identified four different elements in which history has been conceived in human geography, i.e. a series of legacies, an evolutionary motor, a source of agency and grounding for theory. It is time to bring history back into the heart of human geography as an essential part of doing the research on place and its emerging dynamics.

The above discussed concept of geographical information system has been summarised in Fig 3-1. This emerging approach in geography is based on the prevailing ground reality of space and geographical-historical materialism.

Territory vs. Geography

Our globe has many geographical regions based on climatic, physiographic, socio-cultural and other geographic elements. For a considerably long period of time physical geographic conditions have played major role in defining and demonstrating the boundaries of a region. But after 20th century, the emerging global political phenomenon produced many new things in geography, which relate entirely different spatial reality in our earth system. The territorial changes, are most important aspect among these, are becoming vital issues for geography and geographers. The development of communication, nuclear response, global war, command system of super power and worldwide increasing network of market phenomenon have created an environment, which humbles the concept of national boundary. This scenario in society and space incorporated new debate on the issues of territory vs. geography. On this account, before going to any conclusion the ground reality of space needs to be inquired by geographers.

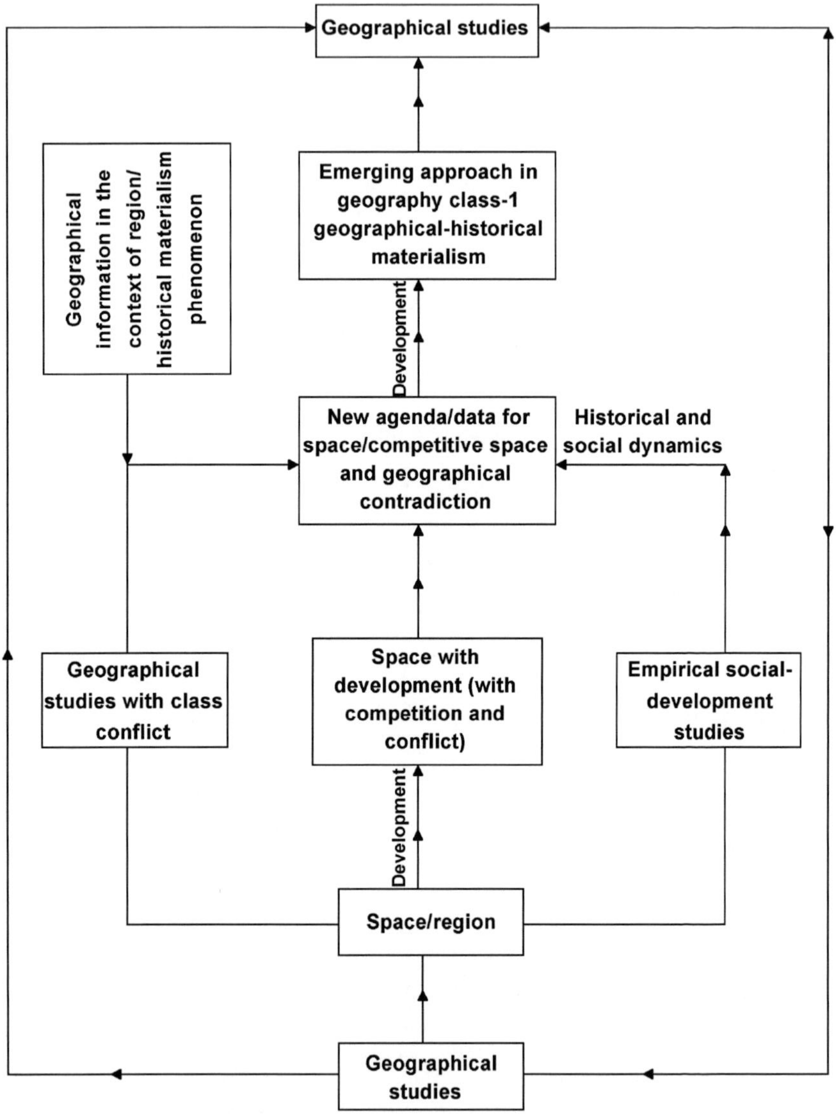

Fig. 3-1: Emerging approaches in geography

The traditional meaning of sovereignty changes as the globalization of finance, production and distributor develops (Williams 1996). Through this prevailing dynamics of development geographical studies requires fresh attention from the researchers. The territorial state became a central issue in the contemporary debate of international economics, and research into the technological revolutions in communications and transportation. Frontiers and boundaries were most popular topic in political geography. However, Pounds (1963) observed a decline of interest in this subject matter. Previously, political geography was largely practiced in Europe and North America region to solve boundary-disputes. Furthermore, many of the early geographers of 20th century were themselves boundary drawers and surveyors in the imperial division of periphery (Taylor and Flint 2004, 161-162). Historically, natural boundary is a product of the strength of the French state in eighteenth century Europe (Pounds 1951; 1954). In contrast, the idea of "national boundaries" is the Germanic reaction to the French expansionist ideas. The functional aspect of territorial boundaries is mostly reflected by state. In modern days, the state is the regulator and facilitator (Clarks and Dear 1984). Johnston (1982, 13) adds the role of the state in producing the physical infrastructure for the smooth running of the economy, roads, railways, power transmission lines, etc. Political geographers have been concerned with these processes as they relate to spatial integration. Deutch (1981) analysed the state in modern era. He described that the entry of the world into the nuclear age in 1945 meant that states could no longer even perform their most basic function—the defence of their people. Brown (1973, 215-216) elaborated the problems for humanity that transcend the territorial state, such as the environmental crisis, the population problem and the widening of the rich-poor gap. He highlighted the growing economic inter-dependence of the world and claimed that national sovereignty is being gradually but steadily sacrificed for affluence. These processes will ultimately erode the prevailing nature of state and its territory. Now the state and its functionaries have realized that their sovereignty is under attack from the global flows of capital within the world-economy and also have to transfer some of their legitimate powers to other institutions. Sassen (1996) concluded that the dilution of state sovereignty has often been the result of process initiated by states themselves.

A territorial restriction on trade is another major factor in the sovereign concept of state. But the emergence of global trade has developed major constraints on the existence and continuity of state and its territory. The first and foremost role of state is to exercise control over the flows across the boundaries. Moreover, in recent time, the dilution of state sovereignty has been the result of process initiated by the state itself. The contemporary world is mostly dominated by large private corporations which undermine the traditional concept of territorial state. Brown (1973) observed that General Motors was economically larger than most states, ranking twenty-third in his combined "league table". He concluded that today the sun does set on the British Empire, but not on the scores of global corporate empires. If the territorial states are facing demise, therefore the consensus would seem to be that large corporations will be their replacement. Under these emerging situations, the geographical explanation of state and society requires fresh attention from the young forward looking geographers. In the wake of increasing pressure of world economy system, the state in its traditional form has faced big problems. Conceptually it is going to disappear in near future. However, the real spatial dynamics at different stages and levels are creating many new challenges and inner dynamics for functioning of state as a strong and living element. The problem with globalization thesis on the demise of state is that it confuses state adaptation to new circumstances with the erosion of state (Taylor 1994). The modern state in its multiplicity is not eternal and will one day disappear when the modern world system reaches its demise. But in the meantime, the inner-state system is integrating with the operation of the world economy. Without multiple states, the economic enterprises would not have their windows of opportunity from state control that have allowed them to expand and prosper. Hence the ambiguous relationship between the territorial state, market and capital is the new reality of space which determines the geographical identity. Deutch (1981) observes, "the state are both indispensable and inadequate today and through at the history of the world economy".

For geographers, all these emerging spatial phenomenon and developmental debates are necessary to incorporate in elaboration for contemporary geographical explanation.

Technology vs. Geography

Technological changes as key for global product on system become the subject of much recent discussion on spatial study. The history of technological change is governed by the production of knowledge. This production of knowledge and its application have changed considerably over time. In this context, the division of labour might be considered a prerequisite for the inventive process; advances in science were not closely linked to technology and economic activity long before the nineteenth century (Usher 1994, 25). Before invention of science, discovery was initially regarded as the sudden and god-gifted from element genius mind. But Gilfillan (1935, 11) rejected this view, claiming that invention is more often the result of incremental problem solving within a relatively familiar framework of ideas and economic relation. Whitehead (1925, 96) advanced this formulation and said that the great invention of the nineteenth century was the invention of the method of invention. The production of knowledge was step by step going to be institutionalised in research and development. The growth in the system of technology could be seen as a major change in the nature of production and spatial dynamics. It also generates new ground for geographical analysis of a developing region. Through global analysis of knowledge it observed that it has tendency to differ over space and mostly technology has also followed the differentiated nature of geography. International differences in technology and the pace of technological change have somewhat clearly elaborated industrial specialization and the overall growth of trade indicating that technological capabilities are nation specific and it has been demonstrated that these capabilities are cumulative. In other words, they persist over time. Consequently industrial specialization and concentration at the sub-national level is, also clear (Cantwell 1991; Ellison and Glaeser 1997). Moreover, the existence of sub-national/regional differences in production techniques remains an open question. A number of studies have documented marked regional variations in labour productivity across a series of industries and at a variety of spatial sector (Careth 1984; Hulton and Schwati 1984; Moonaw 1983; Rigby 1992). Considering the various aspects of technological evolution we observed that technology is a product of human creativity and innovation. Though innovation is the application of new ideas to the economy, the rate of introduction of new technology in an economy and space have generated several debates on development as well as spatial studies. In this context growth, technological development, spatial reorganization, unemployment and process of displacement of society and economy have become vital issues for geography and geographers. The dynamics of long wave of economic activities is summarised in Fig. 3-2.

The above mentioned long wave of economic activity was first introduced by Kleinknecht (1987) who proposed the idea of continuous process of invention. These inventions have made significant impact on society and nature in different way. He also argues that innovation should not be considered a discrete event, because it is merely the prelude to a long series of international improvements that may have a far greater economic impact on the introduction of a new technology.

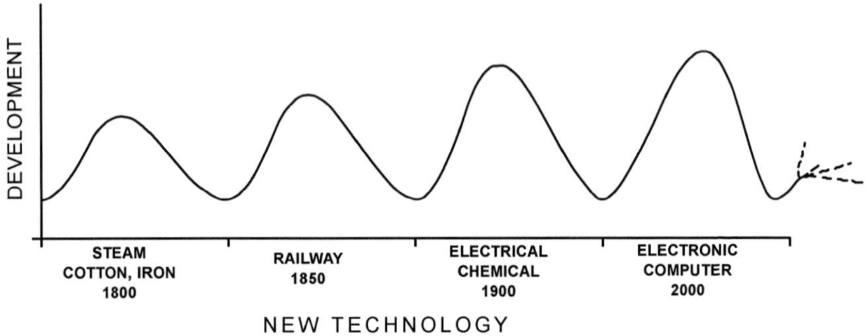

Fig. 3-2: Long wave of economic activity

The above Fig. 2 reveals the various stages of technological invention, i.e. steam power, electricity, electronic to computer and nuclear power, and its impact as a long wave of economic activity. Through spatial analysis of technological changes, it may be concluded that technology opens many new things in society and its trend is not linear and simple. Technological innovation, development in economy and dynamism in society have transformed a simple and stagnant social organization into a more complex, dynamic, antagonistic and conflict generative society which may be termed as developed and diversified society and economy. This emerging socio-economic phenomenon has long impact on the nature and pattern of geographical explanation of space. This trend and elements in geography need to be given greater attention by forward looking geographers. As technological and institutional regimes are produced and reproduced in space they are seen as imbued with distinctive geographical and historical characteristics, and this imparts a strong path of dependence on future trajectories of regional development (Arthur 1989). These elements in spatial studies should be taken into account to develop a new paradigm in geographical explanation

It is observed from the above discussion that the technological innovation of a specific region have a significant role in organizing the geographical space. The process of reorganisation of geographical space is becoming more important in the era of free movement of capital and technology. The role of foreign origin capital and technology in underdeveloped space is more or less untouched by contemporary geographers. The freedom of market expansion, investment of capital and transfer of technology are quite a new phenomenon in contemporary development studies. These new elements in development studies should be incorporated at micro- level spatial analysis by geographers.

Culture vs. Geography

The nature of geographical world has a heterogeneous character in its culture, language, other social structures and climate. The geographical diversification at micro-level may be relevant in case of some events, but the emerging space of globalization has posed serious threats and challenges for the local culture and language. The cultural identity of underdeveloped economy, society and individuals are going to disappear day by day. The process of disappearance of local and undeveloped culture has been initiated by developed countries of Europe and America. Today, global migration and non-territorial diasporas generate new identity agenda and its significance in political process. These identities transgress territorial boundaries, physically and electronically, and state-centric accounts of community. At the same time, identities are increasingly prominent in the analysis of market activities. On the one hand, consumption has always been gendered (Firat 1994). As McDowell (1994) argues, cultural theorists have a growing interest in how increasing globalization, especially of cultural production and consumption, affects people's sense of identity and place at both local and national levels. In this context, geographical perspective of cultural dynamics has become central theme to study the new emerging phenomenon of cultural changes. Some of the important features of these cultural changes and also contemporary challenges are finding ways of understanding the cultural impacts of globalization, rethinking now both the global and the local effects on cultures, finding productive and progressive and non-exclusionary ways of thinking about culture in the contemporary world, and re-conceptualizing cultural identities in increasingly global times.

The above issues are related with contemporary developmental process, and also create now discourse in the process of reorganization of global spatial and social order into different identity and place. The emerging global culture reveals that different places and cultural practices around the world are converging into similar Western culture. Shurmer-Smith and Hannam (1994) examined these cultural dynamics and concluded that the global culture might produce two different processes: (i) the expansion of superior cultural traits and products from advanced countries and their global adoption, i.e. Westernization, Americanization and modernization; and, (ii) The process of hybridization of culture through greater interconnections and time-space compression leading to a new universal cultural practice. Under these phenomenons, the Western cultural values, along with science, technology and capitalist mode of production are being globalized. The manufacture of globalized and universal products and development of global media empires are creating global cultural convergence—people around the world are becoming increasingly similar in terms of consumption, life styles, behaviour and aspirations. This process is very much associated with historical process of exploration, imperialism, colonialism, the creation of a "world market" forced migration and displacement and overall the Western societies' initiated global integration process. It has both positive (as modernization or development) and negative (as cultural imperialism or negation of level value and culture) implications. Under these grained realities, the role of geographer to identify these emerging issues as challenges to society is an important task. This is also a high time to prove the utility of the discipline.

In recent years, we have witnessed the increasing nature of de-centring of culture. The centre of cultural gravity lies at different poles and major cultural centres loose their significance and this process is producing greater cultural diversity in global space. Various cultural segments are now becoming more aware about their culture and its root in their society and space. Consequently, there are now more voices in defence of native culture, which reflects the cultural assertiveness of marginalized groups. Featherstone (1995) analytically observes, "if there is a global culture it would be better to conceive of it not as common culture, but as field in which differences, power struggles and cultural prestige contests are played out. Hence globalization makes us aware of the sheer volume, diversity and many sidedness of culture".

Moreover, the geographer should analyse culture as process that operates at a number of different levels—local, national and global. Therefore rather then conceptualizing a singular cultural landscape geographer should study the ways in which different cultural forms and processes come together in particular location, creating both fragmentation and fusion. In contemporary development process, the Western ideas and cultural practices are considered as a superior and role model for the underdeveloped world. Developmental process also, considers incorporating non-Western economies into global capitalism. On this account, the norms of Western culture have become hegemonic over local values. This process of cultural flow settles down toward local level and the dominant cultures marginalize others on the basis of ethnicity, sexuality, gender, etc. However, the dominant cultures also produce resistances that have potential to create new ways of thinking about culture.

Considering various aspects of global as well as local culture and its emerging conflict and convergence, a new dimension in geographical studies is required. Therefore, geographers should conduct global studies of culture, and try to high light the evolutionary and hybrid nature of culture and demonstrate the influence of marginal cultures on dominant culture. It needs negotiation between local, the national, the global and the historical as well as the emerging dynamics of Diasporas. In this way dominant cultures can also be subjected to critical analysis, rather than taken as the norm without any change. Moreover, the geographical tools for spatial studies have become more vital to meet the requirements of present day culture vs. space analysis. This creates both the challenges and opportunities to young geographers.

Development Space and Geography

The issues of development related with the phenomenon of social transformation in particular regions, where development processes originated, have major impact on regions. The term development is of Western origin and relates with material progress of a people or region. 'Development' is defined as "the process by which traditional society employing primitive techniques and therefore capable of sustaining only a modest level of be capita income is transformed into a modern, high technology, high income economy." Recently its meaning has been enriched by incorporating the advancement in quality of life, i.e. life expectancy, knowledge and comfort level among the peoples. It is measured by the Human Development Index (HDI). However, under the capitalist mode of production, uneven development, both in spatial and social terms has occurred and generate the process of marginalization. It also creates the process of disempowerment and displacement from native resources. Consequently, the natural environment of developing region/space is confronted with many challenges with opportunities. To address these spatial problems, geography and geographers may prove a suitable alternative in this regard. Geography as a holistic science integrating the physical and human dimensions of knowledge looks at the environment-development nexus not from narrow disciplinary perspectives but from systems' perspective. It aims at a course of human progress that keeps the earth liveable while at the same time allows human beings to enrich their life and living from cultural, economic and social perspectives, which in modern terminology is called sustainable development (Misra 2002). Therefore, modern geographic research should be reoriented in relation to the challenges of sustainable development, focusing particularly on resource inventorization, evaluation, environmental impact assessment projects (pre-project and post-project evaluation) and critical unpacking of the discursive practices underpinning the development and environmental conservation policies (Sharma 2007).

The emerging space of global economy accentuated the process of development in a faster way. The expanding economy faces many challenges and process of diversification in its socio-economic and cultural dimensions. In these situations, the analysis of development in a geographical paradigm has become complex and requires a different perspectives for the study of space and its various component. To elaborate these new challenges in space like India, it is most significant and urgent task of present day geography. Table 3-1 reveals the level of development index in different states. The ranking pattern of development levels reveals the nature of inequalities between different states (Table 3-1). On the basis of development index Kerala ranks first and Bihar stands at the 15th place among the Indian states.

In the same way, Table 3-2 depicts the prevailing structure of food consumption of rural population in different Indian state. It observed that the rural population of Orissa (88%) Rajasthan (83%) Uttar Pradesh (83%) West Bengal (81%) have gained more than 80 per cent of recommended calories norm while on other hand the states like Kerala, Andhra Pradesh, Karnataka and Tamil Nadu have only received 60-70 per cent of required value of calorie. Table 3-1 and 3-2 reflect the spatial and structural pattern of uneven development in rural India. In this context, more inequalities in terms of gender related structural pattern of Indian society is apparent (see Table 3-3). It is revealed that on the basis of Overall Development Index (ODI) Punjab has secured first ranking, but its Women's Status Index (WSI) was at the third position. This emerging pattern of development explains that the overall development in Punjab have shown its gender biasness. On other hand, Kerala has shown gender equality in its development process. Through above observations, it is clear that spatial study should not be only confined towards the trend of growth, but also considers the different structural aspects of society. For this purpose, a fresh and creative theory and practice in our discipline is an urgent demand from crisis ridden society.

Hence, as region is now characterized as highly complex. Due to this complexity of space/region a detailed analysis of the most common problems should be undertaken and identified in bigger way by geographers. These are deep rooted poverty, population growth greater than resource invention, scarcity of food, slow growth in per capita income, high and increasing unemployment, low quality of human development, micro- level spatial problems of environmental degradation, slow growth in non-agricultural income, slow shifting of agricultural population to secondary and tertiary sector and other problems of spatial as well as social deprivation, etc.

The emergence of Naxal movement and their intensive impact in 165 districts of India located in the richest resource region (minerals, forests land and water) are becoming day by day a major threat to the security of our country. The reason of conflict in this region is related to the socio-economic and political deprivation of the concern population. These issues should be taken up by geographers because of their training in social and spatial field. Soja has rightly advocated for the need of bigger space for geography in social sciences (1989: 155). But geography as discipline faces problems. Lahiri-Dutt (2005/2009) remarks:

> "The satellite images are digitally processed to provide geographical data that are then stored and retrieved for use in maps through computer software. On the other hand, as the landscape is increasingly understood to be not just a pre-existing world, geographers are now successfully involved in exploring the perceived worlds of the human mind. How one sees the world and what is made of it, are both stimulating subjects of study, leading us from the macro to the micro, from the aggregates to the individual, from space to place. The objects of human activities–cities and towns, roads, factories are no longer the main focus of attention, but human themselves, that is how and why we behave as we do. In other worlds, human geography has indeed become much more human. In India, our application orientation has endured that only the first group of research is patronised and consequently the use of satellite data and geographic information system (GIS) is the branch that is flourishing".

These emerging views on geography remind us that the Indian geographers have conducted so many empirical studies about region without the required understanding of social dynamics and its theory. Consequently, they have been unable to spell out policies even in areas like environmental studies, which is one of the major fields of the discipline. Moreover, the Indian geographers have not done enough theoretical work which can develop an indigenous thought to explain the prevailing spatial conflict in the Indian reality.

While the Western theory has been inadequate in conceptualizing question of Indian society and space; the histories and geographers of "spaces" in India are now becoming more important as a new discipline. In this context, the Indian Geography will become relevant if it gets strength to meet and explain the contemporary issues of spatial conflict and challenges.

Table 3-1: India—Human Development Index by States, 2001

	State	Values	Rank
1.	Andhra Pradesh	0.416	10
2.	Arunachal Pradesh	-	-
3.	Assam	0.386	14
4.	Bihar	0.387	15
5.	Goa	-	-
6.	Gujarat	0.476	6
7.	Haryana	0.506	5
8.	Himachal Pradesh	-	-
9.	Jammu and Kashmir	-	-
10.	Karnataka	0.478	7
11.	Kerala	0.638	1
12.	Madhya Pradesh	0.394	12
13.	Maharashtra	0.523	4
14.	Manipur	-	-
15.	Meghalaya	-	-
16.	Mizoram	-	-
17.	Nagaland	-	-
18.	Orissa	0.404	11
19.	Punjab	0.537	2
20.	Rajasthan	0.424	9
21.	Sikkim	-	-
22.	Tamil Nadu	0.531	3
23.	Tripura	-	-
24.	Uttar Pradesh	0.388	13
25.	West Bengal	-	-

(*Source*: India: Human Development Report, 2006).

Table 3-2: Food consumption by the below poverty-line class, 1999-2000

	State	Calorie /day	Normal consumption[1] (%)	Food expenditure month	% of total expenditure
1.	Andhra Pradesh	1662	69	183.28	70
2.	Assam	1767	74	253.98	70
3.	Bihar	1977	82	219.65	66
4.	Gujarat	1884	70	213.73	67
5.	Haryana	1745	73	219.20	60
6.	Himachal Pradesh	1942	81	227-16	62
7.	Karnatka	1646	69	199.42	64
8.	Kerala	1389	58	226.37	60
9.	Madhya Pradesh	1888	79	199.24	64
10.	Maharashtra	1780	74	196.83	62
11.	Orissa	2117	88	221.50	68
12.	Punjab	1712	71	221.29	61
13.	Rajasthan	2003	83	228.33	66
14.	Tamil Nadu	1543	64	208.68	68
15.	Uttar Pradesh	1990	83	195.04	58
16.	West Bengal	1935	81	253.83	72

Note: 1. Percentage of people below poverty-line class who consume 2400 calories or more per day.

(*Source*: NSS Report, Government of India, 2001).

Table 3-3: Overall development and women's status compared, 2000

	State	ODI	Ranks	WSI	Ranks
1.	Andhra Pradesh	0.074	8	0.073	7
2.	Bihar	-0.523	14	-0.788	14
3.	Gujarat	0.445	2	0.253	5
4.	Haryana	0.189	6	0.071	8
5.	Karnataka	0.000	7	0.147	6
6.	Kerala	0.164	5	0.943	1
7.	Madhya Pradesh	-0.312	10	-0.269	10
8.	Maharashtra	0.297	3	0.341	4
9.	Orissa	-0.471	13	-0.292	11
10.	Punjab	0.831	1	0.381	3
11.	Rajasthan	-0.227	9	-0.216	9
12.	Tamil Nadu	0.150	4	0.466	2
13.	Uttar Pradesh	-0.466	12	-0.763	13
14.	West Bengal	-0.211	11	-0.346	12

(*Source*: Kumar, 2006: 1-14).

The Future Agenda

The emerging challenges of contemporary world have created many opportunities for creative research work in geography. This is high time for geographers to shift their research agenda towards new areas like exploring the relationship between the demand of resources and production of nature as resource, identification of social orders and hierarchies, spatial pattern and its process with historical background, core and peripheries, variety of physical landscapes, use and misuse of resources, mapping of climate change related activities, disaster management, urbanization and its problems, regional growth, rural development and displacement and spatial and social conflict these are some emerging issues which can be solved by geographers in coming days, because they have latest tools like GIS, GPS and geoinformatics. In spite of these things, geographer may prove as an asset to a society by solving the problems of spatial conflict and socio-cultural deprivation, because they have the knowledge and formal skill to study the spatial dynamics. Kapur (2004) described geography as neither an exclusive physical nor pure social science. Its numerous systematic branches of study include geomorphology, climatology and biogeography along with fields like political and social and urban geography. By definition geography does occupy an acetone between the physical and the human realms but there is no rule or law, which states that a science concerned with the synthesis of both is lesser than any one of them. This blurred perception could easily be altered if geographers had fostered an active interaction with social sciences and demonstrated that earth is not, more physical entity but essentially a home of human kind. On this basis, geographers should reorient their work towards human beings—for their justice and welfare.

The above observations indicate that the future agenda in geographical inquiry should be orientated towards following themes (cf. Kapur 2004):

- Just after second world in 1950, the agenda of geographical studies was to initiate the process of regional growth now in 2000 and onward it is for growth with social justice.
- In 1950 it was nation state and national sovereign now in 2000 and onward it is globalization and localization.
- In 1950 it was political boundaries in contemporary world it is political space/region with contradiction.
- In 1950 external threat was major concern for political geographers now these challenges come from socio-economic contradictions from ground reality.
- In 1950 population explosion was big question for Third World countries, now the issues of population displacement through developmental activities is major concern for geographers
- Previously the issues of industrialization were main concern for geography. Now the focus has changed. The main task of geographers shifts towards the analysis of constraints in agricultural land due to fast growing industrialization under the impact global capital.
- The increasing demand for bio-fuel have developed stress on food security especially for less developed countries, and also insatiability for agricultural practices, rational cropping pattern and further development of present economy. These emerging issues have become an urgent agenda for scientific orientation of geography. It should be taken into account by young geographers.

Moreover, geographers should reinvent their commitment for the subject and relate geography with people's agenda. Today, geography and geographers have tremendous potential for solving the contemporary problems. Lastly, discipline of geography has all required elements to become the leader in developing society.

References

Arthur, B. 1989. Competing technologies increasing returns and lock-in by historical events, *Economic Journal* 99: 116-31.

Brown, L.R. 1973. *World without Borders*. New York: Vintage.

Christaller, W. 1966. *Central Place in Southern Germany* (translated by C.W. Baskin). Englewood Cliffs, NJ: Prentice Hall.

Deutch, K.W. 1981. The crisis of the state. *Government and Opposition* 16: 331-43.

Driver, F. 1988. The historicity of Human Geography. *Progress in Human Geography* 12: 497-506.

Featherstone, M. 1995. *Undoing Culture: Globalization, Postmodernism and Identity*. London: Sage.

Firat, A.F. 1994. Gender and consumption: transcending the feminine. In *Gender Issues and Consumer Behavior,* ed. Arnold Costa, 205-228. Provo, UT: Association for Consumer Research.

Gerasimov, I.P. 1983. *Geography and Ecology*. Moscow: Progress Publisher.

Gilfillan, S. 1935. *The Sociology of Invention*. Chicago: Follett.

Johnston, R.J. 1982. *Geography and the State*. London: Macmillan.

Kapur, Anu. 2004. Geography in India: A languishing social science. *Economic and Political Weekly* 39 (37): 187-4195.

Kleinknecht, A. 1987. *Innovation Patterns in Crisis and Prosperity*. London: Macmillan.

Kumar, Momita P. 2006. Regional disparity in the status of women across Indian states. *Indian Journal of Regional Sciences* 38 (2):1-14.

Lahiri-Dutt, Kuntala. 2005/2009. Geography as a marginal science. *Economic and Political Weekly* 40 (7): 689-691. Reprinted in *Indian Geography: Perspectives, Concerns and Issues*, ed. Ravi S. Singh, 242-250. Jaipur/New Delhi: Rawat Publications.

Maddison, A. 1982. *Phases of Capitalist Development*. Oxford and New York: Oxford University Press.

Mander, S. and E. Goldsmith, ed. 1996. *The Case against the Global Economy: And for a Turn to the Local*. San Francisco: Sierra Club.

Massey. D. 1984. *Spatial Division of Labour*. London: Macmillan.

McDowell, L. 1994. The transformation of cultural geography. In *Human Geography: Society, Space and Social Science*, ed. D. Gregory, R. Martin and G. Smith, 146-73. London: Macmillan.

Misra, R.P. 2002. Indian Geography: Problems, challenges and opportunities. In *Indian Geography: Voice of Concern*, ed. A. Kapur, 350-362. New Delhi: Concept Publishing Co.

Mohammed, Ali. 2007. Uneven economic development and deprivation: Its impact on prospects of peace. Moonis Raza Memorial Lecture at the 29th Indian Geography Congress, February 29-March 2, in Udaipur (Rajasthan).

Ollman, B. 1993. *Dialectical Investigations*. London: Routledge.

Pounds, N.J.G. 1954. France and 'les limites naturelles' from the seventeenth to the twentieth centuries. *Annals of the Association of American Geographers* 44: 51-62.

Peet, R. 1977. *Radical Geography*. Chicago: Maroufa Press.

Perkins, J.H. 1997. *Geopolitics and the Green Revolution: Wheat, Genes, and Cold War*. New York: Oxford University Press.

Pounds, N.J.G. 1951. The origin of the idea of natural frontiers in France. *Annals of the Association of American Geographers* 41: 146-157.

Sassen, S. 1996. *Losing Control? Sovereignty in an Age of Globalization*. New York: Colombia University Press.

Sharma, H.S. 2007. Repositioning geographical research: Environment, development and sustainability. Presidential address delivered at the 29th Indian Geography Congress, February 29-March 2, in Udaipur (Rajasthan).

Shurmer-Smith, P. and K. Hannam. 1994. *Worlds of Desire, Realms of Power: A Cultural Geography*. London: Arnold.

Skelton, T. 1996. "Cultures of land" in the Caribbean: A contribution to the debate on development and culture. *The European Journal of Development Research* 8 (2): 71-92.

Soja, Edward W. 1989. *Post-Modern Geographies: Reassertion of Space in Critical Social Theory*. London: Verso.

Taylor, P.J. 1994. The state as container, territoriality in the modern world system. *Progress in Human Geography* 18 (2): 151-162.

—. 1995. Beyond containers: international interstateness, interterritoriality. *Progress in Human Geography* 19 (1): 1-15.

von Thünen, J.H. 1966. *Isolated State*, trans. C.M. Wartenberg, ed. P. Hall, Oxford: Pergamon Press.

Trevor, J Barnes and Eric Sheppard. 2003. *A Companion to Economic Geography*. Oxford: Blackwell Publishing.

United Nations Development Programme 1993. *Human Development Report, 1993*. New York: Oxford University Press.

Usher, A. 1954. *A History of Mechanical Invention* (second edition) Cambridge, MA: Harvard University Press.

Peterson, V. Spike. 2007. Shifting(s), remapping strategies and triad analytics. In *Globalization: Theory and Practice*, ed. G. Young and E. Kofman, 107-122. London: Pinter.

Weber, A. 1929. *Alfred Weber's Theory of the Location of Industries*, trans. C.J. Friedrich. Chicago: Chicago University Press.

Whitehead, A.N. 1925. *Science and the Modern World*. New York: Macmillan.

Williams, M. 1996. Rethinking sovereignty. In *Globalization: Theory and Practice* ed. G. Young and E. Kofman, 109-22. London: Pinter.

CHAPTER FOUR

UNDERSTANDING SYSTEMATIC INTERACTION OF SPATIALLY DISTRIBUTED PHENOMENA

RAMKRISHNA MAITI,
VIDYASAGAR UNIVERSITY

Introduction

Geographers, the spatial scientists, have to select, collect, represent and interpret information (data) in order to make a synthesis to correlate or fit the observation into the systematic inter-working of observed phenomena. Selective attitudes towards geographic information are guided by some intertwined theoretical and methodological believes and approaches commonly practiced by a majority of scholars and researchers. Presently the researches in geography show a trend of diversified approaches juxtaposed each other resembling a post-paradigm phase in the evolutionary history of philosophical growth as a scientific discourse. Although, the postmodern ideas of spatiality offer enormous potential for geography's growth as a spatial science with the use of GIS, the conflicting methodologies hinder the growth of the subject as a unified science and sometimes the social relevance of the researches are lacking. And, hence the importance of geographers as decision-makers is often denied. The job opportunity only in academic field often fails to attract and retain talents in research. The coherence and integration of conflicting ideas are therefore earnestly required to guide excellent research so that geographers can deserve more importance as planners and spatial managers. In the present discussion, the evolution of methodological philosophy and the present trend of the integration of theoretical approach and objective of research with the entire processes of data collection, information extraction and interpretation through temporal and spatial synthesis are exemplified with the author's experience of working in different fields. The potential of future research in postmodern domain and the possibility of coherence through cognitive approach too are assessed with some examples from empirical researches.

Methodological Paradigm and Selective Attitude towards Geographic Information–The Way to Present

The selection of information and the method of analysis depend largely on the approach of study. What is observed depends not only on the context in which a particular phenomenon is set but in the manner is which one is prepared to view it. During the earlier stage of the growth as a discipline, without any distinct methodology, all the available information seemed to be equally important for inclusion in geographical study and research. The information about the countries, oceans, mountains, solar system and other landscapes etc. during the exploration by adventurers, navigators and explorers have equal probability of inclusion in geographical study and thus enriched the geographical encyclopaedia without any selective approach. Without any developed paradigm "all the available facts may seem equally likely candidates for inclusion" (Haggett and Chorley 1967). The professionalisation phase experienced some discrete method of data collection, recording and analysis, e.g. lake level information in relation to climate etc. in a way to subsequent emergence of well defined methodology of determinism for analysing man nature relation. It is the exceptionally marked paradigm phase, the American geography has experienced so far "with the possible exception of the period before 1920 when environmental determinism achieved a measure of intellectual hegemony. American geography has never been a normal, paradigmatic science of the kind postulated by Thomas Kuhn (Abler, Marcus and Olson 1992). The information or data substantiating the dominance of nature over man and interpreting and supporting the processes of determination of man's action and attributes are deliberately selected and the researches necessarily took a particular direction guided by this well defined methodology since 1920. The treatment of information starting from collection, collation analysis and representation etc. and finally drawing conclusion certainly experienced a shift in deductive methodology form earlier inductive methods propounded in pre-paradigm phase.

The chorological approach with its empiricist and implicit exceptionist philosophy, dominated the discipline until the 1940s (Johnston and Sidaway 2004). The information establishing the areal differentiation both in physical attributes [regional theme dominated in Britain (Johnston and Gregory 1984)] and in cultural landscape (regional theme dominat*ed* in *the* USA through the Berkley School (Martin 1998) have enjoyed selected and deliberate importance through the process of regional synthesis by understanding of a place "through knowledge of the general processes which interact to produce unique characteristics". During 1950s and 1960s "spatial theme" superseded the regional theme connected with the re-emergence of systematic study with an increasing emphasis on finding laws of spatial organisation involving distance as a basic influence on human behaviour. The quantitative and precise information about objective geographical phenomena and subsequent analysis with statistical, mathematical and geometric techniques to formulate universal laws or model are some of the great achievements in geography following positivism.

The new disciplinary matrix of spatial science was firmly established in Anglo-American human geography by the end of 1960 and has been sustained since. Over nearly 40 years it has seen many shifts in exemplars, a lot of them linked to methodological, especially technical development in data collection, collation, analysis and display. Such paradigm shifts at the level of exemplar are closer to Kuhn's original presentation with new ways of doing research accepted as superior to those previously used (Johnston and Sidaway 2004). Those include the radical, behavioural and functional approaches. Radical approach emerged as a critical development to the established quantification and has more orientation towards human and social welfare and historical materialism following Marx (Peet 1977) and so information collection and analysis was made accordingly. The same information (stimuli) may yield different human behaviour according to his/her perception followed by culture, tradition and experience and thus geographical study took an orientation towards non-quantifiable subjective information (Kirk 1951, 63). The spatial science disciplinary matrix experience changes in a wider base. The first was the reorientation away from normative modelling, which involved testing observed spatial organisation against a priori-model and towards behavioural studies. The second was the advocacy of "welfare geography" as a response to social condition and a desire to make contemporary human geography more relevant to them. During 1950s and 1960s human geography experienced a new "disciplinary matrix" with twin concentration of spatial organisation of society and human spatial behaviour. The spatial scientists tend to reduce man "to terms in equations" ignoring their individuality and freedom of action. The information are collected and analysed through the application of "power of reason and rationality of scientific methods" based on logic and mathematics involving the belief of "autonomous, sovereign and self-consciously directed individuals able to make same moral and other judgements" for finding out universal truth in an orderly sequential world. D. Gregory prescribed that geography should concentrate on "social organisation of space" rather than "spatial organisation of society" practiced in quantitative domain. As a criticism to this positivist spatial science, the concept of humanistic geography came into play in mid-1970s with challenging the ideas of "power of rationality and reason", monolithic order, universal truth and of "rational man". Another challenge to the spatial science disciplinary matrix comes from radical geography which is concerned with the failure of positivist spatial science to tackle and solve pressing societal problems. The core of radical disciplinary matrix is to uncover the mechanism that drives society, providing accounts of how people act and how the empirical world is organised. Since 1990 geography witnessed turbulence in methodological belief in a rapidly changing economic, environmental, political, technical and cultural scenario as a result of increased interaction through globalisation (Johnston and Sidaway 2004). In the 20th century, it was collative geography trying to discipline randomly collected data, arranging and rearranging them so as to make them meaningful using subjective approach followed by quantitative methods (Munsi 2000).

Through globalisation we are moving towards a situation in which our world is being turned into a village, cutting across the barrier of space and time, and geographers will have to adjust their expertise accordingly. The world village concept now redefined as "Space-Ship-Earth" brings home the danger *of in* a closed space of overstepping the boundary of sustainable man-environment relationship in the terrible consumer orientation of few surrounded by the abysmal poverty of the overwhelming majority (Munsi 2004).

Researches at Present: Information Collection, Collation and Representation

Role of geographers to generate proto-type data through field survey

In geography, the information on the physical attributes is mainly collected through intensive field survey. The measurement, tabulation and representation of the data follow some standard methodology in order to understand the general process of interaction and the resultant effects reflected in sensible form. The same form may even be changed with minor deviation in any of the interacting factors. Presently geographers try to measure that change occurred in temporal (short-time) span by repeated field survey and to identify and prioritize the factors according to their intensity of effects.

In the present discussion, all the above methodological objectives are exemplified through the empirical study on the temporal evolution of mining pit on Himalayan Slope at Tindharia, Darjeeling. The synthesis is drawn by understanding the interaction among physical and human processes though repeated field survey during 1998-2006. It follows the deductive methods to substantiate the extent and magnitude of destructive effects of mining on slope, soil and water etc. The data collected from repeated field survey are tabulated (Table 4-1) and can also be represented graphically to show temporal evolution (change) both in profile and area.

The diagrams, graphs and tables help to appreciate the season wise relative importance of physical as well as human processes. During dry season human actions are more active through mining and associated dissection, excavation and drilling etc. But during rainy season, as the shear strength of the slope forming materials reduces, the safety factor reduces to less than one and thus slide occurs to attain temporary equilibrium slope.

Each of such changes in plan of the slide scar on mining pit is due to the effect of rain induced saturation and subsequent instability. Major changes occur during February, 2000 to June, 2004. This understanding through proto-type survey helps in cognition into the processes active seasonally for evolution of the slope, surrounding Mining Pits and can be expressed through cognitive model (Fig. 4-1).

Survey with structured questionnaire and generation of primary data

The information about economic and social attributes is collected through surveys with structured questionnaire. It, again, can be exemplified with the prototype study of illegal mining on hill slope. The collection of information about the illegal action, specially about 'the families of direct involvement' in illegal mining, is not easy to access and so, after a continuous effort, it was possible to survey all the 85 families settling at Tindharia Cricket Colony, Kurseong (Darjeeling) with the active help from local students of Kurseong College being guided by the author, the then teacher of that college. One indirect inference was made to identify the families "directly involved in mining". The families having almost no expenditure for fuel and satisfying almost 100% of total fuel need from locally available coal can be identified as the families of direct involvement. Seventeen of such families were identified out of a total of 85 and are characterized by low income and poverty (Table 4-2). The poor families have no alternative but to extract coal either for domestic use or for income by selling at local or neighbouring market. After realizing this information, the study took a *humanistic* bend searching for scientific recommendation for mining to cause less harm to environment rather than the suggestion of complete closure from ecological and legal point (Maiti 2004).

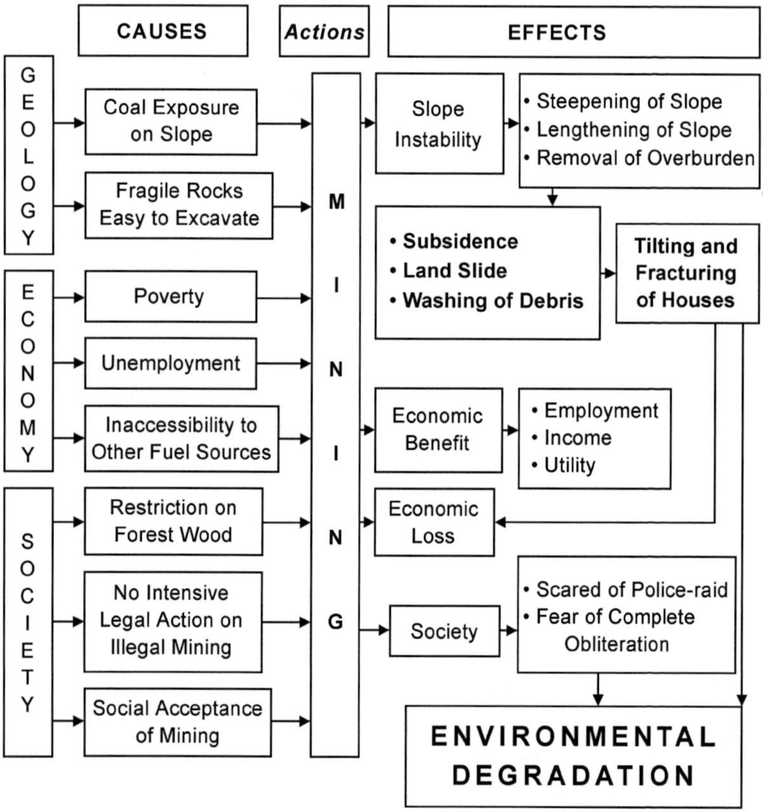

Fig. 4-1: Cognitive model expressing effect of mining (with cause and effect relationship) involving selective attributes of geology, economy and society

Data generation and verification through laboratory analysis and laboratory experiment

The data required for geographical research is also generated through tests in laboratory. The samples of rocks, soils, and water are collected with proper care and are tested in laboratory to harness required information (Fig. 4-2) which is precise in nature and quite essential to achieve cognition about the concerned phenomena and their relation and ultimately guides to the formation of universally applicable laws, theory and models. The process of generation of *Trans links* and *Cis links* in association with slope and texture is understood through a laboratory experiment on 1m x 1m area with 50cm thick prototype soil and controlled rain. The stages of development of channels (Branching) are thus studied and mapped accordingly (Fig. 4-3A, B & C).

Temporal synthesis–dealings with limited information

Non-availability of data is a compulsory outcome of selective approach towards geographic information. After adoption of a paradigm, a standard theoretical belief and methodological practice, the selective information or data is necessary. One of the curious effects of paradigm adoption is to replace a period of data abundance with one of data scarcity. In the case of physical science such problems are commonly accompanied by design and construction of measuring apparatus of increasing precession and complexity (Hagget and Chorley 1967).

The social sciences particularly geography generates primary data by field survey on physical, economic and cultural phenomena (Table 4-1, 4-2 and 4-3) to compensate the non-availability of information and to collect required information well suited to the existing methodology. Sometimes geographers have to depend on information from official and other sources and the non-availability of such information for required temporal coverage may stand in the way of smooth progress of research and sometimes may yield unsatisfactory result. One example of Magnitude-Frequency analysis of flood at Silabati river basin, Paschim Medinipur, West Bengal, at its lower catchments may be cited in this connection (Ghosh and Maiti 2006).

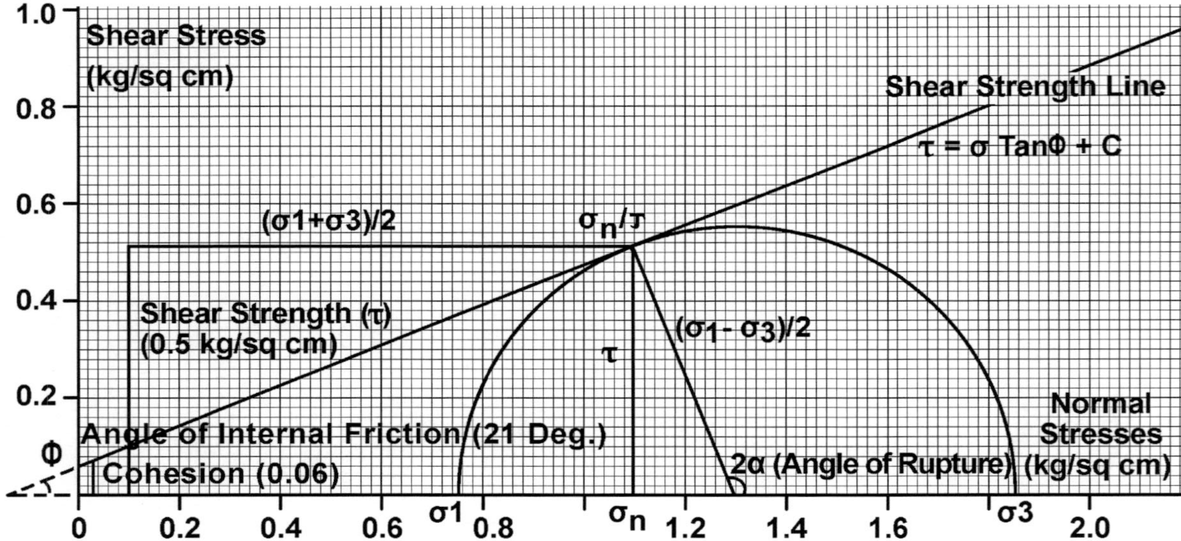

Fig. 4-2: Graphical representation of the results of the tri-axial test.

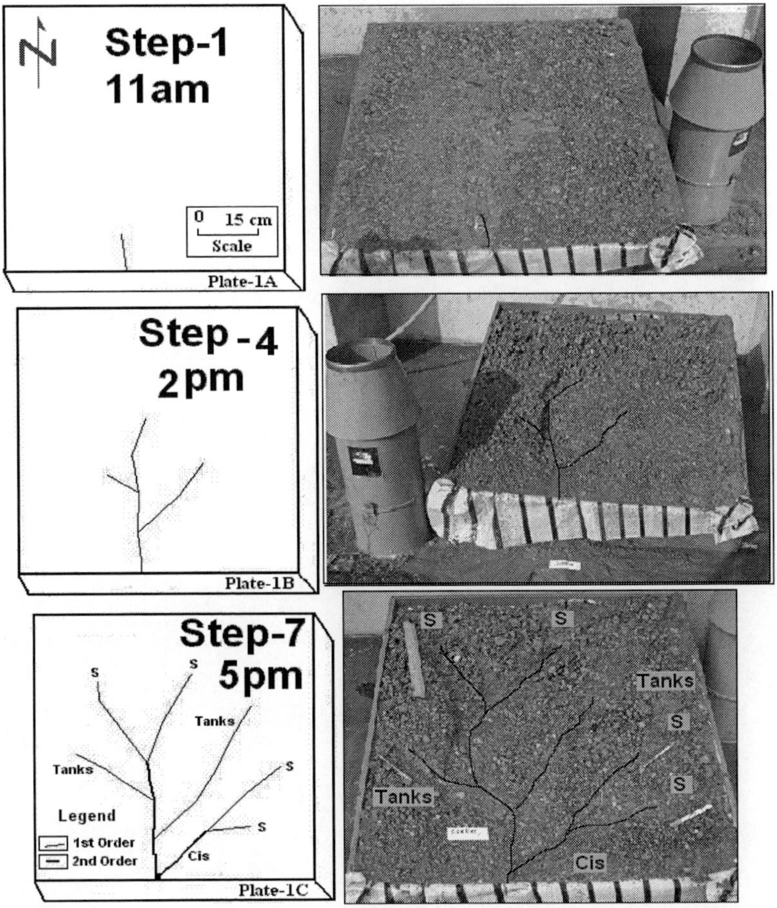

Fig. 4-3: Temporal evolution of branching pattern of a gully in the controlled laboratory environment.

Table 4-1: Temporal evolution of mining pits: the primary data generated through field investigation (The result of repeated field survey at mining scar on slope below Tindharia cricket Colony)

	Feb,2000	Oct,2000	April,2001	Oct,2001	May,2002	Jan,2003	Feb,2005	Jan,2006
Max.length (m)	12.5	15	21.5	24	32.4	39	39.5	40.2
Max. width (m)	10	12	21	21.6	22.5	25	25	25
Max. Depth(m)	5.75	3.75	5	3	3.5	1.5	1.5	1.5
Min. Depth (m)	1.25	0.75	0.5	1	0.75	0.5	1	1
Avg. Depth(m)	3.5	2.25	2.75	2	2.12	1	1.25	1.25
Area affected(m^2)	72	79.92	184.5	300.37	378	401	407	410
Vol.displaced (m^3)	253	179.9	507.38	600.74	801.20	401	508.7	512.5
Angle on the scar face (Avg) in Degree	45	23	35	22	20	21	20	20.5
Slope on the back wall in Degree.	60	30	55	25	35	-	-	-
Height of back wall (m)	3.5	0.25	4.2	0.30	0.2	-	-	-
Concavity of the scar	5.76	5.33	8.58	12.52	11.67	10.28	10.30	10.38
Remark	Mining hole	Slide during Rain	Re-excavation and Mining	Slide cum slump during rain	Mining and re-opening of pit	Pit abandoned due to huge slide in rainy season	Pit abandoned	Pit abandoned

Table 4-2: The directly involved families in mining
(Identified through indirect inference).

Sl. No.	Quarter No.	No. of family members	Annual Income in Rs.	Profession	Division of Annual Expenditure			Use of fuel				Source	
					Basic Need	Education	Fuel	Coal	Gas	Ker.	Elec.	Coal	Gas.
1.	77	5	18000	Purahit	6000	2000	Nil	100%	-	-	-	Local	-
2.	Private	6	42000	Ex Railway service	24000	12000	Nil	100%	-	-	-	Local	-
3.	73	6	18000	Labour	10000	40000	Nil	100%	-	-	-	Local	-
4.	Private	4	10800	Labour	8400	2000	Nil	100%	-	-	-	Local	-
5.	128	5	36000	Business	22000	2000	Nil	90%	-	10%	-	Local	-
6.	73	11	48000	Govt. service	40000	8000	Nil	90%	-	10%	-	Local	-
7.	Private	5	60000	Photo grapher	48000	2000	Nil	100%	-	-	-	Local	-
8.	100	4	12000	Labour	10000	2000	Nil	100%	-	-	-	Local	-
9.	Private	9	14400	Driver	12000	2000	Nil	100%	-	-	-	Local	-
10.	-	4	12000	-	9000	1400	Nil	90%	-	-	-	Local	-
11.	-	4	42000	Business	28000	10000	Nil	95%	-	-	-	Local	-
12.	97	10	42000	Business	30000	2000	Nil	80%	-	20%	-	Local	-
13.	26	7	30000	Ex employee	21000	3000	Nil	100%	-	-	-	Local	--

14.	-	6	46000	-	27000	10000	Nil	80%	-	20%	-	Local	-
15.	108/8	5	60000	Railway employee	35000	5000	Nil	100%	-	-	-	Local	-
16.	43	5	24000	Labour	15000	4000	Nil	100%	-	-	-	Local	-
17.	117	6	36000	Railway employee	30000	2000	Nil	100%	-	10%	-	Local	-

(*Source*: Maiti, 2004)

The flood discharge data for last 13 years were available (previously) and the analysis shows that the flood of 1993 with1130.35 cumecs discharge is the worst of all and may be expected at a recurrence interval of 15 years i.e. in 2008 following Gumbel (1954) (Table 4-4). But according to Mockus (1960) a minimum of 16 years' data is required for such analysis to yield maximum satisfactory results. Subsequently the same analysis with 38 years' data shows the flood of 1993 may have a recurrence interval of 9.5 years. So information of sufficient quantity with acceptable quality is needed for smooth progress of social science research and for yielding acceptable results.

Table 4-3: The result of a tri-axial compression test

Sl. No.	Parameters	Result
1.	Major Principal Stress(Kg./ Sq cm)	1.83
2.	Minor Principal Stress(Kg./ Sq cm)	0.76
3.	ormal Stress(Kg./ Sq cm)	1.10
4.	Angle of Rupture (Degree)	56
5.	Angle of Internal Friction (Degree)	21
6.	Cohesion (Kg./ Sq cm)	0.06

(*Source*: Maiti & Mondal 2006)

Spatial synthesis through mapping

Map may be considered as the spatial synthesis of collected information and can be exemplified from a prototype study in the Indian part of Sundarban. The study on the causes of embankment breaching at Sundarban (West Bengal) reveals that internal channels, which could once accommodate huge tidal water, are now incapacitated completely due to gradual choking and complete separation from the parent streams by erecting strong cross embankment. The study on the mechanism of choking and its temporal sequence are made through the structured questionnaire survey. The roads of present day across the choked channels are the then cross embankments and indicates the stages of occupancy by human settlement. The sequence of the longitudinal embankments are also the indicator of the stages of shift of land use within the segments between cross embankments, from fishery to cultivation and ultimately to homestead. The queries about the time of reclamation and the time of initiation of different land uses yield coercive answers and are plotted accordingly on the *mouza* map to get a spatial synthesis revealing the mechanism and stages of choking of the channels clearly (Fig. 4-4).

Table 4-4: The flood-frequency analysis with limited available information

Year	Peak Discharge Banka In Cumecs	Discharge In Cumecs	Ranking Order	Recurrence Interval (Year)
1993	1130.35	1133.01	1	15
1994	958.93	1130.35	2	7.5
1995	1093.41	1119.74	3	5
1996	1133.01	1098.72	4	3.75
1997	1007.38	1093.41	5	3
1998	948.89	1007.38	6	2.5
1999	1098.72	1006.83	7	2.14
2000	1119.74	993.00	8	1.88
2001	912.60	992.78	9	1.67
2002	943.90	973.78	10	1.5
2003	992.78	958.93	11	1.36
2004	973.78	948.89	12	1.25
2005	993.00	943.90	13	1.15
2006	1006.83	912.60	14	1.07

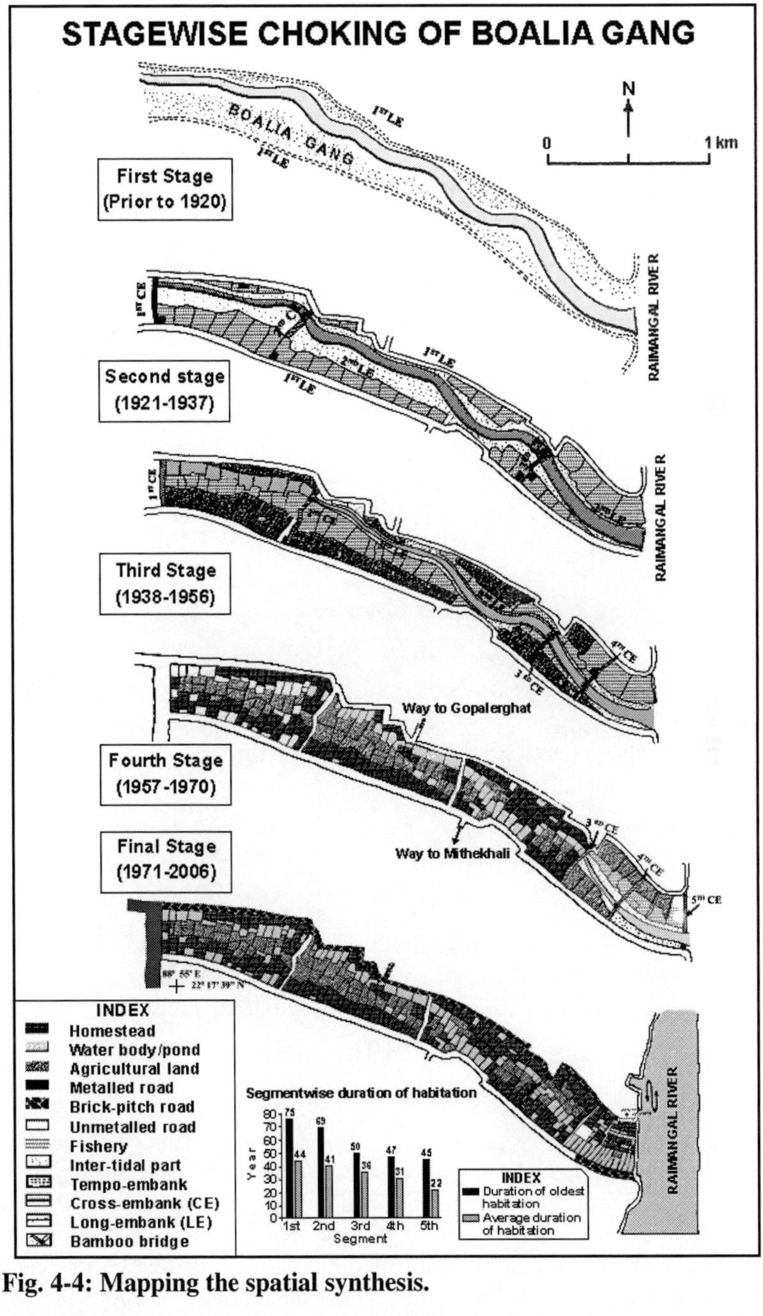

Fig. 4-4: Mapping the spatial synthesis.

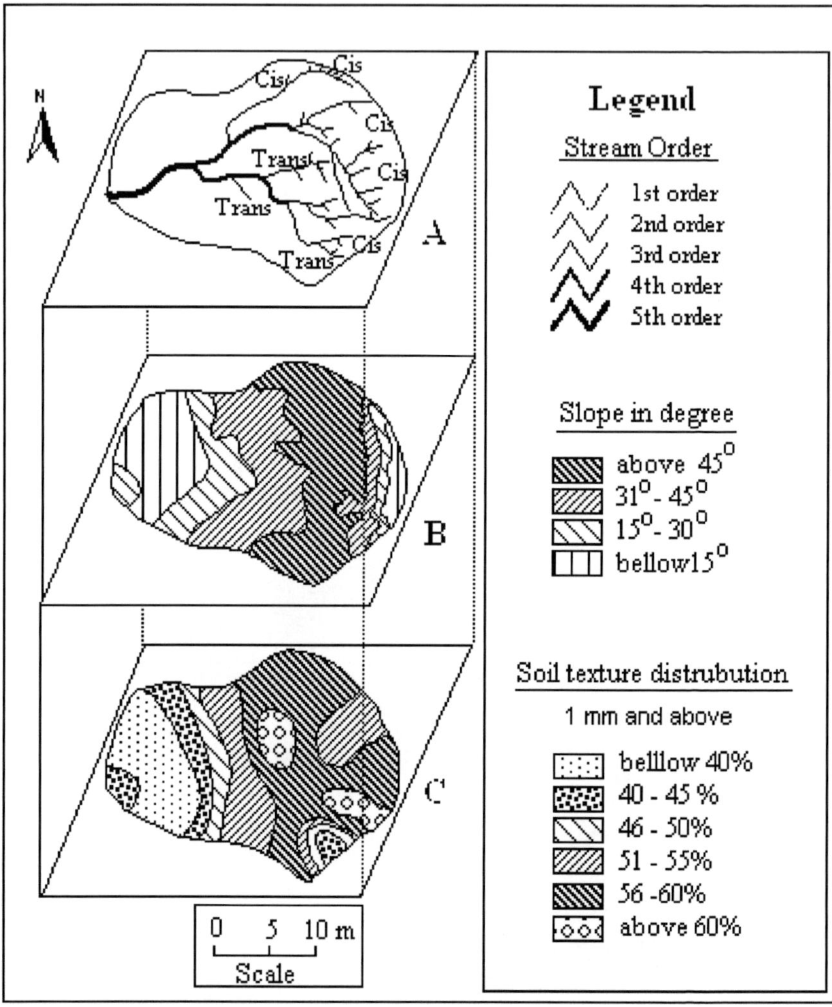

Fig. 4-5: The synthesis of spatial analysis using GIS techniques.
(*Source*: Field survey, 2006-08)

Introduction of GIS techniques enhanced the possibility of precise spatial synthesis, spatial integration and correlation through overlaying a number of geographical phenomena. The classification scheme of links in a drainage system after Abraham is applied on a precisely surveyed gully-basin of Rangamati area, Paschim Medinipur. The distribution of *Trans* and *Cis* links are developed on the spatial distribution of slope and soil texture and a spatial co-ordination is thus depicted through GIS Techniques (Fig. 4-5A, B & C)

Coherence of Conflicting Methods through Cognitive Model

In the 21st century, geography must be turned into 'cognitive geography' a science endowed with a true understanding or clear insight into the working of phenomena, which now characterises the writing of Humanistic Geography contrasted with positivism. Cognitive geography will possibly include not only studying a natural phenomenon but also the study of manifestations of man's activity comprising the physical and biotic environment together with those phenomena altered or created by human endeavour (Munsi 2000; 2004). Thus the conflicting methodologies of present time can get some coherence through cognitive approach. The re-emergence of 'spatial difference via cultural turn; the strong remaining of spatial science' etc. essentially give rise to a substantially fragmented discipline. If geography is to prosper, the basis for coherence (necessary in post-modern discipline) has to be realised (Johnston and Sidaway 2004).

The inter-working of phenomena both physical and human can well be displayed through a role, relation, or an equation i.e. a cognitive model, developed after synthesis of information acquired by subjective and (or) objective understanding of the interactions. A model is a theory, law or a hypothesis or a simplified structuring of reality which presents supposedly significant features or relationship in a generalised form. Models are highly subjective approximation in that, they do not include all associated observations or measurements, but, as such, they are valuable in observing incidental detail and in allowing fundamental aspects of reality to appear (Haggett and Chorley 1967). The acquisitive nature of the models provides a framework for defining, collection, ordering and organising the data and also have a role of fertility in allowing the maximum number of information to be squeezed out of the data.

These interactions both in the form of morphological and cascading system may yield a generalised relation and it is the duty of the geographers to understand it and represent it subsequently through maps or models. The use of GIS software makes this correlation and their representation easier through composite maps. The former example of mining from hill-slope and introduction of instability to slope and soil and nature of acquiring equilibrium by sliding can be sited again for the construction of cognitive model displaying the interactions among human actions of mining affecting geological, economic and social processes and their subsequent results (Fig. 4-1).

Postmodernism and Future Research in Geography

In the fast changing world, geographers might be required to produce meaningful geographies which can convince the future generation about the content and the form. This can be possible if geographers seriously connect the subject matter with the immediate context of social structure. The social upheavals in the tribal, backward, underdeveloped, drought-prone and other lagging regions of the country have enough scope of geographical theory (Ahmad 2004).

The spatial difference in development may be another emerging field for geographers. The ideographic concept as reinforced in modernism is still relevant in socio-spatial dialectic (Banerjee Guha 2004). In postmodern idea of spatiality and the context of environmental crisis, the environmental degradation may be studied as a result of spatial inequality in social justice (Chattopadhya 2004).

The underdeveloped and deprived regions bear the reflection of the systematic interaction of physical and social features and to some extent these lagging regions are handicapped in at least any of these factors. Interaction of human with nature may be performed in two stages of the development of spatial identity—the first one includes the space consciousness or understanding the situation as it is inherited and in the second, activities are planned. The new space system is created as a result of the activities in the second stage (Chattopadhya 2004).

At this juncture, there is a scope for convergence in the physical and cultural geography. The nature and extent of such difficulties and their spatial distribution and differences may be linked with the social deprivation (cf. Fig. 4-6) and possible solution may be proposed in the combined framework of geographical research to make the study socially relevant.

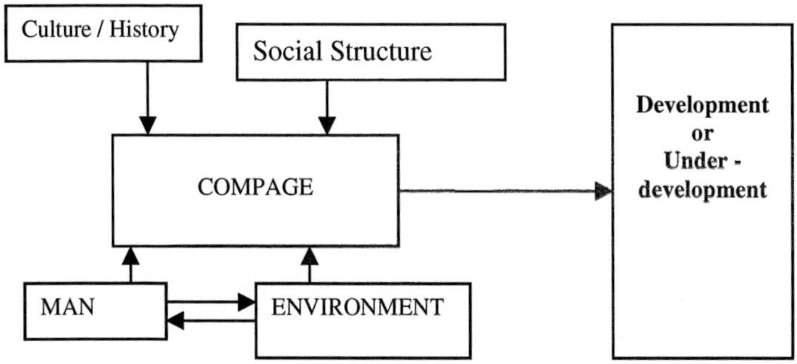

Fig. 4-6: Development as the outcome of systematic interaction of man and environment.

Conclusion

Since 1970s geography experienced a number of conflicting methodologies which are juxtaposed each other giving rise to utter confusion to the geographers specially the young geographers regarding the direction of research. The present tried above to bring a coherence in all the conflicting ideas and approaches through collecting information by field studies with proper accessories and questionnaire to make a spatial synthesis through mapping by understanding the systematic interaction among natural and human processes. The postmodern views of spatiality reinforce the ideographic concepts of spatial difference in development, and spatial difference in social justice. Growing consciousness of environmental degradation may be addressed better by geographers through the study of spatial inequality and unequal spatial distribution of opportunity and justice. Cognition of the processes underlying underdevelopment, poverty and starvation may make geography socially relevant and expression of the systematic interaction through cognitive model both in mathematical and verbal language may bring a coherence among all the conflicting approaches and thus can stimulate the development of geography as a unified discourse.

Acknowledgement

The author expresses his deep indebtedness to Mr. Tapan Kumar Das and Mr. Pravat Kumar Shit for their great contribution but for which this paper would have not been possible. He is also thankful acknowledges the help received from Mr. Sujit Mondal and Miss Suparna Choudhury in the present work.

References

Abler, R.F., M.G. Mercus and J.M. Olson. 1992. *Geography's Inner World: Pervasive Themes in Contemporary American Geography.* New Burnswick, NJ: Rutgers University Press.

Ahmad, A. 2004. Rethinking geography in the post modern world. In *Space, Society and Geography*, ed. S. Banerjee Guha, 46-60, Jaipur: Rawat Publications.

Ahnert, Frank. 1977. Some comments on the quantitative formulation of geomorphologic processes in a theoretical model. *Earth Surface Processes* 2 (2-3): 191-201.

Banerjee Guha, S. 2004. Spatial hermeneutics of social processes. In *Space, Society and Geography*, ed. S. Banerjee Guha, 15-33, Jaipur: Rawat Publications.

Basu, S. R. and R.K. Maiti. 2001. Unscientific mining and degradation of slopes in the Darjeeling Himalayas. In *Changing Environmental Scenario of the Indian Subcontinent*, ed. S. R. Basu, 390- 399, Kolkata: ACB Publications.

Berry B.L.J.1964. Approaches to regional analysis: A synthesis. *Annals of the Association of American Geographers* 54: 2-11.

—. 2002. Paradigm Lost. *Urban Geography* 23: 441-45.

Brudsen, D. 1979. Mass movement. In *Process in Geomorphology*, ed. C. Embelton and J. Thornes, 130-186. New York: John Wiley & Sons.

Chattopadhya, S. 2004. Integrating environment in social geographical studies. In *Space, Society and Geography*, ed. S. Banerjee Guha, 129-139. Jaipur: Rawat Publications.

Foucault, M. 1986. Of other spaces. *Diacritics* 16: 22-27 (Translated from the French by Jay Miskowiec).

Gumbel, E.J. 1954. Statistical theory of extreme values and some practical applications. *Applied Mathematics*, Series 33, Washington, DC: U.S.Bureau of Standards.

Haggett. P. and R.J. Chorley. 1967. Models, paradigm and the new geography. In *Models in Geography*, ed. R.J. Chorley and P. Haggett, 19-42, London: Methuen.

Harvey, D. 1969. Conceptual and measurement problems in the cognitive-behavioural approach to location theory. In *Behavioural Problems in Geography*, ed. K.R.Cox and R.G. Gollege, 35-67, Evanston, III: North Western Studies in Geography, No.17.

Hartshorne, R. 1939. *The Nature of Geography: A Critical Survey of Current Thought in the Light of the Past.* Lancaster: Association of American Geographers.

Johnston, R.J. and S. Gregory. 1984. The United Kingdom. In *Geography since the Second World War: An International Survey*, ed. R.J. Johnston and P. Claval, 107-131. London: Croom Helm.

Johnston, R.J., D. Gregory, G. Pratt and M. Watts. 2004. *Dictionary of Human Geography* (4th Edition). Oxford: Blackwel.

Johnston, R.J. and J.D. Sidaway. 2004. *Geography and Geographers: Anglo-American Human Geography since 1945*. London: Arnold Publishers.

Kirk, W. 1951. Historical geography and the concept of the behavioural environment. *Indian Geographical Journal* 25: 152-60.

—. 1963. Problems of geography. *Geography* 48: 357-71.

Maiti, R.K. 2004. Mining and environment-a few case studies in Eastern Himalaya. Ph. D. thesis, University of Calcutta.

Mockus, V. 1960. Selecting a flood-frequency method. *ASAE Trans.* 3: 48-51.

Munsi, S.K. 2000. Trends in geographical thought: Past, present and a peep into the future. In *Contemporary Dimensions in Geography*, ed. N. Prasad and R. Basu, 19-26. Burdwan: Academic Staff College, University of Burdwan.

—. 2004. Critical issues in social geographical research. In *Space, Society and Geography*, ed. S. Banerjee Guha, 37-45. Jaipur: Rawat Publications.

Peet, R. 1977. The development of radical geography in United States. *Progress in Human Geography* 1 (3): 64-78.

Pred, A.R. 1967. Behaviour and location: Foundation for a geographic and dynamic location theory, Part II. Lund: C.W.K. Gleerup.

Schaefer, F.K. 1953. Exceptionalism in geography, a methodological examination. *Annals of the Association of American Geographers* 43: 226-49.

Soja, E.W. 1997. *Post-modern Geographies: The Reassertion of Space in Critical Social Theory* (Indian Edition). Jaipur: Rawat Publications.

Tuan, Yi Fu. 1976. Humanistic geography. *Annals of the Association of American Geographers* 66: 266-76.

PART II:

RESHAPING GEOGRAPHY EDUCATION

"...geography is first of all knowledge gained..."
—Carl Ortwin Sauer (1889-1975) in 'The Education of a Geographer' (1956).

CHAPTER FIVE

THE STATE OF GEOGRAPHY IN INDIAN SCHOOLS: REFLECTION AND ACTION

SARFARAZ ALAM,
BANARAS HINDU UNIVERSITY

Introduction

> "The strange fact of the matter is that the role of geography in the school curriculum is at once anomalous and ubiquitous. Geography lacks a clear identity …. Nonetheless, by its very nature, geography is integral to all human inquiry. It is difficult, or even impossible, to separate what is geographic from what is not. In this sense, then, geography is everywhere in the school curriculum. The major problem, both for geographers and geographic educators, and for all curriculum planners and teachers, is to find ways to acknowledge and act on this reality."
> —Malcolm Douglass (cited by McDougall 2000)

Geography is one of the ancient and honoured disciplines. The history of the discipline is one of the most fascinating one which has been unfolding for hundreds of years as we have been trying to understand not only ourselves and our surroundings but also distant people, places and environments. As a subject, geography has always had a great educational, pragmatic (utilitarian), existential, ethical and symbolic significance not only for the kings (rulers) but also for the commoners (ruled).

Geography stimulates our curiosity about the diverse people, places and environments of the earth's surface and thereby enhancing our spatial and ecological understanding about it. As every parent knows, children begin life as uninhibited, unabashed explorers of the unknown. In other words, they begin life as little geographer. It is because of this that geography is considered as one of the three most useful disciplines through which education should be imparted particularly in the primary schools; the other two subjects being language and mathematics.

Geography also has utilitarian value in the contemporary world. Geographical knowledge and skills are used in business, politics, environmental and resource management, spatial planning and development, etc. The discipline enables us to understand not only the intrinsic nature of our planet but also help us to know where we live on this planet. Geography education also provides ethical and moral reasons for cooperation amongst humans for the preservation of their only home—the planet earth and the protection of the whole humanity. Its symbolic significance lies in the fact that the study of geography helps us form images about different parts of the earth. It is because of these and many other functions that the educational value of geography and its place as a key subject in school curriculum in India can not be underemphasized.

School geography in its modern avatar in India is a gift of the British educational system. In fact, the discipline's alleged connivance with imperialism and colonialism helped school geography to flourish during the British period. But, the study of the subject in the schools witnessed a decline since the Secondary Education Commission (1952-53). Even then, it had a fairly reasonable run till the publication of the report of the NCFSE 2000 which strongly advocated for the integrated teaching of Social Studies in which geography became an insignificant part. The NCF 2005 attempted to rectify this problem. However, being motivated by the subaltern ideology it had its own limitations. In an attempt to make school geography popular, student-friendly and interesting, it completely dropped World Regional Geography from school syllabus.

Keeping this historical development of the school geography in the background, this paper focuses on the state of the subject in India—its major problems and concerns. In the process, it sets out some agenda for the young geographers towards the strengthening the place of geography in school curricula. The paper mainly focuses on the geography praxis in the upper primary and lower secondary school curricula.

Evolution of Geography as a School Subject

Geography education in India has ancient roots. There are ample evidences of elementary lessons on geography and cosmography being imparted in the ancient Indian *gurukuls* and *pathsalas* as well as in the medieval Indian *madrasas*.[1] However, the study of geography in its present form was initiated by the Britishers in India. The reason was, of course, far from an interest in the promotion of the academic aspects of the discipline. "Geography gained a strong hold on the school curriculum in the nineteenth century due to growing recognition of its utilisation value by the middle and upper classes. Suiting the vocational needs of future merchants, clerk, statesmen and strategists, school geography also served to facilitate increased exploitation of the physical environment and legitimate nationalism and imperialism." (Huckle 1985, 296). Imperial geography as a discipline allegedly was in close connivance with "the generals, politicians and corporate chiefs" (Harvey 1984, 7). Thus, the promotion of geography in the British India owed itself to the fact that it served the interests of the imperialists and those Indians who were beneficiaries of the British imperialism. What made things brighter for the discipline was its value in terms of employment as surveyors that attracted members of the middle class who aspired for jobs in the British Indian administration. All these factors resulted in publication of good, simple, textbooks of geography written in Hindi even in the 1920s (Mukerji 1991, 135). One could also find geography texts books in other vernacular languages such as Bengali and Marathi that primarily aimed at serving the local elite and middle class of those languages.

In 1950, UNESCO organised an International Educational Seminar on the Teaching of Geography as a Means of Developing International Understanding in Sainte-Anne-de-Bellevue, Canada in the year 1950.[2] It brought out few publications on the teaching of geography in schools. The efforts of the UNESCO created awareness amongst geographers and geography educationists all over the world. Its impact was also felt in India. Professor S.P. Chatterjee (1964, 269-271) prepared and presented a report on the "Fifty Years of Science in India: Progress of Geography" in the 51st Indian Science Congress held in Calcutta. The last but one chapter of the report was on the progress of school geography education. Again, when the International Geographical Union Congress was organised in 1968 in New Delhi, in the Presidential Report Professor Chatterjee devoted few pages on school geography education (Chatterjee 1968). Similarly, in the first Indian Geographical Congress which was held in Delhi in the year 1972, there was one section on the teaching of geography and the largest numbers of papers in this section were on geography teaching in schools. While geographers were pretty enthusiastic about their subjects and wrote important articles and books on geography education, other educationists and education planners were totally insensitive to the school geography.

As a matter of fact, the report of the Secondary Education Commission (1952-53, 81) was the first attempt to marginalise geography. The report states:

> "It is not often realized that the complaint of overcrowding is largely due to multiplicity of subjects prescribed as separate entities without bringing their organic interrelationships. Thus it is psychologically preferable to present subjects centring around the study of social environment and human relations under the comprehensive heading of 'social studies' than to teach a number of separate subject like history, geography, civics and economics in a watertight compartments as 'segregated wholes' lacking coordination and compassion rather knit them in a 'completely whole' assigning their due places in the universe of educational thought and set up."

By 1960s the adverse effect of marginalised position of geography in school curricula was felt. It is important to remember that the Government of Mysore (now Karnataka) deplored the position of geography and history in the school curriculum and at the same time made a strong case for their more important role in school education; in the meeting of the Standing Committee on School Education in 6th and 7th August 1965 (National Council for educational Research and Training, NCERT, 1965):

> "The present position of Social Studies in the School Curriculum leads one nowhere. It has been found by experience that the pupils who come out of S.L.L.C. examination do not posses knowledge of even the rudimentary of geography and history of India. It is very necessary that a strong foundation is laid in both history and geography of India if a sense of patriotism in our younger generation is required to grow. A special Committee of experts may be appointed to examine this question and formulate a syllabus to remedy the falling standards in the study of history and geography of India".

The chairman of the committee virtually ignored the suggestion of the Mysore government by giving a vague answer. Kothari Commission (1966) tried to rectify the problems of geography, history and civics teachings in the schools. But the Commission advocated for the study of geography within the overall framework of social study.[3] The Commission recommended that geography should start in the higher primary school (that is from the 5th class onward). In the higher primary stage, the Commission recommended that, "Social studies may continue as an integrated course if competent teachers and the requisite facilities are available; otherwise the study of history, geography and civics should be taken up as separate disciplines." (Kothari Commission 1966, 189)

At the lower primary stage, the integrated approach is desirable. Instead of giving the pupils miscellaneous and unrelated bits of information in history, geography and civics, it is far better to provide a coordinated programme of social studies centring round the study of man and his environment. In the upper classes of the primary schools, the content of social studies may still be organized as an integral whole in connection with the treatment of certain topics, but the pupils should be gradually introduced to an appreciation of history, geography and civics as separate subjects. In the secondary schools, these subjects will be treated as separate disciplines and form the basis of specialized studies in social sciences at the higher secondary stage. (Kothari Commission 1966, 201)

Kothari Commission was an important landmark in the history of education in general and school education in particular. All other commissions and committees dealing with school education considered Kothari Commission as the bench mark. The framework of geography teaching in school continued to be adopted in India till recommendations of the NCFSE 2000. This curricular framework proved to be extremely controversial for more than one reasons. Five years later, the NCF 2005 gave its recommendations. Both these recommendations have proved to be extremely harmful to geography. The next part of the paper will critically examine the recent trends in school geography education based on these two curricular frameworks.

Recent Trends in School Geography in India

As mentioned above the recommendations of the NCFSE 2000 and NCF 2005 have far reaching repercussions on school geography. In this part of the paper, an attempt has been made to examine some of the recent developments in the school geography, with special emphasis on NCFSE 2000 and the NCF 2005. Geography has always been a greatly admired subject in the Indian schools. It is being taught as an independent and separate subject since the second half of the nineteenth century, particularly from the upper primary class onwards. However, the NCFSE 2000 had recommended for an integrated teaching of geography as part of social studies whereby geography was be taught in combination with history, political science, economics and sociology. Integration is a noble idea. But in actual practice it is a very difficult exercise. Rather, integrated teaching of geography became a highly controversial issue. Integration does not mean producing mix of knowledge of various disciplines or a useless amalgam of facts coming from different disciplinary traditions. It is the integration of the various components of knowledge, skills brought about by well developed disciplines each of which makes its own contribution on the basis of its rigours. This combined social study subject is supposed to be taught by each and every teacher. There is no special need for subjects' knowledge and skills and teacher training required for teaching social sciences. While in principle it may look a good idea to give an integrated picture of social reality, in practice, it turns out to be a very impractical and disintegrated amalgam of information about society.

In an integrated curriculum, geography is the main sufferer for it is difficult for a non-geographer to teach geography. There are several reasons for this. Geography has in it quite a heavy science component and it becomes difficult to fit it completely with social sciences. In many cases geography is taught by teachers who may not have studied geography in colleges. They often find it difficult to teach geography. Moreover, being trained in social science discipline, instructions in geography conform to the norms of social sciences. Geography is taught to serve teaching requirements of history, political science and other social studies subjects. This kills the very objective of geography teaching. It has been found from the experiences of the many countries that the teaching of geography as a part of social studies has done harm to educational objectives of schools. And it is not surprising that in many countries curriculum framers have started refocusing on the significance of geography as a school subject and its compulsory and independent teaching.

True to the fear, social studies books were written in such a way that topics from the disciplines of history, geography, civics, economics and sociology were compiled without any order and logic. Therefore, the so-called integrated social studies books became one of the major issues of discontent amongst a large number of educationists and social scientists. As a result, the NCF 2005 reverted to the pre-NCFSE 2005 system of separate teachings history, geography and civics.

It is true that the NCF 2005 helped restored an independent and separate teaching of geography. But, on the other hand, the Report called for more focused study of 'local' geography at the cost of the study of the 'world geography'. It is true that the teaching of a discipline by taking examples from the immediate surroundings of children help them easily comprehend the concepts and skills of the respective discipline. But it does not mean that children should not be taught about the other geographical regions of the world, particularly in the era of globalisation. It is worth noting that an understanding of world geography helps children develop a world perspective. In other words, they do not develop ethnocentric attitude. The question is when children of most of the countries of the world learn about India; then why our children should be denied their right to learn in the classrooms geography of other countries of the world. To illustrate this point, cross-country comparisons in the level of geographical literacy can be mentioned here. In the United States, geography is not a compulsory subject in the school. It is taught as a part of broad social studies. On the other hand, geography has always been a major school subject in India. It is only natural that among young Americans the level of geographical literacy has been found to be lower compared to Indian students (Sukhwal 1984, 228-229). In today's phase of globalisation, there is a strong case for the studies of the world regions. The ever-increasing interconnectedness and interdependencies among the countries of the world, the increase in travel facilities in an ever shrinking world, the growing complexities of international problems, increasing disparities in levels of development at all geographical scales, and continuing degradation and depletion of natural resources, have all made the study of world regional geography far more relevant than ever.

There is a discernible development trend towards systematic studies in geography and away from regional geography which has long been the core and culmination of the subject. Regional geography has always been considered as the core of the subject. Regional geography is taught very badly in India. It is partially because of the fact that the teaching of regional geography is a quite a difficult task. The teacher needs to have basic knowledge of natural and social sciences including humanities in order to be an effective teacher of regional geography. It is being neglected. Rather than improving teacher quality through appropriate training, the regional geography has been done away with. This is an unfortunate development.

In recent years, with the introduction of environmental study and geology as a separate and compulsory paper in the schools there is pressure from some quarters of curricular planners and educationists on the discipline of geography to prove its usefulness and in the school curricula (Alam 2003; 2008). So much so that the Bihar School Examination Board has removed geography from the school curricula and in its place environmental study has been introduced (The Times of India, 2004). Now only few chapters of geography are taught in the environmental study discipline. This has been done in spite of the strong objections from the geography community of Bihar. This is a dangerous trend. This step of the Bihar Government creates a strong doubt on the intellectual competence and academic integrity of those educationists and educational planners who are responsible for this bizarre decision to marginalize and eliminate one of the most ancient and honoured school subject.

One of the major criticisms of school geography books has been that there is lots of emphasis on factual information about the people, place and environments of the earth. As a result, its teaching and learning become dull and boring wherein there is emphasis on rote-learning. The focus is on memorization rather than understanding and skill development. The teaching is characterized by a one way communication and students become passive listeners of useless factual information about the earth. However, the NCFSE 2000 and the NCF 2005 have recommended for concept and skill oriented teaching of geography. As a consequence, there is discernible movement from the realm of facts into that of concepts and principles in the school curricula in NCERT geography textbooks. This is a welcome step as this will enhance students' interests in geography and consequently help improve the profile of the discipline (see Chapter 6, of this book, for a detailed analysis of this issue).

The Present Status and Concerns

Even though geography continues to remain a school subject since Independence, its position and status have always been a major point of controversy. The discipline no longer enjoys the status of "the queenly subject" as was the case during the British period. Now, it has to struggle for its survival as its very presence is being questioned in the school curriculum. It has gained popular respect among the masses but lacks recognition and respectability amongst academicians and government. Writing on the problems of geography in India, in general, Bhardwaj (1987, 329) suggests that: "Some problems are, indeed, systemic to geography {meaning thereby these problems are universal to geography, others exist due to the peculiarities of the way geography developed in India with the baggage of its colonial past and the Indian socio-cultural milieu". Probably, the contemporary praxis of school geography in India has evolved under conditions mentioned above.

The place of geography in the school curriculum

Over all, geography has been given a marginal position in the school curricula. Geography is treated differently at different stages of school curricula. At the primary level, geography is generally not taught. Rather some topics and themes of geography are taught with environmental studies. It is introduced as a subject only in the upper primary school. In the upper primary school and the secondary school, it is taught as a part of social studies. In the higher secondary school it has been elevated to a position of an optional subject and introduced as a separate subject in its own right.

Geography syllabus

Even though geography is studied within a social science framework, the actual geographical content of courses in school includes both physical and human aspects. In general, three main branches of geography are taught—regional, systematic and practical geography (i.e., tools and techniques). As indicated above, earlier regional geography used to constitute perhaps the most important part of school geography curriculum. But since NCF 2005, regional geography has lost out to systematic geography. There were some major problems with teaching-learning of regional geography identified by Kumar (1996, 246-253). The treatment of the subject is very encyclopaedic in form. There is focus on facts and figures about starting from geology and landforms to populations, society, polity and economy of the regions. Children were expected to cram facts and figures related to the region. The presentation of lesson is also very uninteresting and boring. Therefore, the learning process is generally very dry and a burden on children. The recommendations of the NCF 2005 in this regard have been very discouraging and reflect its nonchalant attitude for geography. Rather than improving teaching and learning practices of regional geography, the NCF 2005 has totally removed regional geography from the school curricula.

Systematic geography textbooks focus on patterns of different natural and cultural phenomena and features without bothering about the underlying causes and processes which produce the existing patterns. No value based questions are raised (e.g., what ought to be?). Practical geography is third important component of geography teaching. It is concerned with tools, techniques and methods of collecting (field work), presenting (cartography) and analysis data (statistics). It is perhaps, the most neglected component of geography teaching in the schools. Geography practical is introduced in the senior secondary schools. Some of the problems of practical geography teaching have been highlighted by Deshpande (First Indian Geographical Congress 1972, 6-7). These problems persist even today. The practical stands a part from the rest of the courses. It is non-applicative and functionally useless except when one undertakes a research. The field studies, though a very significant part of the practical, do not seen to be purposefully and imaginatively planned and honestly executed. Students score very high marks in practical geography but hardly develop proper understanding of it. There is less emphasis on map reading and map interpretation. Field work is completely ignored. Practical work including field work is a much ill planned area of study both at the school and the higher stage.

School geography books

As mentioned above, the three branches of geography that were generally taught in schools, the NCERT as well as some private publishers used to publish books of regional, systematic and practical geography. Notwithstanding, with merits of those books, there were many problems.

A study by Kumar (1996, 246-253) suggests that geography books of schools are grossly overloaded with facts. Virtually all regional geography books were very encyclopaedic in forms. These were so written as heaps of trashes without any meaning. Children were expected to cram the facts and figures about people, places and environment of the earth. The learning process is generally dry and a burden on children. The presentation of lesson is generally also very uninteresting and boring. Practical works including field work were much ill planned areas of study at the school stage. There was hardly any emphasis on map reading and map interpretation. Some other weaknesses of school text books of geography in India have been noted by Misra (1983, 7-8):

1. The do not start with and explains the principles; they start with description of the distribution. The student is never able to answer the question *why*. They know the distribution of phenomena but do not know the principles underlying the distribution of phenomena.
2. The texts books are not related to the level of students. Often hey reflect the level of the writer, though
3. The texts books are not written by the experts in the field are printed rather poorly. The rationalisation of the textbooks ought to have improved matters to a large extent.
4. Poor quality printing keeps the prices down no doubt but makes the book unattractive to the children.

Teaching methods

Teaching style in schools is very formal and encourages rote learning. Teaching is driven by the content of the books. Books are considered as syllabus. In geography, there use to be one book in each class of upper primary school and secondary school which increases to three books in one academic year in the senior secondary school. Generally, teachers do not teach students from sources other than the book prescribed. In the absence of up-to-date teaching aids like maps and globes, lectures become boring. There is little emphasis if any at all, on field work. One important reason for this is lack of funds for school geography.

Field teaching

Field work is a powerful teaching method. Geographers and geography teachers consider fieldwork as a vital instrument for understanding our world through direct experience, for gathering basic data about this world, and as a fundamental method for enacting geographical education. Field is considered as laboratory for geography teaching. But, in India the use of field method of teaching is virtually absent. Very few schools have resources to conduct field work. Students are not taken even to their surroundings, not to speak of distant places.

Teaching aids

In a way, geography is considered as a costly subject. For its effective classroom teaching, teaching aids like maps, globe and three-dimensional models are required. Not many schools have these aids in their repertoire. On the other hand, one will not be surprised to see wall maps, depicting countries of the world prepared during 1980s, hanging in schools, if at all they have any. This is in spite of the fact that so many changes have taken place over these years.

Involvement of school teachers

School teachers feel that they are being forced into the role of retailers (merchants) of information. They are asked to cover so much ground in so much detail that only stereotyped methods which impart information quickly and accurately enable the job to be done. "The university professors frame curricula, write textbooks, and conduct refresher courses; and the teachers who teach the students only carry out the policies framed by the university professors. It is obvious that teachers lose their interest in the educational system and feel themselves outsiders" (Sukhwal 1984, 223).

Research on school geography education

Geography education in India has remained a neglected field, least of all in the area of school geography education. For unknown reasons, interests of university teachers and researchers have rapidly declined in recent decades on the issues of school geography. Since 1970s, very few papers on school geography teaching are presented in seminars and conferences in India. It is very hard to find papers and articles on school geography teaching in geography journals and magazines published from India. Only few Ph.D. scholars have worked on geography education.[4] So much so that some geography departments even discourage students who wish to work in this area perhaps either due to lack of capability to supervise researches on geography; or because of disciplinary orthodoxy. They have been of the view that researches on geography education should not be undertaken in the departments of geography, instead education departments should do research in this field.). Education faculties have few geographers. Education department is mostly filled with faculty members from economic, psychology, philosophy, sociology or public administration backgrounds. Naturally, they are neither competent nor interested to conduct researches on geography education. What is even more heart aching to note that in the past two decade many university teachers both from India and abroad have written research papers and given lectures in conferences and seminars on the status, problems and prospects of geography in higher education but none of them have found it necessary to write or say even a few lines on school geography. All the Surveys of Research in Geography published after S.P. Chatterjee's *Fifty Years of Science in India: Progress in Geography* have been absolutely insensitive to school geography education. It is also very sad indeed to note that even the NCERT's work in this regard has been primarily notional and patchy.

Geography teachers

For the prospective primary school teachers, the essential qualifications vary from state to state. But generally the prospective teachers should have passed Matriculation or Intermediate with BTC training. The essential qualifications laid down for the prospective teacher of upper primary and secondary teacher is B.A./B.Sc. with B. Ed. The qualification laid down for a senior secondary school is M.A./M.Sc. in geography with B. Ed. degree. There may be exceptions to essential qualifications for each category of teachers. In many cases, teachers do training course many years after joining their jobs.

Knowledge of the subject and knowledge of teaching strategies are both important qualifications for an appropriate school teacher of any subject. In many instances, a teacher with good knowledge of geography fails to become a good geographer. On the other hand, a teacher, who is well-trained to handle teaching profession, does not possess adequate knowledge of geography. It is generally seen that geography is not always taught by the right kind of teachers. In many cases, it is taught by either by an untrained geography graduate or trained non-geographer graduate. Generally, in the absence of geography teachers, social studies teachers specializing in history, political science or economics are asked to teach geography in schools. In both situations, geography and the helpless students, few of whom could have become prospective geographers, are the real losers.

Teaching times for geography

There are many educationists who give geography equal value at par with mathematics and language in the school education purely on the basis of its educational value. But the time devoted to geography teaching is quite below par than these subjects. For instance, at the upper primary school stage, geography has just one half of time budget of mathematic and language (both Hindi and English are given six classes each); and at the secondary school stage, the share decreases to one third.

Geography has to share with social studies for class teaching hour. Of the total time allotted to social studies classes, geography gets one half in the upper primary school stage, which decreases to nearly one third in the secondary school stage. To be exact, of the total six classes per week allotted to social studies in the upper primary schools, three classes is given to geography, two classes to history and one class to civics (now political science). At the upper primary stage, two classes are given to each history and geography while rest two classes are equally divided between political science and economics.

Examination and evaluation

In the upper primary stage out of 100 percent marks, questions of 40 % marks are asked from the discipline of geography and history while questions worth 20 % are asked from political science (civics). In the secondary school the distribution of marks in the examination is as follow: questions from geography and history constitute 30% each, political science 20%, while the rest 20 per cent marks are equally shared between economics and disaster mitigation.

The manner in which the questions in the examination are set leaves little scope for testing the quality of the examinees' mind. Geography teaches us how to explain a geographical phenomenon logically. It is the reasoning and how one approaches the questions that should be focused in the evaluation process. However, in examinations, questions are asked in such a way that they fail to elicit reasoning and thinking from students. Hence, it is not at all surprising o find study which shows that students just cram the text of their books or the answers to questions given at the end of their lesson, without understanding the basic principles (Kumar 1996, 248). We rob geography for life when we focus solely on results and encourage students to recite facts without a commensurate emphasis on encouraging them to think beyond the realms of facts.

Towards Young Geographers' Agenda

Harvey (1984, 7) once remarked that "geography is by far too important a subject to be left for geographers only". Using the same analogy it can be said that "school geography is far too important to be left for teachers of school geography only". There is need for the active involvement of all stakeholders of geography—be it school teachers, college or university teachers, geographical societies, researchers, students and all those who love geography. A joint endeavour from all is required for the cause of the one of the most important and useful disciplines in child education. Their efforts and energy could be directed for improving the profile of the discipline amongst academic community, decision makers and society. More specifically, the following measures should be adopted which might help to improve the health of the discipline.

Reform the curriculum

Curriculum making is an exercise in contest. There is disagreement on what should be the focus of school geography. It is very interesting to see the practices of school geography being problematised differently by different scholars. This is in spite of the fact that the Central Board of Secondary Education (CBSE) considers a non-dualistic identity of geography in school curriculum, wherein both physical and human elements are taught (Kapur 2004, 4189). Those who have earth sciences leaning criticise geography in school for its overwhelmingly "human" nature. They are strongly against geography being practiced as a social science discipline. They are of the view that geography should be taught as an earth science discipline along with geology and environmental science. (Kaur and Chaudhri 2003, 618-619) There are scholars with social science background. They criticise school geography practice because of its close proximity to physical sciences. They are of the view that even though the spectrum of geography education in school encompasses physical, human, regional environmental geographies and maps; its overwhelming "physical" nature is quite explicit not only through its content but also through its ontological posture (Sunny 2006, 270-278). Actually, the truth is that both physical and human elements are treated as if they exist in isolation. The study of law of nature in geography rarely takes into account human factors. Similarly, human geography is treated as if natural world does not exist. In fact, most of the malaise of geography in school stem from the very fact that it is taught as a part of social studies or earth science. Geography should be introduced as a separate and independent subject. When taught as a part of social studies or earth sciences, students just study geography for the sake of studying geography but fail to gain knowledge of genuine geography education.

Scholars holding these anti-polar positions, for some unknown reasons and interests, do not advocate for the 'true' geographical education. They have, perhaps, become prisoners of their own training as geographers and do not try to remove their self–styled intellectual veils. Those trained in geography as a branch of the earth sciences in the university strongly advocate for geography as an earth science discipline. On the contrary, geographers trained as social scientists consider geography essentially a part of social sciences. They have to understand that school geography either as an earth science or as a social science subject is nothing more than a curricular compromise. Perhaps curricular planners apprehend of geography's pre-eminent status as an integrative discipline which when practiced in true spirit will make some established subjects of primary schools redundant.

Geography is an integrative subject. Its hall mark is regional synthesis. If geographers do not provide the synthesis of phenomena in parts of the earth's space, who else would? (Gosal 1999, 2). Therefore, rather than prioritising either the physical or the human elements, the major focus of school geography text should be on the areal synthesis of both natural and human phenomena. By doing so, geography's unique yet a very significant role in school curriculum could be ensured. The actual educational role of geography can only be achieved if it is neither taught as a part of social sciences nor as a part of natural sciences. On the other hand, it should be taught in such a way that the elements of both physical and social sciences can be studied within the regional framework, at least in the primary school.

Curricular goals of school geography should be rational, objective and universal. One of its goals should be to bridge the dichotomy of natural and social sciences. Due emphasis should be given to knowledge, skills, attitudes and values about people, places and environments at local levels, regional, national to the global. The curricula objectives of geography, in particular, should focus on the appreciation of physiographic, social and cultural diversities; on the problems of environmental degradation and regional disparity in development. Unity in diversity, respect for regional differences and similarities and ways of lives of people should be given due place in school geography education.

Relevant syllabus

Identifying the content of geography education for school is a very challenging task. It is extremely difficult to accept the view that there is a single corpus of geographical knowledge which should form the basis of all school syllabuses. The difficulty is due to all-embracing nature of its subject matter and diversity of approaches and perspectives applied by its practitioners. There are disagreements among geography educationists over what should be specific geographical knowledge, skills and value to be imparted t various stages in the school.

We should be aware of the reality that the geography of the earth is changing very fast. Therefore, the discipline of geography must be changed to suit the purpose. Geography, like natural sciences, is a living subject with ever-widening boundaries. Therefore, text books need to be constantly upgraded. Therefore, selection of contents, skills, tools, techniques and value system at various stages of school education should be rationally introduced. It does not mean changing everything all the time. Change for the sake of change should be avoided.

Training of the school teachers and their trainers

Geography should be given an important place in the school curricula. Being an integrating subject, it deals with physical as well as human phenomena on the earth's surface. In the absence of geology, astronomy in the school subjects, their topics are taught in geography classes (Kaur and Chaudhri 2003, 618). If it is taught by a social sciences teacher, it will be most unfortunate for the students as well as the discipline of geography. Good teachers are scarce commodities. This is particularly true in the case of geography. It is because geography is a synthesis discipline. It draws heavily from natural and social sciences as well as humanities. Therefore, geography teaching demands good knowledge of these disciplines also. Many students develop likings for the subject because of some outstanding teachers of the subject. Yet, our quest for quality teachers should be always present. There is need for quality training and regular orientation courses for all the teachers.

New technologies for teaching spatial concepts

Spatial competence is considered as the fundamental aim of teaching in geography, which should be obtained with regard to both: an understanding of spatial conditions in the world (local, regional, national and global; and the active and participating wielding of geographical knowledge with regard to spatial problems (Birkenhauer 2002, 272). GIS and RS and other spatial technologies are opening new vistas for geography educationist as these are extremely useful not only for collecting, presenting and analysing spatial information but are also of immense value in the geography classrooms for their role in enhancing spatial competence. Google Earth, NASA's World Wind, Wikimapia and other free virtual globes have potential to revolutionise geography teachings because they generate not only powerful spatial image of the earth with great manoeuvrability, but are easily accessible. It is therefore encouraging to note that the US Academy of Sciences has recently called for the introduction of "spatial thinking", including GIS, into school curricula, and for governmental research agencies to launch research into the nature of the cognitive processes involved in such thinking. (*Nature* 2006, 763)

McDougall (2000, 232), a professor of international relations, has emphasized the need for reorienting geography teachers in new technologies: "Geography teachers should try to convey how notions of space and time have changed as a function of technology. From the first irrigation systems to the internet the human race has reinvented the world. But just as students can not handle calculus until they have mastered algebra, so they can cannot deconstruct human conventions of space and time until they know what got deconstructed in the first place."

Need for quality school books

Quality books are those, which provide sufficient information; stimulate readers' interests, easy to understand and thought provoking. As mentioned above, really good teachers of geography are scarce commodity in India and therefore books should be so written as would make it possible for a teacher of average quality to impart proper geographical knowledge and skills in content and quality. The need for good books in geography becomes indispensable when geography is taught as part of social studies with history and civics (political science) in school after the recommendation of the NCFSE 2000 and revised NCF 2005. Thus in the new circumstance, it is important to have quality books in geography so that a social studies teacher not trained in geography could teach geography by simply reading geography books. This is not to suggest that trained geography teachers are not required.

Text books should be written keeping in mind the following three aspects—the degree of complexity, the method of presentation and choice of content (including details). Further, for each level there should be different mixture of these (Misra 1983, 7-8). Reading geography books and studying maps *should* evoke beautiful visions of distant lands and provide geographical knowledge in a fascinating way (Hudson 1994, 226). Privatisation of school text books will definitely not solve the problems. On the contrary, professionalisation of book writings might go a long way towards this end. All the stake holders of school geography education should be involved in the process of books writings.

Re-focusing on the regional geography

Regional geography has been the traditional forte of geography. It has traditionally been considered as the core or bedrock of the subject but has been totally ignored. One of the main complains about school geography is that it puts unnecessary burden on students by forcing them to learn in a dull and boring fashion. It is true as well as imperative that the teaching of the subject should be freed from the excessive factual burden which put unnecessary burden on its students. It is a bizarre situation indeed. Just because the teaching of regional geography is boring and dull does not give us right to remove them from the school geography syllabus. In fact, it is not a problem of regional geography. We have approached regional geography with a taken-for-granted attitude without any serious thought on its pedagogy. Regional geography text books are written very badly presented. Regional geography is very badly taught. Writing and teaching about region of the world is a difficult art to acquire. If we remove regional geography from the school, the value of regional geography will not diminish. But, what will certainly diminish is the power of thinking and imagining the world as mosaic of places, people and environment from the mind of the future citizens of India. Therefore, regional geography needs to be refocused in view of increased connectivity across geographical space.

Harvey (2001, 210) quotes Martha Nussbaum (one of the main proponent of cosmopolitan ideal in the USA) to emphasize the critical importance of geography education in the contemporary world: she complains how "the United States is unable to look at itself through the lens of the other, and, as consequence, (is) equally ignorant of itself" precisely because the population is so "appallingly ignorant of the rest of the world". In order to conduct any adequate global dialogue, she continues, "We need knowledge not only of the geography and ecology of other nations—something that would already entail much revision of our curricula—but also a great deal about other people, so that in talking with them we may be capable of respecting their traditions and commitments. Cosmopolitan education would supply the background necessary for this deliberation." Harvey further quotes her by saying "cosmopolitanism without a 'sound' and 'proper' understanding of geography and anthropology is…", she implies an empty ideal.

Teaching tools and technologies

Geography teachers should not only be made available but should be trained and encouraged to make use of teaching aids like maps, globes, three-dimensional models of geographical features and other modern tools and technologies for classroom teaching purposes. In new circumstances, remote sensing, GIS and computer based Google Earth have become important technologies through which geography education can be more effectively imparted in the schools. There is need to use teachings aids including multimedia.

Reinvent and reintroduce field teaching

Class-room learning should be linked to the day-to-day lived experience of children. Field work as a teaching aid is an essential requisite for a good geographical teaching._Therefore, it needs to be re-invented and reintroduced as a teaching strategy. Students learn best by observing the geographical features and patterns on the earth's surface as well as various processes which produce these features and patterns. Field teaching means proceeding from known to unknown, from real to abstract and from large scale to small scale. Geography teaching should always start from the feet of the students and progressively move outward. In other words, field teaching should follow the principle of concentric model

Restructuring practical geography

Practical geography is the most neglected component of school geography. This is in spite of the fact that its utility as a learning method is extremely rich. There are basically two major problems: there is less emphasis on practical geography and whatever of it is taught is generally detached from the theory components. There is an immediate need to weave practical geography syllabus with the teaching of various substantive geography theory papers, even for the purpose of examination, making it functional, productive and full of interest (Deshpande 1972, 7). Side by side, practical geography should be linked with field work and field teaching so that students are presented with opportunities to learn directly from the field.

Involvement of school teachers

Traditionally, school teachers were kept out from every steps of curriculum building of school geography. As Sukhwal (1984, 223) states: "The university professors frame curricula, write text books, and conduct refresher courses; and the teachers who teach the students only carry out the policies framed by the university professors." However, in recent years the curriculum development has become more democratic and participatory in nature. In course of the development of the last two curriculum frameworks (2000 and 2005), almost all concerned sectors and sections of society people, including school teachers, educationists, scholars, non-governmental and research organisations were involved at various stages of curriculum development. These efforts of the NCERT will go long way in democratising and rationalizing the curriculum building. The efforts should not stop here. There is need of the involvement of professionals and experts from other countries who have long experience in geography curriculum development.

Geographers for geographical education

There is a pressing need for a geographical society specifically to look after geography education in general and the school geography education in particular. The society can be formed on the pattern of *The Geographical Association* of the United Kingdom or *The National Council of Geographical Education* of the USA. The society should undertake research on geography education praxis from primary school stage to that of the university. At the same time, it should act as a pressure group to fight for a rightful place of geography in the society, university, colleges and schools. It should also focus to enhance the image of the discipline by interacting with the wider society.

Conclusion

The very fact of the continued existence of geography as a school subject clearly underlines its importance in the school curriculum. Geography exists as a school subject because it performs certain functions which are not performed by other subjects. The importance of geography in everyday life is intensely felt by the people from most walks of life. Apart from its utilitarian and symbolic significance, geography education helps us shape our spatial and ecological behaviour. It helps us shaping our spatial behaviour by teaching us to make sense of the world around us and imparts the order to the spaces of which the world is comprised, as well as by helping us to navigate through varied and diverse places of the world. Geography education also helps us acquire knowledge about the ecological and environmental systems of the planet earth. This knowledge is a precondition to appreciate the value of living in harmony with the nature. At the same time, in view of globalisation and the fast changes taking place in the people, places and the environmental conditions of the planet earth, the teaching of geography is in schools has become more important than ever.

It is also required that the needs and desire of the rapidly changing world must be reflected in school geography curricula. It must keep pace with the changing requirements of society. What to teach? How to teach what? These questions must be constantly posed by curriculum planners as well as geography teachers. For these to happen, the school teachers need to be trained and retrained through specifically designed orientation and refresher courses.

It is a natural right of every human being to look beyond their immediate surroundings to see how other places, people and environment and compare them with their own places, people and environment. We must embark for a curriculum change that places geography in its rightful place alongside language as mathematics as indispensable part of education right from primary stage. For this to happen, there is need for a paradigm shift in our educational system—which is what the agenda should be for every young geographer—to ensure rightful place of geography in the school curricula and then work hard to change the way we teach and learn geography.

Young geographers are required to develop an understanding of the politics behind the curriculum construction (for details of politics behind school curriculum, (for details, see Johnston 1986, 144-167). They have to constantly engage themselves in complex yet intriguing political battles externally with wider academic community, to defend and justify the important place of geography within the academic division of knowledge. At the same time, they have to keep on critically debating the nature of geography in terms of its Strength, Weakness, Threat and Opportunities (SWOT); and ceaselessly putting efforts towards genuine teaching and socially and intellectually useful and productive research to enhance its profile in the society in general and academic community in particular. The key will be hard work and dedication because as Hart (1982, 3) in his presidential address to the Association of American Geographers reminded us by saying that: "We will be known by our products, not by our protective garb, nor by our procedures, nor by our protestations of virtue unrewarded."

At the same time we also need to acknowledge the fact that: "There is no greater gulf in knowledge today than that between the public's image of geography and the reality of what geographers actually do. That won't change until we tell our own story and entice the world to listen." (Dobson 2006, 483). Therefore, we need to be as articulate as we can. And above all, to make these things happen, we need to our keep our passion and commitment for one of the most ancient, honoured and useful disciplines alive and kicking.

Notes

[1] British records show that indigenous education was widespread in the 18th century, with a school (i.e., *pathsalas, gurukuls* or *madrasas*) for every temple, mosque or village in most regions of the country. The subjects taught included Reading, Writing, Arithmetic, Theology, Law, Astronomy, Metaphysics, Ethics, Medical Science and Religion.

[2] Some suggestions on the Teaching of geography (Paris: UNESCO, 1950); Handbook of Suggestions on the Teaching of geography, (Paris: UNESCO, 1951); Source Book for Geography Teaching (London: UNESCO/Longman's Green & Co., 1965)

[3] The Kothari Commission also suggested for the teaching of earth sciences in the secondary schools -- "Earth sciences should be introduced in the secondary schools, geology and geography being taught as an integrated subject. There are also many areas in chemistry, physics and biology to which certain topics in the study of earth sciences can be naturally related (p. 199).

[4] The author found most of such works spread out in Maharashtra, Gujarat, Tamilnadu and Faculty of Education of Banaras Hindu University.

References

Alam, Sarfaraz. 2003. The role of geography in environmental education. *The Primary Teacher* 28 (4): 5-13.

—. 2008. Environmental study as an independent subject in Indian schools? *The Geography Teacher* 5 (1): 26-28.

Bhardwaj, Surinder M. 1987. Toward an Indian geography agenda. *National Geographical Journal of India* 33 (3): 327-334.

Birkenhauer, Joseph. 2002. Proposal for a geography curriculum '2000+' for Germany. *International Research in Geographical and Environmental Education* 11 (3): 2002.

Chatterjee, S.P. 1964 *Fifty Years of Science in India: Progress of Geography*. Calcutta: Indian Science Congress Association.

Dobson, Jerome E. 2006.Why can't geographers write their own story? *Geographical Review* 96 (3): 480-483.

Geography vs. ecology leads to fire in Bihar. *The Times of India*, January 22, Patna edition. Available at http://timesofindia.indiatimes.com/articleshow/438067.cms (Accessed 9 February 2007).

Gosal, Gurdev Singh. 1999. Fourth Survey of Research in Geography in India: An Overview. In *Fourth Survey of Research in Geography 1976*-1982, ed. Gurdev Singh Gosal, 1-7. New Delhi: Manak Publications Pvt. Ltd.

Hart, J.F. 1082. The highest form of the geographer's art. *Annals of the Association of American Geographers* 72 (1): 1-29.

Harvey, David. 1984. On the history and present condition of geography: an historical materialist manifesto. *Professional Geographer* 36 (1): 1-11.

—. 2001. *Spaces of Capital: Towards a Critical Geography*. Edinburgh: University Press.

Huckle, John. 1985. Geography and Schooling. In *The Future of Geography,* ed. R. J. Johnston, 291-306. London: Methuen & Co. Ltd.

Hudson, B. 1994. Geography in the colonial schools: The classroom experience in West Indian literature. *Geography* 79 (4): 322-329.

Johnston, R.J. 1986. *On Human Geography*. Oxford: Basil Blackwell.

Kapur, Anu. 2004. Geography in India: A languishing social science. *Economic and Political Weekly* 39 (37): 4187-4195.

Kaur, Parampreet and Naveen Chaudhri. 2003. The status of geoscience education in school curriculum. *Current Science* 84 (5): 618-619.

Kumar, Pramila. 1996. Teaching-learning process in geography. In *New Frontiers in Indian Geography*, ed. L. R. Singh, 246-253. Allahabad: R. N. Dubey Foundation.

McDougall, Walter A. 2000. You can't argue with geography, Footnotes: *The Newsletter of FPRI Marvin Wachman Fund for International Relations* 6 (5). Available at: http://www.fpri.org/footnotes/065.200009.mcdougall.cantarguegeography.html (Accessed 24 August 2008).

—. 2003. Why geography matters…but is so little learned. *Orbis: A Journal of World Affairs*. 17 (2): 217-233.

Misra, R.P. 1983. Introduction. In *Contributions to Indian Geography: Concepts and Approaches*, Vol. 1, ed. R. P. Mishra, 1-10. New Delhi: Heritage Publishers,

Mukerji, A.B. 1991/2009. What ails Indian Geography? Some statements. In *Emerging Trends in Indian Geography*, ed. Jaymala Diddle, 135-155. Jaipur: Rawat Publications. Reprinted in *Indian Geography: Perspectives, Concerns and Issues*, ed. Ravi S. Singh, 223-41. Jaipur/New Delhi: Rawat Publications.

National Curriculum Framework for School Education 2000. New Delhi: NCERT.

National Curriculum Framework. 2005. New Delhi: NCERT.

NCERT. 1965. *Social Studies: A Draft Syllabus for Class I to IX*. New Delhi: NCERT.

Report of the Education Commission, 1964-66. New Delhi: Ministry of Education, Government of India, 1966.

Report of the Secondary Education Commission 1952-1953. New Delhi: Ministry of Education, Government of India.

Sukhwal, Bhiru L. 1984. Geography in Indian secondary schools. *The National Geographical Journal of India* 30 (4): 223-230.

Sunny, Y. 2006/2009. Analyzing current practices in geography education. *Economic and Political Weekly* 41 (3): 270-278. Reprinted in *Indian Geography: Perspectives, Concerns and Issues*, ed. Ravi S. Singh, 257-278. Jaipur/New Delhi: Rawat Publications.

The Section on the Teaching of Geography. In *Abstracts of Papers and Rapporteur's Statements. First Indian Geography Congress*, 1972. New Delhi.

Think Global. *Nature* 439 (7078): 763.

CHAPTER SIX

LEGITIMISATION OF KNOWLEDGE: POLITICAL CONNOTATIONS IN GEOGRAPHY

YEMUNA SUNNY,
EKLAVYA

Text books conventionally define a unique literature, in style of writing as well as in theoretical and political positions that they hold. The present article examines text books and especially looks into how geography texts deal with an all encompassing relationship of human-nature that shapes the world historically, politically, aesthetically etc.

Learning is enriched through reading; it helps us to go beyond our experiences of everyday life, adds new dimensions to our observations. Very often it takes the reader to time-space experiences that s/he might not otherwise visit. These provide historical and spatial depths to interpretations. But are these potentials lost through the particular nature of text book writing?

What are the premises in which cryptic and formula—like text writing becomes established and where elaborations and explanations are avoided? What pedagogic and socio-political implications do these enhance in interpretations of the world?

These issues are examined through experiences of some workshops and through reviews of geography school textbooks. This essay begins with a reflection on the rich enhancement of literature and how text book traditions depart from those. The following section examines some nuances of human-nature relations as portrayed in school geography text books. It is interesting that explicit influences of capitalist relations can be traced to both the style and the content of the texts.

In the final section Eklavya[1] text books, that were created on the basis of pedagogic and social sciences critiques of existing practices in schools, are examined. Both content and style of Eklavya texts books are seen to do away with the alienation of conventional texts. Hence these had demonstrates possibilities of creating meaningful learning contexts.

Nevertheless, a critical appraisal of human-nature relations as seen in these texts reveal some aspects to be considered for further enrichment. These are (1)the role of social organization without which human relationships with nature are not materialized (2) the role of spatial demarcations (like regions) in analyzing human-nature relations.

As Objectives Blur…

Some time back a "reading workshop" was organized by Eklavya in Hoshangabad[2] to gather interesting reading materials for students. This opened up many reflections encouraged by the fact that the workshop was not tightly packed or streamlined for very strict objectives.

To read what you desire, what you like is often an act of "robbing time" or an act performed when in leisure. Many of us would remember how we have fooled teachers by reading a book of our liking camouflaged by the text that the teacher wants you to read in class. And it was a voluntary act to continue reading it on the bus way back home and again do "hide and read" when you were supposed to do your homework. But whenever stories or plays were part of the course, you would read it all as soon as possible; you would be far ahead of the teacher.

This dichotomy between what you want to read and what the system (whose job is to make you a well- read person) wants you to read is like a divide between differing perceived intentions of reading. These are—to read to fulfil a concrete destination of moulding oneself as worker; and to read simply to know, whose destinations are less focused, subtler and perhaps more lasting.

It is the second sort of reading that was encouraged in the workshop. And what were the realisations? Some of my realisations surprised me quite starkly. I realised that the feeling of "hide and read" had never left me; it had travelled with me from school and stay with me even as I work for Eklavya today. I realise that the school is indeed a strong institution and the moulding of me into a worker (as designed by schooling) had been successful even as I had always been some sort of a critique of the designs and processes of classrooms and schools.

I wouldn't normally pick up any book of my choice and read them at length on my office desk because I believed that I was there to "work", so my choices would have to be Geography or Education. All other fruits of "sin" were to be issued and read at home. Capitalism has made me a good worker, even though my last years of school were spent in an "enlightened" experimental school and my first and only employment has been Eklavya.

The first day of the workshop, I picked up children's books and then I thought of various children I loved and thought it would be a pleasure for them to indulge in these books. But by the end of the day, I realised that given the flexibility of the workshop, why should I unconsciously stick to those spaces of our (Eklavya's) traditional work—of designing and selecting materials that would be pleasurable or "correct" for children?

Let me search for books that fancy my mind, which provides pleasure for me, I thought. So those were the books that I searched and read the rest of the days. And there came my second realisation. Most of us were workers trying to promote reading, through school and out of school and through our own publications etc. But to indulge in the act of reading ourselves sufficiently (I have no measure of what "sufficient" is) is something most of us are not doing. What I mean is the indulgence to read for our self, not necessarily those books that the institutional frame might deem desirable.

I read Farooqui's book on the role of opium in creating the metropolis of Bombay and Rifkin's book on transition of society's concept of time with the transitions on work and the production mode. And so on and so forth.... So there was a wide range of things that one enjoyed to read. And it was not that you were to tie them all up in the usual academic exercise to write a piece yourself—I'm here referring to the lack of compulsion at times.

I was sharing with a friend a beautiful experience of a novel. This was a Malayalam novel by Anand (Sachidanand) titled "Govardhante Yatrakal". Parts of it remain with me vividly even after several years. There are interesting conversations that Govardhan makes with Galileo, with Umrao Jaan and with Kabir. These leave visual impacts on the mind, as though it was not a book, but a movie—the narratives were very illustrative.

Interestingly, this book had lead me to search out and read "Umrao Jaan", which was quite tough reading for me, given a liberal usage of Urdu. But it is indicative that you pick up tough readings when the process follows your own searches.

These also lead me to reflect on the role of aesthetics. One is the sort of aesthetics offered by a nicely designed book like Scholastic's "A grain of rice" with a rich appeal offered by paintings. But the appeal of the content or matter is equally important and such matters make one take pleasure in books.

Books like "Govardhan" or the one a friend was reading out (K. L. Misra's "Lal Taaga Furr") with not a single illustration, the narrative proved itself capable of both—a rich content that makes the reader create the visuals in the mind.

It is true that most of the interesting readings that we do in life, whether they are literary, political or any other are mostly done outside the legitimised spaces of learning (there could of course be exceptions). Schools and colleges have many jobs to do, so there is no time to pause. This situation enforces a harnessing of the mind to go for quick answers, *the* answer and *the* conclusion. It cannot wait to probe into more than one explanation or interpretation. And paradoxically, it never places on the learner expectations of seeking an answer. The schooled learner extends a bowl for ready-made food.

In the workshop, we did not read to answer any questions, so we obviously got no ready-made answers. That was certainly a good shift of position from the usual routine of schooling and also of workshops of organisations and institutions.

Many felt the need to fine-tune the age groups for which the reading materials would be suitable. Some broad categorizations could be fine, but going beyond that could imply being stuck to the regimented sort of learning process we are all used to. When we have to design a textbook for a particular grade, we tend to be very careful in pitching it right for an average student of that grade. But when we are selecting a large number of books (for any broad age-group), it means that a certain flexibility of choice by the reader is part of the design. Any person can choose or not choose to read a particular book selected by us. It could even be that someone reads a book half way through and decides not to pursue it further. So, unlike the process of making a textbook, fine-tuning and feedback from learners become almost irrelevant in a situation where the idea is to enhance reading from a large variety of books.

I would want to reflect more on the potentials of reading from a wide canvas. It allows you to pick and choose and allows time and space to delve into the matter. It shapes your mind with the richness of variety. It also makes one want to explore the possible richness of interconnections among say, literature, paintings, geography, history etc. There is a lot that can and need to be explored.

Many were trying to carve the space of wide reading into the body of legitimacy (questions like how such readings can be part of schooling) even though that space is contradictory to the rest of the body. The two differently perceived functions of reading seem to clash. Given the nature of moulding the workforce, schooling takes on a definitive shape. Any attempt to alter its nature gets perceived as a threat to potential employment. That is why children get discouraged from libraries by parents as examinations get near. The popular question arises as to how knowing a lot of interesting things and the shaping of the mind through that process can help in utilitarian terms?

Yet people thought aloud on how we could take the learning of the workshop into the education system. There is no ready-made answer. I would respond to that question through another question—how can we dream of a world where knowing and learning can be delinked from the dominant link created with capitalism? Only learning and knowing can help us to find answers to such a question, at least to try to find answers.

And the Text Books?

Many people argue that schools with their textbooks represent only a fraction of the entire socialization and learning processes that people go through. Then why should the contents of text books matter?

Perhaps true, more true for some people than for others. As schooling becomes mechanically intense with great stress on examinations, the text books become very significant. And it becomes the knowledge that is legitimised and which also extend legitimacy to the employment sector.

If we look at text books of Science, Geography or any other discipline for that matter, we get no indication schooling encourages any depth of analysis. On the contrary, discouragement of analysis is clearly reflected in them. What is encouraged is the memorization of large amounts of information in capsule forms, the sort of information that remains at surface levels, in what is termed "general knowledge", considered useful for employment (through competitive tests).

Madan Sarup analyses such relationships as follows:

> "...any understanding of the complex interrelationships between education and the economy must be based on an analysis of the nature of work and of monopoly capitalism in the twentieth century...Many features from the factory system, that resemble the time and motion of 'Taylorism' have been introduced into the schools. There has been a systematic fragmentation of different sections of education and then within each sector. A parallel process in industry is the increasing division of labour. There is the increase of competition, in 'pacing' the rate of expected learning. And as the period of formal schooling has increased in terms of years, its content has gradually been trivialized. Again this reflects a parallel process that has been taking place in industry: deskilling...these features are encouraged in schools because they sustain hierarchies. These are some of the mechanisms of social control by capital" (Sarup 1978, 190-191).

The school synchronises with such patterns of economy. And that is the reason why school text books acquire a peculiar nature of content and expression.

In another context, the observations made by Freebody become important when we see that text books are almost the sole resource through which education is organized: "The approach of texts taken here was to consider them as crafted, representational objects which evoke a readership of a particular kind, which make representational claims on some aspect of reality and experience, and which entail purposeful choices and distortions as part of these representational processes" (Freebody 2003, 203).

The point of view with which text books are made is related to the point of view of schooling. Should schooling cater to knowing with depth and understanding or cater to a selected "coverage" approach?

Human-Nature Relations in Geography Text Books

The cryptic language and presentation of text books go along with an important qualification—the perspectives in writing of the texts. Do we find multiple perspectives and stand points? Or do we find only a dominant view? This becomes important in the premises of social sciences because it analyses social relations. With the prime role of the text book in schooling and the legitimacy that it carves out from a body of knowledge, it is important to critically review the contents of what transpires to be the official knowledge.

Human-nature relation which is at the heart of Geography's ontology and epistemology and is an influential area of human operation is examined here. It is to be kept in mind that those relations develop along with spatial, political and historical organization of society.

The Point of View

It is mainly the text books of State Council of Educational Research and Training (SCERT), Madhya Pradesh that are examined here in detail. However, these are not in any significant way different from those of other states and of National Council of Educational Research and Training (NCERT). So the following observations apply to the overall status of Geography text books in our country.

A striking feature observed in the texts is that it assumes a confrontation of two entities called "human" and "nature". The human takes too much from nature, an assumption that is very similar to Arvill's (1983) expressions of *assault* and *attack* on environment.

And according to the text, this situation can be remedied by a changed posture from humans: "Man has to learn to live in perfect harmony with the physical and biological environment so that the earth continues to be habitable for future generations as well"[3]. As in Toffler's new image of nature, "instead of conceiving ourselves as engaged in a bloody war with nature, we are moving toward a fresh view that emphasises symbiosis or harmony with the earth" (Toffler 1981, 307).

The texts attempt to look into the issue of degradation of the environment, for instance, why does human intervention tend to become destructive? The cause is pointed out as population growth. This simplistic arithmetic assumes that greater dependence on the environment is caused because more people would need food etc.

Obviously this explanation ignores the varied relations that people exercise with the environment. It ignores the social structures through which the implications on environment operate differently within society. The same numbers of people would not utilize the same amount of resources; their socio- economic placements need to be known.

It is stated that technological efforts like green revolution would combat the problems of food for a large population. Thus more intense human interaction to alter nature (as in the case of green revolution) is aspired to combat the environmental problem (brought forth through population growth).

But this contradicts the tone of human restraint seen elsewhere in the text with a clear articulation of human limitations and the belief that environment determines human actions: "We know that all man's activities are according to his natural environment. His selection of industries, his agricultural products ... even his religions and ideals are guided by his natural environment"[4]. These contradictory postures make it difficult to find out the environmental problem, its nature, causes etc.

For a long time, Geography had debated on who dominates the human-nature relation. Whereas the Deterministic School of thought stressed the upper hand of environment, the Possibilistic School pointed out the dominant role of humans in shaping Geography.

In such frameworks, the nature of their relationship was not getting sufficiently addressed. But as geographic analysis diverges much from such naive positions, the complexities of "human-nature" get revealed.

When the environmental problematic is explained in terms of sustainability of nature and the control of human reproduction (as in the text books), contradictory and simplistic positions are created that veil a closer examination of the problem. If it is to be recognized that relations of production transform the environment, it can be seen that nature is often transformed in the image of capital.

This qualitative transformation of nature gets illustrated through biotechnology enhanced through seed industry, the transformation of river ecosystem through dam construction or of forest through monoculture etc., but it is to be noted that ecological crisis as we know it today is not only a crisis for humanity, but also a crisis for capitalist reproduction itself (D'Souza 2003, 27).

Political Connotations

It is noted that in our textbooks, a uniform entity called society (human) is shown to interact with another called nature. The description of society as being homogenous leads to an assumption that humans exercise and can exercise only one sort of relation with nature. But this can easily be unwrapped by examining the different ways in which societies historically organized for livelihood and production.

The simple and egalitarian social relations seen in tribal societies have much to do with the ways in which their livelihood is organized in relation with nature. But settled agriculture is spatially and hierarchically organized for shaping the environment. And this is done through the structuring of labour and the accumulation of wealth in hierarchically different strata of society.

In capitalist industrial societies, both technology and human labour are shaped to maximize profit. The emergence of class society is the social organization of capitalist production. Here nature is conceptually transformed into resources for production through factory organization that distances labour and capital.

That is why homogenous (synonymous to equal) society can be discussed only in the context of tribal societies. In other societies, caste/strata and class are part of the ways in which the relationships with nature are organised.

Hence, when text books discuss environmental problematic in terms of population control and sustainability of resources, it speaks from hierarchical positions of the production system and the social organisation. This trend which has political connotations is further examined:

> "Ideas about environment, population and resources are not neutral. They are political in origin and have political effects. Once, for example, connotations of absolute limits come to surround the concepts of resource, scarcity and subsistence, then an absolute limit is set on population. And

> the political implications of a term like overpopulation can be devastating. Somebody somewhere, is redundant and there is not enough to go round. Am I redundant? Of course not. Are you redundant? Of course not. So who is redundant? Of course it must be them…it is only right that they, who contribute so little to society, ought to bear the brunt of the burden…" (Harvey 1996, 148-149).

Economically and socially marginalized sections of society are made to bear the guilt of growing population and the coercion of population through state policies.

The frame of homogeneous "human" and "nature" ignores socio-political dimensions of human interventions. For instance, the innumerable struggles of peoples thrown out of their relationship with nature (through development projects, urbanization etc) does not become a point to be discussed. The process by which people are displaced from their living spaces and relationships is the outcome of a particular sort of relationship of humans with nature. Marcuse (1972) captures this as "the ever-more-effective domination of man by man through the domination of nature" (in Gregory 1978, 45).

It is therefore necessary that our understanding move beyond explanations focused on population as the threat to environment. We are experiencing a phase where the World Bank and other agencies advocate environmental sensitivity, often employing terms like "sustainability" and "harmony with nature". Such terms can have varying implications.

For instance, Bridge (1998) in examining mining capital, writes on the active use of discourse that reflects an offensive attitude towards environmental problems—indicating a new posture of capitalists as protectors of nature. Opposition to mining has emerged not only from environmentalists but from capital interests that are "implicated in the economic restructuring of former mining regions (e.g. real estate, the high-tech sector, tourism) and who are alarmed at mining's potential for ecological expropriation through the devaluation of property" (Beck in Bridge 1998, 221).

Therefore different interests portray different discourse on nature, each explicating a different concern on environmental risk. Nature gets defined as stocks of resource, which is to be managed, and in the process the concerns and discourse are trapped within the capitalist frame: "Sustainable development becomes … a question of efficiently managing environmental stocks and flows, a project for which capital is not only eminently suited, but which it alone can undertake" (Beck, op. cit., 228)

As Harvey remarks it is very hard not to be in favour of "sustainable" practices as the term has a positive connotation; politics and policies give it a status of being environmentally sensitive. He critiques the concept of "sustainable development" in the following way:

> "The duality of ecological and social projects here takes some interesting twists for while it is true that debt repayment, as ecologists argue, is at the root of many ecological problems, it is precisely the threat of debt default that forces international finance to recognise that all debate about ecoscarcity, natural limits, over population and sustainability is a debate about the preservation of a particular social order rather than a debate about the preservation of nature per se". (Harvey 1996, 148)

New Frames for Geography Text Books

One of the noted efforts of educational innovations in the country was brought forth through "Eklavya" in the 1980s whereby major points of change were seen in the role and nature of text books. In different ways both science and social science texts of Eklavya attempted to develop contextualized learning. To a remarkable extent, the pedagogic bottlenecks usually seen in schooling through abstract ideas and concise texts were overcome.

The present section focuses on Eklavya text books that depart from the frameworks mentioned in the preceding section. A critical appraisal would be attempted because the need is to look for possibilities and their refinement in education so that an all encompassing relation like that of human-nature could be analysed keeping in mind their theoretical and political implications.

In Geography, the theoretical difference between the dominant school text books and that of Eklavya[5] was that a concerted effort was seen in the latter to integrate the physical and the human. This was a major shift from the separate status of these in text books that made any understanding of relationships very challenging, in fact a non-issue.

The integration of the physical and the human is done largely through case studies. For instance, take the example of water inflow and retaining capacity that get influenced by the landforms, slope, texture of soil etc. Human interventions are to a certain extent guided by these natural premises. Three case studies focus on such processes—Pahwadi, a village in Satpuras, Kotgaon in the Narmada plains and Balampur in the Malwa plateau.

Human-nature relationships as developed through agriculture, irrigation etc. are examined in these cases. The high production in agriculture in the plains enhanced through water availability draws in surplus labour (migrant) from the Satpuras during harvest. Thus people move into the plains seasonally for wages.

Another set of case studies focus on regions/countries as representations of torrid, temperate and frigid zones. These are Indonesia, Japan and Tundra region. These regions can be differentiated on the basis of natural environments; variations are seen in temperature, rainfall, natural vegetation etc. Human interventions like intensive paddy cultivation through contour stepping are prominent features in Indonesia and industries are prominent in Japan etc. In Tundra region, permafrost does not enable much of natural vegetation and cultivation, yet humans make a living through certain interventions.

In both the sets of case studies we see that a physical/natural criterion is selected as the basis for classification of regions. Thus landforms in the first and heat zones in the second become the primary bases for categorization. The logic of selecting such a base needs to be examined closely if we perceive that human-nature relation is at the ontological and epistemological root of geographic understanding.

It is true that in all the studies human interventions are examined in detail. But the categories of regions and comparisons encouraged among them project an assumption that human-nature relations develop on the basis of the physical characteristics of places. This can be contested because each type of landform or heat zone do not produce the same type of geography (here geography implies a process, product and cause of human-nature relation).

All plains (mountains or plateaus) do not produce similar human interventions. Similarly each heat zone cannot be generalized for human-nature relations. The similarity lies on the natural processes only; they cannot be extended to the human interventions. For example, it was mentioned that from Satpura Mountains labourers migrate to the plains during harvest. But today harvesters and threshers have altered the scenario. The migrant labour could possibly be going elsewhere or could have made their migrations relatively more permanent or could have done away with migrations altogether by creating more interventions in the Satpuras. All these are possibilities; one can find out what really happened. There cannot be a generalisation that uplands are areas of out-migration. The tea/coffee estates of Western Ghats and Assam are areas of in-migration of labour.

Likewise, each heat zone cannot be demarcated for a particular type of human intervention that differentiates it from other heat zones. In fact, many natural differences are transcended through human interventions. This is aptly described below:

> "Night and day, the seasons, life cycles in the animal and plant world and the biological processes which regulate human reproduction and the body, are typical encounters with various kinds of temporality. But each of these stands to be modified or even transcended as we harness sources of energy to turn night into day, as we use an international division of labour to put fresh produce into our shops at all times of the year, as we speed up the lifecycles of chickens and pigs through genetic engineering and as human life expectancy rises with improved living standards and medical knowledge" (Harvey 1996, 211).

By adopting natural criteria as the base for examining a region, it gets trapped in environmental determinism which is historically a major point of critique in geography. Theoretically this has been debated and abandoned even though school texts are seen to still adhere to deterministic assumptions (*see* Sunny 2006/2009). If categories are required, they need to be shaped out of human-nature relations because the present categories are structured to emphasise environmental determinism. If relations become the criteria, then analysis gets opened to reflect on social organizations that necessarily evolve along with human interventions. Then geography would not grapple with separate entities called human and nature; the epistemology would rest on their relation itself.

So what constitutes an area or region demarcated for geographic study? What should be the logic or criteria? How and why should demarcations be important if the focus is on the processes of socio-spatial change? These are some of the questions that are kept in mind in this section.

Changing Geographies

Historically irrigation was a major intervention (much before green revolution) through which agricultural production was increased. It is true that major irrigation systems were built in plain areas be they the deltas of South India, the plains of Western India or the Gangetic plains. What sort of socio-spatial organizations developed with such agricultural production? Some examples can be examined here.

The anicuts of the deltas of South India and the inundation canals of Western Uttar Pradesh gave way to new irrigation systems during colonial times: "the intention of the British rulers of India was to replace rain fed subsistence agriculture by impoverished villagers with high input commercial agriculture by a substantial yeoman peasant class" (Hardiman 2008, 56).

Colonial interventions of irrigation were experienced in Western Uttar Pradesh and Punjab, but decisions were also influenced by the opposition of people in the canal regions to taxations etc. These developments reorganized the geographies of colonial India and produced new regions.

For instance, Eastern Uttar Pradesh had been traditionally more agriculturally developed than Western Uttar Pradesh. Natural endowments of water and fertile soil in the Gangetic plain had enhanced these developments. But the social geography of the east was also marked by *zamindars* ('landlords') who took much rent from the tenants, a situation not so prominent in the west. After the colonial intervention through irrigation canals in the Ganga-Doab region, a reversal of agricultural prospects was seen. The relative autonomy of farmers in the west also favourably affected the process. Thus geography of Western Uttar Pradesh got redefined through labour intensive agriculture and high outputs.

In colonial Punjab, canal irrigation brought forth the emergence of a hydraulic society. It saw social organization getting shaped in a particular way because land was assigned to only one caste in a village and that became the undisputed dominant caste there, "The middle level caste rank established village societies by encouraging subordinate peasants, artisans and agricultural labourers to migrate with them. Old hierarchies of class and caste were thus reestablished in the new areas…" (Hardiman 2008, 54). He also notes that "Society in the canal colonies became as hierarchical and exploitative as anywhere else in rural India—a chance to make a clean break with the past had been abandoned" (*ibid.*, 56)

The proposed legislation by the British to increase tax on canal water brought in a wave of protests in 1907. It is interesting that whereas the British usually crushed such protests in the colonies, in Punjab the Viceroy vetoed the legislation on taxation which was already passed by the Punjab assembly. The reason for this attitude was that the British relied on Punjab for recruitment of soldiers into the Indian army and they did not want to take risks that could shake the loyalty of the Punjab troops (see Hardiman 2008).

It can therefore be seen that a large number of factors shape and reshape the geographies of regions. It is such examination of multiple processes that help understand geography. A process like colonialism that spread over the entire Indian subcontinent or a natural premise like plain or river valley do not necessarily produce the same sort of geography. The variations in socio-spatial sphere are not only after effects but at the same time the causes for continuity and change in geography.

Recently in a workshop for text book writing held in Trivandrum, attempts were made to examine Geography in the light of such concerns. One such experience is discussed here to examine the rationale for demarcation of an area to study human- nature relationship.

Kuttanad in Southern Kerala was chosen for study, an area marked physiographically by its elevation below sea level and its location on the banks of Vembanad Lake. A natural relation between the fresh water lake and the sea is seen through infiltration of saline water into the lake in those months of the year when fresh water inflow from the rivers become relatively less. Both biotic (like breeding of fish) and the annual cycles of agriculture and other activities of humans were shaped by these processes.

Kuttanad gets geographically marked as a region because of several interventions in agriculture that made it one of the most prominent paddy cultivating areas of Kerala.

In the predominantly water filled Kuttanad, land used to be reclaimed for cultivation. One of the major challenges to cultivation (contradictory to other paddy cultivating areas) is that water has to be removed to enable cultivation. The earlier interventions were made through water wheels that employed a number of people, often in shifts, so that the work of draining water continued into the night as well. Later, kerosene pumps replaced the wheels.

Quite early in the nineteenth century, Travancore had merged with the world market through trade in cash crops. This had created shortage of food grains in the region and one of the steps taken by the Travancore princely rule was the encouragement of paddy cultivation in Kuttanad. Cultivation here was much more labour intensive than elsewhere, given the challenges of removing water.

The social organization of agricultural production had evolved through feudal relationships that demanded a slave-like labour and loyalty from the tenants. Caste and gender bifurcations helped evolve a hierarchical relation that met the challenges of cultivating Kuttand and kept the controls and wealth within the premises of an elite class.

The infiltration of saline water had made cultivation impossible in the region during half the year. During the other months it was susceptible to floods. This twin problem was sought to be technologically intervened in the 1950s. One intervention was a spillway that would divert the "excess" water flowing down rivers (into Vembanad Lake) into the sea. This was designed to control floods in Kuttanad. Another intervention was a bund that checked the saline infiltration. Along with other interventions of modernity like chemical control of pests, chemical enrichment of soil, hybrid seeds etc., Kuttanad became what was termed as the "rice bowl" of

Kerala. It became a region of high productivity of paddy through modern agriculture.

Social organizations also underwent changes. Even though the landlord-tenant relationship continued relatively a long time with its slave-like labour, modern education and political thought influenced the social fabric. Social movements sought to question and challenge the caste basis of socio-economic life. Class struggles that emerged through communist ideology also took roots here. Moreover, Kuttanad became an area where land ceiling was successfully materialized through government legislation.

Agricultural production did rise considerably, making Kuttanad a region of surplus paddy production and establishing market relations with other areas. But it is also to be noted that the traditionally rich sources of protein like fish and duck etc. were affected by the chemicals that accumulated in waters of Kuttanad that were stagnant due to the technological interventions of the spillway and the bund. The natural regenerating qualities of the soil were heavily influenced by the chemicals. The limits of agriculture which delved into overdoses of chemicals were reflected through questions on sustainability of high yields. The threat to ecological regeneration also became a concern.

The main economic activity of people living in the region is still paddy cultivation, but with changes social organization—the larger land holdings and labour intensive tenant relations gave way to smaller holdings worked on through wage labourers. As agriculture was influenced by modernity, the social makeup transited through modern occupations that were not necessarily agriculture based. Education also paved mobility to young people in search of employment. The eroding of caste and gender based socio-economic structure opened up new economic avenues for people both within and outside the region.

Demarcation of Regions

The demarcation of regions had always been an epistemological practice in geography. It is true that geography has gone through various paths of questioning, rejecting and reinstating the regional frame of enquiry. There is need to reflect both theoretically and epistemologically on how spatial and regional bases are to be perceived.

It is seen that "space" is borne of multiple social and political processes and therefore there is nothing permanent about demarcations of space.

The pre-modern spatial organizations were dramatically modified through modernity. The tightly bound and hierarchical social organization of production eroded to reorganise through a less hierarchical order that opened up Kuttanad and integrated it with a widespread modernizing process that was both technological and social. Post-modern changes emphasise the continuing importance of spatial processes. It would be interesting to explore socio-spatial modifications that continue to shape the region. For instance, intensification of the following processes is observed in the region:

1. Increasing construction of roads and bridges (the area is traditionally dependent on water transport).
2. Lesser number of people depending economically on agriculture.
3. Out-migration of people for employment to other parts of the country and world.
4. New economic spaces of tourism that commodifies nature through house boats and resorts.

These need to be further examined to understand geographic developments of the region, also vis-à-vis the processes of economic globalization-liberalization. How have these processes changed the geography of Kuttanad? Do we still see that Kuttanad can be demarcated as a region on the basis of human-nature relations? Or would we make new demarcations which take into account the current processes of geographic change? Only a detailed examination would help us to assess the situation.

Thus nature of changing geographies tells us that the concept of regions and their demarcations need to be necessarily flexible.

Conclusion

Reading is an important part of learning and the education system aspires to make people well-read. But there is a very tight selection and scheduling of what should be read in school. As examinations and competitions become the soul of schooling, legitimization of knowledge operates within a narrow spectrum accompanied by high doses of compulsion. This robs reading of the pleasure that knowledge gives the reader.

The style of writing of text books are marked by fast-pacing and tight packing which not only brings forth a peculiar language into text books, it also encourages learning where analysis and reflections are not priorities.

In analyzing the content it is seen that only the dominant side of issues is represented in the text books. This is problematic as far as social science education is concerned. As social sciences largely relies on interpretations, it is essential that text books be broad based and state different points of view on any issue.

Moreover, socio-spatial organizations that are part and parcel of human-nature relations are not even mentioned; human and nature are treated as two entities without internal differentiations.

All these bring very explicit political connotations into schooling because the fast pacing (in style) and analysis of human-nature relations (in content) are seen to be nuanced with the rationale of capitalist relations.

Eklavya's text books demonstrate significant departures from conventional school texts. In Geography, the major change was the attempt to bring the physical and the human into a single frame of analysis. And this is done by studying specific regions whose demarcations are made on the basis of physical/natural premises. If we critically appraise this base of categorization of regions, it is seen that this enhances environmental determinism—something that has been challenged and abandoned theoretically in Geography. How are demarcations to be perceived? Are they permanent or are they indications of socio- spatial changes over space and time?

Regions indicate space and scale of human operations, it can indicate being part of a larger flow of relations, it can also indicate geography of struggle. All these functions have clear connotations of geographic change that are spatial, social and scalar. The issues and concerns of geography rest on these processes and not on the physical demarcation per se.

It is a study in social sciences that Geography entails through space (or region). This implies space that is not simply physical but something created through human- nature interactions. To demonstrate the point, the article delves into examination of a region (Kuttanad) and it is seen to be the cause as well as the effect of geographic processes. Pre-modern Kuttanad shaped agricultural production through certain social formations and relations. It is seen that this Geography later became the springboard for struggles that aspired for a less hierarchical social order.

The pedagogic perspectives in designing a text book are discussed in the article through the nature of language and content that the books use. Contextualised writing like that of Eklavya does demonstrate major breakthroughs from the conventional texts that employ cryptic writings and formula. They are also commendable for open endedness and balanced social science perspectives. But even as these are enhanced, there is need to critically examine the frame works in which geography is handled.

All these have implications on the ethos of mass education and the aspirations of education for equality. Phule and Ambedkar had argued for *accessibility* to legitimised knowledge which was traditionally represented by Brahminical knowledge and replaced by the knowledge systems of modern education. They realized that through modern education, the age old exclusions of caste system could be questioned. Modern education, especially its functions of opportunities became a political arena.

Whereas Phule and Ambedkar argued for inclusion of marginalized castes into the system of legitimacy, Gandhi argued for the legitimization of that knowledge which was traditionally designated as that of the lower caste groups; and this included almost all life skills and physical labour. But Gandhi's "basic education" which gave importance to manual work and skills of the hands remained only as symbolisms, not obtaining the importance it called for. Thus what is taught and who are included/excluded become important and interconnected concerns on the question of education for equality.

Levitas (1974) observes that if competition governs the school system to equip students for social climbing, it becomes a challenge to equality (through schooling): "...education would only serve the distribution of opportunities. And this contradicts the demand for social revolution whose depth and spread if borne in mind would show that a "vital source of its power is education" (Levitas 1974, 198).

It is argued that the content of education becomes an important influence in the concern of education for equality. For instance, when schools legitimize the capitalist version of human- nature relationship, it places the onus of environment deterioration on the poor and marginalized people. The processes by which displacement and marginalisation occur are not analysed. And when texts use the frame of environmental determinism, it is unable to explain the social organizations (and change) that are part of human-nature relations. Thus, even as attempts are made to create new conceptual frameworks for text books, they need to be continually evaluated on perspectives and epistemology.

It is also concluded that natural processes like climate, landforms, soil etc or macro socio-economic-cultural ones like colonialism, modernity, capitalism, economic globalization etc. do not produce similar geographies. Multiple processes, interactions and struggles produce socio-spatial variations that are flexible and are undergoing changes. And Geography bases itself on such enquiries that can be defined as an interweaving of the spatial and the social. Geography text books still have not gathered these epistemologies to create learning tools and perspectives to help students understand the world.

Notes

[1] In Eklavya, a voluntary organization working in education for more than 2 decades and based in Madhya Pradesh.
[2] In Madhya Pradesh
[3] M.P, SCERT, 2004, IX; 3
[4] M.P, SCERT, 2004, IX; 3
[5] Social Science text books for grades 6, 7 and 8, made by Eklavya and implemented by Government of M. P. for some years.

References

Arvill, R. 1983. *Man and Environment*. Harmondsworth: Penguin.

Bridge, G. 1998. Excavating nature: environmental narratives and discursive regulation in the mining industry. In *An Unruly World? Globalisation, Governance and Geography,* ed. A. Herod, G. O'Tuathail and S.M Roberts, 219-238. London and New York: Routledge.

Claval, P. 1993. *An introduction to regional geography*. Oxford/ Massachusetts: Blackwell Publishers.

D'Souza, R. 2003. Environmental discourses and environmental politics. In *The Value of Nature*: *Ecological Politics in India,* ed. S. Kothari, I. Ahmad, and H. Reifeld, 23-38. New Delhi: Rainbow Publishers.

Freebody, P. 2003. *Qualitative Research in Education*: *Interaction and Practice*. New Delhi/ London/ California: Sage.

Gregory, D. 1978. *Ideology, Science and Human Geography*. London: Hutchinson and Co.

Hardiman, D. 2008. The politics of water in colonial India: the emergence of control. In *Water First*: *Issues and Challenges for Nations and Communities in South Asia,* ed. K. Lahiri-Dutt and R.J. Wasson, 47-57. Los Angeles/ London/ New Delhi/ Singapore: Sage.

Harvey, D. 1996. *Justice, Nature and Geography of Difference*. Oxford: Blackwell.

Jaffrelot, C. 2005. *Analysing and Fighting Caste*: *Dr. Ambedkar and Untouchability*. Delhi: Permanent Black.

Levitas, M. 1974. *Marxist Perspectives in the Sociology of Education*. London: Routledge and Kegan Paul Ltd.

Rammohan, K.T. 2006. *Tales of Rice*: *Kuttanad, Southwest India*. Trivandrum: Centre for Development Studies.

Sarup, M. 1978. *Marxism and Education*. London: Routledge and Kegan Paul Ltd.

Sunny, Y. 2006/2009. Analyzing current practices in geography education. *Economic and Political Weekly* 41 (3): 270-278. Reprinted in *Indian Geography: Perspectives, Concerns and Issues*, ed. Ravi S. Singh, 257-278. Jaipur/New Delhi: Rawat Publications.

Toffler, A. 1980. *Third Wave*. London: Macmillan for Pan Books.

Textbooks and Documents

National Curriculum Framework, 2005, NCERT, New Delhi, 2005.

Our Environment, Textbook in Geography for class VII, NCERT, New Delhi, 2007.

The Earth, our Habitat, Textbook in Geography for class VI, NCERT, New Delhi, 2007.

Hamari Duniya, Class V, NCERT, New Delhi, 2004.

Bhoogol, NCERT, New Delhi:Class 6 (2002, 2004), Class 7 (2004), Class 8 (2004), Class 9 (2002), Class 10 (2003), Class 11(2002).

Madhya Pradesh SCERT, *Social Science*, Madhya Pradesh Textbook Corporation, Bhopal: Class 5 (2004), Class 6 (2000), Class 7 (2002), Class 8 (2004), Class 9 (2004), Class 10 (2004).

Social Science, Eklavya, Bhopal: Class 6 (1998), Class 7 (2001), Class 8 (2004).

CHAPTER SEVEN

BRING GEOGRAPHY ALIVE AT SCHOOL LEVEL VS. REDISCOVERING GEOGRAPHY AT HIGHER EDUCATIONAL LEVEL IN INDIA

P. K. RYNGNGA,
NORTH-EASTERN HILL UNIVERSITY

Introduction

Quiz after quiz has shown that children today do not know where some places are. How often have you heard this lament about "geographic ignorance" or "geographic illiteracy," as it is commonly called?

Now, take that complaint and turn it around. What does it say about geography? It says geography means knowing *where* places are. That's what some geographers call "place-name geography." It's vital, but it's the least of what we expect budding geographers to learn.

Geography is more than most people think. Geography is to space what history is to time. It is a spatial way of thinking, a science with distinctive methods and tools, a body of knowledge about places, and set of information technologies that have been around for centuries. Geography is about understanding people and places and how real-world places function in a viscerally organic sense. It's about understanding spatial distributions and interpreting what they mean. Geography is a dimensional science and humanity based on spatial logic in which locations, flows, and spatial associations are considered to be primary evidence of earth processes, both physical and cultural. Its hallmarks are spatial analysis, place-based research (e.g., regional studies, area studies, urban studies), and scientific integration.

The familiar litany also implies that geography is just for kids, something you learn in elementary school or high school if you're lucky, and use for the rest of your life without any need for new learning. Does geography really matter for grownups? Of course it does. Geographic knowledge, understanding, and skills matter, for instance, is important in formulating foreign policy, designing and using GIS, and just about everything else in society that involves locations, movements, and flows.

As seen today, there is a wide gap of geographic knowledge between school going children and the higher educational level. When making such a statement it does not mean that the same measuring scale is used. But to elaborate on this, it simply means that the weightage of geography as a subject at school level is becoming lighter and lighter these days. This is so because geography is now forming only a small segment of a syllabus to bring about one subject i.e., Social Science/Social Studies. This practice has really killed the thirst for obtaining more knowledge and information and in fact it has limit students to recognize geography as a subject. But on the other hand, at higher level the subject of geography has widened so much particularly with the advent of technologies- the use of computers and the combination of Remote Sensing (RS) and Geographic Information System (GIS). The success and extensive use of these tools have penetrated to other streams and disciplines of studies in which these other streams have also adopted and borrowed the techniques of geography.

An Excellent Invention Worldwide

Geography was founded at least 2,500 years ago and advanced by Greek, Roman, and Chinese scholars throughout the Classical Age. Prior to the Renaissance, geography and astronomy—interrelated, spatially oriented disciplines—were pre-eminent sciences. For a thousand years, geography was recognized and valued. "How excellent inventions are geography, arithmetic, astrology, and the rest!" wrote Saint Augustine in *The City of God* in the early 5th century.

Then came the Great Interruption of the middle ages, and geography became a fantasy. For a thousand years, its real body of knowledge was preserved by Irish monks and advanced by Arab and Persian scholars. Its rediscovery by European scholars was central to the enlightenment of the Renaissance. It thrived from c. 1450 to 1948 based on explorations, from c. 1600 to 1900 based on the information needs of westward expansion, and from c. 1915 to 1947 based on geopolitics. Its accomplishments include the following:

a) The first proposal of continental drift was published by geographer Abraham Ortelius in 1596. Its most famous champion, Alfred Wegener, was a climatologist (climatology is a subspecialty of geography).

b) Biogeographer Alfred Russell Wallace co-discovered evolution in 1859. By today's rules of precedence, he would be considered the principal discoverer because he published first, but Charles Darwin's friends made sure his paper was read ahead of Wallace's at a meeting of the Royal Society.
c) President Woodrow Wilson's geographer Isaiah Bowman was the author of America's globalization policy, which proclaimed that America could lead the world through political and economic means rather than military conquest. Bowman went on to become President Franklin D. Roosevelt's geographer, as well, and one of the top six architects of the United Nations.
d) Geographer Carl Sauer led the way toward new understanding of the ancient Americas and the vast populations they once contained. His first inklings were published in the 1930s; Geographer Bill Denevan discovered massive supporting evidence in 1961; Geographers widely accepted their findings by the 1970s; and science journalist Charles Mann's *1491* announced those findings to the public, to great acclaim, in 2005.
e) Geographer Roger Tomlinson is universally recognized as the father of GIS in the early 1960s. Duane Marble (2006) and other geographers were instrumental in laying the groundwork for GIS and have been heavily involved ever since. Geographer John K. Wright of the American Geographical Society published the earliest known expression of points, lines, and areas—concepts now central to GIS—and pioneered quantitative techniques, such as dissymmetric interpolation, that serve vital GIS functions today.

Development of Geography in India

Before World War II, writing in geography was in their infancy. Geographers were content with the plain description of facts. The major focus on which geographical writings appeared was in the field of agriculture, and the commodity approach was applied. If one look at the development of the subject in India, one can see that the primary institute in the sub continent has been Aligarh University and the foundation stone 'Butler Department of Geography' of 1927 can still be seen in the existing department. However, the real conferment of postgraduate degree started in 1932. Today there are 270 universities in India and a very large number of universities and the post graduate colleges impart teaching in geography.

After the Second World War, the country has seen the second phase of development of Geography. There was the freedom movement going on in the country, which culminated in the Independence of the country in 1947. There was a realization among the geographers that there was abject poverty in the country, in spite of abundance resources. Therefore, it was felt that along with the rising population, a lopsided development of towns, was taking place. Fortunately, this was the time when few Indian geographers who had done their doctorate abroad were available. A notable contribution in land use studies was made by S.P. Chaterjee of Calcutta who prepared land use map of Hoogly district and 24 Parganas and presented at the International Geographical Congress. Subsequently he worked on a book "50 years of Geography in India". Chaterjee's efforts led to the establishment of a very big organization "National Mapping Organization", later in 1978 renamed as National Atlas and Thematic Mapping Organisation (NATMO), in Calcutta.

In the fifties and sixties, geographers turned their attention towards Applied Geography working on "Problems of Land Use", and "Morphology and Development of Towns". Thus two major schools of geography emerged in India—one at Aligarh Muslim University with major concern in agricultural land use, agricultural efficiency, standard nutrition units, potential production unit, population and health care bearing on Medical geography and problems of environment. The other school of Urban Geography had its base in Banaras Hindu University, where urban studies particularly of small and medium towns, reached its high water mark under the guidance of R.L. Singh and his colleagues.

The strong wave from the western world greatly had impact the Indian geography through the application of statistical methods leading to precision of result. A new branch therefore emerged in Geography under the title of "Statistical Geography" and gradually gained the status of Advanced Statistical Geography. This has gone to the limit that no research papers/PhD thesis was submitted worth the name, if they did not apply some statistical techniques.

Geography has completed one full circle in its development in India. The relationship between man and land, man and air, man and water or man and the total environment has become the major focus. With the increase in population in India, the environment has come under great stress. The urban population in India is 30 percent and by 2025 it may go up to 45 percent, thereby the quality of life in cities would deteriorate. Slums and squatter settlements would increase. There would be more air and water pollution, which would lead to scarcity of drinking water. The official figure of people living under poverty is around 300 million people.

Therefore, poverty becomes a social issue that has to be looked at. Growing population brings about unprecedented negative ailments. Thus India needs planned utilization of resources for sustainable development, which may take care of protection of the environment; as well as development of resources.

Indian geographers are also focusing their attention on rural development, reduction of poverty, and problem of urban poor. Many Indian universities have focused attention on these problems by holding national and international conferences and inviting research paper on the above mentioned themes.

Lately, geography is enjoying resurgence in the history of its development due to the phenomenal success of GIS and the need for better understanding of foreign lands and peoples in this age of globalization and geopolitical turmoil. Peers in other disciplines now respect, and some are adopting, the hallmarks—spatial analysis, place-based research, and scientific integration of geography. With this resurgence, many existing geography departments in Indian universities are adding faculty and new graduate degrees.

Changing Quality of Education in School Geography

Geographic education as a core paper has been eliminated by the School Boards (both at the State and the Central Board of School Education) from school curricula in India during the 90's except the Indian Council of School Education. In elementary school, it has been lost in mishmash called "social studies" that neglects physical geography and spatial thinking. Most matriculating students do not even know it is possible to earn a college degree in geography.

Ironically, as the world enters a new age of massive change highlighted by technological innovation and international cooperation—and faces the pressing need to balance the effects of a century of worldwide exploitation with a new emphasis on conservation—geography is in danger of becoming a second-class subject in Indian schools. The status of geography in Indian educational system with geography particularly has been heavily affected by shifting sands of educational reform in the late 20th century. Geography suffered from uninspired instruction that was often restricted to memorizing place names, rivers, and capitals of countries and states at school level. Often devoid of context or enthusiasm, geography eventually was lumped with history, economics and civics into the social studies curriculum.

In spite of its broad adoption over the past decade by states and districts as a core academic subject, geography faces significant challenges in finding a place in a standards-based school curriculum that is increasingly focused on accountability for achievement. The recent syllabus revision considerable emphasis on merging of geography into Social Science/Social Studies has had the unintended consequence of jeopardize the role of geography in education. In many schools, geography is taught not as a discipline but as a set of skills and concepts that focus on the local community.

Rejuvenated Geography in Higher Education

Recent gains in geography's position in Indian institutions of higher education suggest that the discipline's domestic situation is improving. Geography emerged in India as a formal university discipline in some universities only with its own departmental standing around the turn of the twentieth century when the modern disciplinary matrix was being forged. During the first half of the twentieth century, geography carved out an increasingly significant place in Indian colleges and universities. The country experiences only a small percentage of increase in the number of students taking college-level geography after Independence. This period also saw significant weaknesses in geography's institutionalization. Geography never found an independent foothold in India's research universities. Even though there were weaknesses in the discipline, geography shared significantly in the baby-boom-driven expansion of Indian higher education in the 1960s and early 1970s. New universities and departments were founded and most existing departments expanded substantially. This was partly a response to the demand for programs that could train college teachers. In addition, in some institutions, geography departments were favoured because they could provide general education courses at a relatively low cost—thus helping to support expensive specialized research programs in other parts of the university.

In many respects the mushrooming interest in GIS in the country both in private and public sectors has opened the door of intensifying the importance of the subject which is a special case of geography's growing extradisciplinary impact, but the magnitude of this development and its impact beyond the academy suggest the need for separate treatment. Just four decades ago, the discipline's principal contributions in the mapping arena were in cartographic design, thematic mapping, and cartography. That all began to change with the development of GIS and related technologies—first in the 1960s, but with limited broader impact until the early 1980s. No single discipline can lay claim to GIS, but the tie to geography is strong. Geographers played a key role in the initial development of GIS, its purpose is to yield insights into a core geographical concern (the relationships among phenomena on the surface of the earth), and its end product is often a map (a type of representation closely associated with geography).

The development of GIS represented far more than an incremental modification of the way data is analyzed and displayed; it opened up entirely new avenues of inquiry, analysis, and representation. GIS offers a means of tying data digitally to a specific location so that its nature and impact could be assessed in relation to other data. Recognition of the analytical and applied value of the undertaking has precipitated what some have termed a revolution in the way contemporary society collects, stores, and disseminates information. Its uses are wide and deep. From ecological restoration to congressional redistricting, from military strategizing to transportation planning, from emergency service deployment to the modelling of the impacts of climate change, GIS has become an indispensable tool for a wide array of practitioners and analysts in the public and private sectors.

Among those in colleges and universities who were part of the GIS revolution, attention initially centred on programming and display issues. As the technology became more sophisticated and its utility became more widely recognized, these matters were taken up more and more by the private sector and by individuals working for government agencies. By the mid-to-late 1980s those in institutions of higher education were increasingly turning their attention to the nature and impacts of different geographic concepts used in GIS, the cognitive underpinnings of GIS, and the impacts of GIS on society. Many research universities devote significant attention and resources to this field—and its presence is growing at other colleges and universities as well.

The precise institutional situation of GIS differs greatly from campus to campus. In most places, GIS courses form part of the curriculum in more than one discipline—but which disciplines are involved varies. In institutions with geography departments, it is quite common for that department to be the principal home of GIS and the principal surveyor of GIS courses. Any forward-looking institution of higher education can scarcely avoid serious engagement with GIS. Not only are employment opportunities for students mushrooming in geo-technologies; GIS has become so ubiquitous that new arenas of research and practice are opening up with dizzying speed.

Under the circumstances, curricular offerings in the geographic technology arena tend to be popular among today's students—something that certainly has not escaped the attention of administrators despite their relatively expensive costs. When considered alongside the scholarly and applied impetuses for devoting attention and resources to GIS, the case for an expanded emphasis on the science and applications of geo-technologies in colleges and universities can be (and often has been) quite compelling. Because GIS have an independent existence from geography in some places, such expansions do not necessarily work to the institutional advantage of geography. But the close link between geography and GIS lends strength to the argument that the expansion of one should be accompanied by the expansion of the other.

The Expanding Job Market for Geographers

A further catalyst for geography's enhanced institutional position in Indian colleges and universities is the expanding job market for individuals with geographic training. This does not mean that the term geographer is often found in job descriptions. But seen in utilitarian terms, geography has long had an employment advantage over some social sciences and humanities because there are clear practical applications for some of the basic skills that are a part of geographic training: map making, location analysis, terrain analysis, environmental assessment, etc. Moreover, contemporary geographical training also typically involves a set of critical thinking, writing, and presentation skills which, when combined with substantive knowledge of particular places or geographic phenomena, can be very appealing to employers. It is thus no surprise that geography majors find their way to positions ranging from foreign policy analyst to travel agent, from forest conservation monitor to weather broadcaster, from map maker to planner, and from elementary school teacher to surveyor.

Geography's heightened general profile in recent years has led to expanding opportunities for geographers in a variety of arenas, but the turn toward geographic technologies has been a particular catalyst for growth. In fact, the world as a whole currently identifies geo-technologies as one of the three most important emerging and evolving employment arenas (along with biotechnology and nanotechnology).

Burgeoning employment possibilities have led many institutions of higher learning to develop courses and laboratories that teach students how to use and apply the latest GIS-related tools. But because the tools are evolving so rapidly, it is important to introduce students to more than software packages; they need to understand the nature of the tools they are using and the conceptual foundations on which they are built. This has led many institutions to develop a curriculum emphasizing the conceptual underpinnings of GIS. Such initiatives can push the curriculum in the direction of geography, for that is the discipline that most clearly focuses on the acquisition, analysis, and display of spatial information or data. Any layer in a GIS involves decisions about data prioritization and spatial representation that are rooted in geographical principles and concepts. Without an understanding of these principles and concepts, the nature and import of the decisions being made can easily be ignored. Recognition of the importance of geographical understanding for GIS users has surely spurred the growth of a number of geography programs over the past fifteen years. Moreover, many job openings in the geo-technology arena are not simply for individuals with technical skills. They are in substantive areas (environmental analysis, transportation planning, marketing, planning, and weather forecasting) where geo-technologies are being heavily employed. Under the circumstances, students who have been exposed to substantive issues and problems along with geo-technologies are at an advantage.

Even outside the geo-technologies arena, larger social developments are making the understandings and skills of geographers ever more relevant to the world of paid work. The past fifteen years has seen significant growth in the fields of environment management and planning. At the same time, the globalization of business has led to increased opportunities for individuals with foreign area expertise, and the expansion of globally or regionally focused social, political, and economic programs has opened up a wide array of possibilities for those with substantial international experience and understanding. Geography's ties to all these arenas make it a potentially attractive focus for investment at a time when universities are under pressure to adapt to changing employment possibilities for their students.

General Awareness of Geography's Relevance in India

There is much to suggest that geography has increasingly entered the public conscience. In recent years, headlines have been dominated by widely publicized discussions of climate change, debates over new geopolitical maps, competing claims about the benefits and drawbacks of globalization and reports of extraordinarily devastating natural disasters. Each of these matters focuses attention on core geographical concerns: the human-environment interface, the nature and significance of differences from place to place, and the interlinked fortunes of places. Moreover, geographic representations and analysis (mapping, spatial analysis, etc.) are often a part of their public consideration.

Indians have long been aware that developments in other places have an impact on the country, of course, but there has arguably been a significant increase in the level of concern about global connectivity in an era dominated by news about terrorism, the outsourcing of jobs, growing gaps between rich and poor, and major natural disasters. At the same time, the personal computer revolution and the development of the Internet has brought the wider world into people's offices and homes in a way that was unimaginable just two decades ago. These developments have inevitably made many people feel that geography, or at least geographical understanding, is essential in the contemporary world.

Of course much of the popular concern over geography is based on the assumption that it is a subject to be addressed at the pre-college level—a clear reflection of people's limited understanding of the discipline. But the surge in interest in geopolitics, globalization, and environmental change is surely fuelling a more general interest in geography—including among college and university students. Fuelling the growing level of public interest in geography is a gradual rise in the public visibility of geographers themselves. Geographers have a long way to go before their public presence approaches that of political scientists, economists, historians, or biologists, but increasingly high profile contributions by geographers—identified as such—undermine the sense that the discipline involves little more than place-name memorization. The institutional impacts of this perceptual shift are difficult to discern—particularly since academic administrators sometimes come from institutions that do not have geography programs. Nonetheless, growing general awareness of geography's relevance has arguably served both to strengthen the discipline's position at colleges and universities where it has long been present and to foster growing interest and passion in geography at institutions of higher education where it has no formal presence.

Scholarly Awareness of Geography's Potential in India

The work of certain geographers has long had an impact outside the discipline. William Morris Davis was a major figure in the earth sciences in the early twentieth century and Carl Sauer had a significant influence on anthropology in the mid-twentieth century. Nonetheless, before the 1960s such examples were more the exception than the rule. The situation began to change in the 1960s and 1970s—first when geography's quantitative revolution brought the discipline closer to the intellectual currents shaping a number of neighbouring disciplines and then when the works of other western geographers with a structuralist or humanist bent of mind began to influence Indian geographers. The writings of Richard Chorley, Arthur Strahler, and C. Warren Thornthwaite paved way for Indian physical geographers to appreciate the potential of a quantitatively grounded physical geography. The work of Peter Haggett, Torsten Hägerstrand, and Brian Berry opened the eyes of social scientists to the potential of spatial analysis. By the mid-1970s David Harvey's Marxian analyses of urban geographic change made him a major figure in urban studies circles.

Geography's growing presence across the Indian education over the past decade did not quickly translate into an enhanced institutional position in India. It took time for momentum to develop and for Indian geographers to be seen as more than individual aberrations. The pace of extra disciplinary recognition began to accelerate in the mid-to-late 1980s, however, as geographical scholarship began to achieve greater prominence and scholars from other disciplines began to draw more explicitly on geography in their work. Additional fuel for geography's extra disciplinary impact came from the development of an increasingly high-profile feminist geography, which in part reflected the growing number of women joining the discipline. The work of Susan Hanson, Doreen Massey, Geraldine Pratt, and Linda McDowell, among others, helped shape the larger feminist project. Finally, the development of RS &GIS played a role that impact the pioneering work of not only the geographers but other allied disciplines as well became increasingly visible by the mid-to-late 1980s; the use of the term "geographic" in GIS highlighted the fundamentally geographical character of this new, powerful approach to understanding and representing the hanging character of the earth's surface. As the 1980s gave way to the 1990s, larger intellectual currents helped to foster interest in geography among the practitioners of neighbouring disciplines. Growing questions about the large-scale normative assumptions of everything from rational choice perspectives in political science to mechanistic process approaches in geology directed attention to differences from place to place: a core concern of geography. In addition, GIS provide scholars to look at a wide array of questions like the ability to analyze and represent spatial differences in new and exciting ways. In a related vein, the turn toward postmodernism in the humanities and parts of the social sciences led to a wave of interest in the role of context in the production of ideas and meaning.

What Can We Do to Bring Back Geography Alive at School Level?

Restoring geography at school level should be the best of our interest as a citizen of India and as geographers. Nowadays, there is a conspicuous reluctance to say the G word particularly at school level. *Spatial* or *geospatial* are even more acceptable at the higher educational level than *geography* or *geographic*. When geographers discover anything new, they are likely to be identified with the closest thematic discipline rather than geography, even if geographic methods were absolutely essential to the discovery. However, the importance of geography should be emphasized by adopting the following measures:

1. Tell people what geography really is: The vast majority of geographers define their discipline based on spatial perspective and methods rather than content. Tell everybody that geography is to space what history is to time. No one would seriously suggest that a university should exist without a history department. Why should any exist without geography?

2. Make administrators accountable: If your favourite university does not have a geography department, ask why. If in a school when you find that geography is combined with another discipline, ask administrators to explain the reasoning behind its placement or even insist the school board through the state government,/authorities to bring geography as separate subject and not as a squeezed curriculum. Together, we must send the clear message that every discipline that helps explains our world matters to every one of us. We and those who support us throughout society must draw a line that no politician or administrator can cross without pain. Every individual who crosses that line must face accountability. The hardest part, of course, will be to set aside our own discipline's short-term interests in favour of the greater good, but solidarity is our best hope to staunch the purge of disciplines. Together, we must oppose every threat, starting with the one that has already occurred.

3. Lobby for a rational legislative agenda: Funding for geographic education, development, and research must be increased by at least two orders of magnitude, partly to solve the labour shortage in GIS and partly to educate the general public. These funds are needed to fulfil six modest principles:

a) Every elementary and high school student must have the opportunity to learn basic geography and experience GIS technology.
b) At a minimum, every freshman should reach college knowing that geography is a viable major with solid career prospects after graduation.
c) Every college student must have access to a full geographic curriculum—thematic, regional, methodological, and technological—within the set of college destinations among which he or she normally would choose.
d) Scholarships must be available to support the best and brightest students who choose to pursue undergraduate and graduate degrees in geography.
e) Research grants must be available to encourage substantially increased geographic research, including fieldwork, both foreign and domestic, by faculty and students.
f) Development grants must be available to upgrade or create geography faculties throughout the nation; to ensure topical, regional, methodological, and technological coverage; to upgrade GIS facilities; and to promote community outreach.

Conclusion

The trends discussed in the latter part of this article seem so encouraging to those who believe that geography has an essential role to play in higher education and society. It is important to remember, however, that geography starts from a comparatively marginal position. This suggests the need for some caution in drawing conclusions from this piece. Two points deserve particular emphasis. First, since geography's presence in India's institutions of higher education is comparatively small, positive trends can take on an appearance of significance that is greater than they deserve. Second, since geography's institutional position is somewhat fragile, it would not take much to reverse the current upward trajectory.

A second issue arises out of the complex relationship between geo-technologies and the discipline of geography. Even as geo-technology is, in many respects, at the leading edge of geography's forward movement in India, some have expressed concern that it could either spin off as a subject unto itself or so dwarf geography's traditional conceptual foundations and approaches that the discipline will lose its status as one of the liberal arts and sciences While neither of these scenarios may be likely, there are certainly instances in which technically oriented programs have moved significantly away from general geography course offerings or have developed a techniques curriculum that is almost exclusively oriented toward vocational training. Such strategies may enjoy success for a time, but unless efforts are made to bring in broader geographical concepts and theories, geo-technology programs can become distanced from the intellectual currents shaping the sciences and social sciences. Conversely, there are instances in which GIS is being pushed away by geographers uncomfortable with the technological turn or concerned about diverting attention from qualitative or social theoretic matters. Such actions risk alienating the GIS community from geography while undermining an important catalyst for geography's growth. The challenge is to build bridges between GIS and other parts of geography. Since some of the most innovative work in geography is taking place at the intersection between these arenas there is room for optimism on this point.

However, for all the potential pitfalls, the last word must necessarily be positive. Over the past decade geography has moved from a position of substantial vulnerability to one of increasing centrality in the India's higher education landscape. It has expanded its presence at institutions with long-standing programs, and it is finding a foothold at a growing number of institutions that long neglected the discipline. If current trends continue, there is hope not just that geography will grow in prominence, but that the students and faculties at Indian colleges and universities will be increasingly enriched by what the discipline can contribute to their institutions' intellectual life and pedagogic mission.

Acknowledgement

The author would like to thank the Head of Department of Geography, North Eastern Hill University, Shillong for his valuable help when I needed the most. I also extend my sincere thanks to Mary Ann Jyrwa, lecturer PGT, Shillong for sparing most of her valuable time for discussions. Special thanks to the Librarian North Eastern Hill University, Librarian North Eastern Council, Shillong for allowing me to stay beyond time in the library so as to enable me to complete my work.

References and Select Bibliography

Ahmad, A. ed. 1996. *Progress in Indian Geography 1992-1996*. New Delhi: Indian National Science Academy.

Ahmad, A. 1997. Fifty years of geography in India and abroad. Presidential address delivered at the 20th Indian Geography Congress, December 28-31, in Thiruvananthapuram (Kerala).

Agnew, J. 2005. *Hegemony: The New Shape of Global Power.* Philadelphia: Temple University Press.

Association of American Geographers 2005. *Guide to Geography Programs in the Americas 2004-2005.* Washington, DC: Association of American Geographers.

Bagchi, K. 1977. Relevance in geographic study and research: A case study for restructuring of geographical syllabi from secondary to post-graduate level. *Geographical Review of India* 39 (4): 295-301.

Bhardwaj, S.M. 1987. Toward an Indian Geography agenda. *National Geographical Journal of India* 33 (3): 327-334

Bockenhauer, M. 1993. The National Geographic Society's teaching geography project. *Journal of Geography* 92 (3): 121-124.

Bonnett, A. 2003. Geography as the world discipline: connecting popular and academic geographical imaginations. *Area* 35 (1): 55-63.

Chakraborty, S.C. 1987. Contemporary directions in geography and our options in India. *Transactions, Institute of Indian Geographers* 9 (2): 1-8.

Coppock, J.T. and D.W. Rhind 1991. The history of GIS. In *Geographical Information Systems: Principles and Applications*, Vol. 1, ed. D.J. Maguire, M.F. Goodchild and D.W. Rhind, 21-43. Harlow, Essex: Longman Scientific & Technical.

Clark, W. 2003. *Immigrants and the American Dream: Remaking the Middle Class.* New York: Guilford Press.

Das, H.P. 1972. The teaching of geography as a science subject in schools. *Geographical Review of India* 34 (2): 93-97.

Dayal, P. 2002. The agenda for Indian Geography. In *Indian Geography: Voice of Concern*, ed. A. Kapur, 313-330. New Delhi: Concept Publishing Co.

De Blij, H.J. 2005. *Why Geography Matters: Three Challenges Facing America: Climate Change, the Rise of China, and Global Terrorism.* London and New York: Oxford University Press.

Deshpande, C.D. 1991. Indian Geography in 2001 A.D. In *Emerging Trends in Indian Geography*, ed. Jayamala Deddee, 156-165. Jaipur: Rawat Publications.

Diamond, J. 2005. *Collapse: How Societies Choose to Fail or Succeed.* New York: Viking.

Dickason, D.G. 1992. Geographic research on India: contribution from the United States. *National Geographical Journal of India* 38 (1-4): 147-163.

Dobson, J.E. and R.C. Durfee 1995. ORNL and the geographic information system revolution. *Oakridge National Laboratory Review*

28 (1). Available at: http://www.ornl.gov/info/ornlreview/rev28-1/text/gis.htm (Accessed 6 January 2008).

Foucault, M. 1980. Questions on geography. In *Power/Knowledge: Selected Interviews And Other Writings, 1972-1977*, ed. C. Gordon, New York: Random House.

National Geographic Survey 1988. *Geography: An International Gallup Survey*. Princeton, NJ: Gallup Organization, Inc.

Geography Education Standards Project 1994. *Geography for Life: National Geography Standards 1994*. Washington, DC: National Geographic Research and Exploration.

Giddens, A. 1984. *The Constitution of Society: Outline of the Theory of Structuration*. Berkeley: University of California Press.

Gewin, V. 2004. Mapping opportunities. *Nature* 427: 375–376.

Gober, P. *et al.* 1995. Employment conditions in geography. *The Professional Geographer* 47(3): 317-346.

Gregory, D. 1994. *Geographical Imaginations*. Oxford: Blackwell Publishers.

Hanson, S. 2004. Who are 'we'? An important question for geography's future. *Annals of the Association of American Geographers* 94 (4): 715-722.

Harvey, F., M. Kwan and M. Pavolovskaya 2005. Critical GIS: Editorial introduction. *Cartographica* (Special Issue) 40 (4): 1-3.

Hossler, D. and L.H. Litten 1993. *Mapping the Higher Education Landscape*. New York: College Entrance Examination Board.

Kapur, A. 2002. *Voice of Concern: Indian Geography*. New Delhi: Concept Publishing Co.

Krugman, P. 1991. *Geography and Trade*. (Cambridge, MA: MIT Press.

Kwan, M. 2004. Beyond difference: From canonical geography to hybrid geographies. *Annals of the Association of American Geographers* 94 (4): 756-763.

Marble, D. 2006. Defining the components of the geospatial workforce—who are we? *ESRI ARC News* 27 (4). Available at: http://www.esri.com/arcnews/Winter06articles/winter06gif.htm (Accessed 6 January 2008).

Marcus, M.G. 1979. Coming full circle: Physical geography in the twentieth century. *Annals of the Association of American Geographers* 69(4): 521-532.

Mark, D.M. 2003. Geographic information science: defining the field. In *Foundations of Geographic Information Science*, ed. M. Duckham, M. F. Goodchild and M. F. Worboys, 3-18. New York: Taylor and Francis.

Mark, D. M., N. Chrisman, A. U. Frank, P. H. McHaffie and J. Pickles 1997. The GIS history project. Available at: http://www.ncgia.buffalo.edu/gishist/bar_harbor.html (Accessed 26 September 2007).
Marston, R. A. 2005. Geographical Research for Public Policy, *AAG Newsletter* 40 (11): 3 & 6.
Martin, G. J. 2005. *All Possible Worlds: A History of Geographical Ideas* (4th ed.). Oxford and New York: Oxford University Press.
Massey, D. 1994. *Space, Place, and Gender.* Minneapolis: University of Minnesota Press.
Murphy, A.B. 2006. Enhancing geography's role in public debate. *Annals of the Association of American Geographers* 96 (1): 1-13.
Murphy, A. B., H. J. de Blij, B. L. Turner, II, Ruth Wilson Gilmore, and Derek Gregory 2005. The role of geography in public debate. *Progress in Human Geography* 29 (2): 167-73.
National Research Council 1997. *Rediscovering Geography: New Relevance for Science and Society.* Washington, DC: National Academy Press.
Pandit, K. 2004. Geography's human resources over the past half-century. *The Professional Geographer* 56 (1):12-21.
Pred, A. 1990. *Making Histories and Constructing Human Geographies: The Local Transformation of Practice, Power Relations, and Consciousness.* Boulder: Westview Press.
Richardson, D. and P. Solis 2004. Confronted by insurmountable opportunities: Geography in society at the AAG's centennial. *The Professional Geographer* 56 (1): 4-11.
Sayer, A. 1992. *Method and Social Science: A Realist Approach* (2nd ed.). London: Routledge.
Schulten, S. 2001. *The Geographic Imagination in America, 1880-1950.* Chicago: University of Chicago Press.
Schuurman, N. 2004. *GIS: A Short Introduction.* Oxford: Blackwell Publishing.
Sharma, H.N. ed. 2004. *Progress in Indian Geography 2000-2004.* New Delhi: Indian National Science Academy.
Shearer, Christopher 2003. Geography education: learning the "why of where". *Journal of National Association of Elementary School Principals* 82 (3): 32-35.
Smith, N. 1987. Academic war over the field of geography: the elimination of geography at Harvard, 1947-1951. *Annals of the Association of American Geographers* 77 (1): 155-172.

Turner, B.L., II 2005. Geography's profile in public debate "Inside the Beltway" and the national academies. *The Professional Geographer* 57 (3): 462-67.

PART III:

RESURRECTING PHYSICAL GEOGRAPHY

"If we do not find anything pleasant, at least we shall find some thing new".
—Voltaire (1694-1778) in *Candide* (1759)

CHAPTER EIGHT

MICROGEOMORPHOLOGY: AN UNFINISHED JOURNEY

SUDIP DEY,
TRIPURA UNIVERSITY

Prelude

The natural earth displays various colours in different spectral resolutions as all the natural features of the earth from nanometre to kilometre scales have their own geophysical signatures, which is one of the core interests of landform scientists. In the forwarding note of the proceedings of the 2nd Conference of Indian Institute of Geomorphology, Prof. S.N. Sinha remarked that for a long time geomorphology was considered as a subject that deals just the academic interest of landform and its origin. Development of applied geomorphology before the '50s of the last century prepared the platform of application of geomorphic knowledges. Since then the interests of the landform scientists were also directed towards the impact of human interventions on the surface of this earth. During the mid-20th century, geomorphologists started to deal with the environmental issues. This trend surely enriched this subject by developing its interdisciplinary nature as many scholars, from the other disciplines, like geologists, environmental scientist, hydrologists, civil engineers etc. were attracted to this field which encouraged inclusion of environmental and disaster management in geomorphic studies later. Very recently in the editorial note of the famous journal *Geomorphology*, Lang and Glade (2008) remarked that the landform researches have developed due to the advancement of methodologies and technical applications. In fact it has been observed that the approach of landform science is more systematic and scientific during the recent years than few decades back. Now the landform scientists are concentrating on the study of hazards and geomorphic disasters by the help of advanced field techniques, laboratory analysis and statistical and modelling tools. This global trend has also influenced the geomorphic studies in this country. Since India is considered as the natural laboratory of geomorphology due to its physiographic diversities and geomorphic problems, Geomorphology in India is now an applied science which is not only discussing the geo-hazards, but also actively playing role in planning for sustainability. The dynamic nature of this subject always attracted the present author who believes that studying of landform is one of the most scientific ways to understand the environment and impact of human interventions on it. This made him decide to enter in the "realm of Geomorphology" to continue his learning of nature.

Background: Current Geomorphic Issues and Learning Geomorphology

Since the present author started his research on coastal environmental changes, the most attractive issue to him was sea level change in Holocene epoch and its regional impact on the coastal landform. Though most of the previous works done by the well known scholars like Niyogi (1970), Chatterjee (1972) Umitsu (1987) and Merh (1987) highlighted long term changes of coast, this author always views that the direct impact of seasonal variation particularly in tropical coastal areas cannot be ignored for the assessment of its geomorphic changes. Thus concentrated on sea level fluctuations by the seasonal changes of Digha area in West Bengal and tried to establish that it plays an important role for the gradual changes of coastal morphology (Dey 1999). After that I started to search the evidences of long term shoreline change of West Bengal. In this study we worked with the geological evidences to detect the actual changes of shoreline in this area over the last 2000 years (Dey *et al.* 2002). In this work we studied each individual feature like sea beach, sand dunes, spill basins, mud flat, etc. The impact of shifting of shoreline upon human civilisation was also studied in the process (Dey and De 2002; Dey 2005a). We also observed that individual features of the coastal area like sand dunes preserve the evidences of early shoreline changes and sea level conditions (Dey *et al.* 2005). A couple of years back I published my work on spatial characters of landform in a small area like Digha-Shankarpur coastal tracks by sea level changes (Dey 2006). From 1999 to 2004 this author and his fellow researches worked on different geomorphic features of mesotidal coast which play very important role to understand that even small parts of land have their own expression according to process efficiency.

In Search of New Dimension in Landform Science

Studies on geometric forms of small features in coastal area

We were not satisfied with the general expressions of geomorphology that we were doing that time. Thus we were seriously thinking for a different window. I have always a view that though little has been discussed in the scientific literatures, the small features or micro-features are integral parts of landforms and cannot be ignored in the discipline of geomorphology since influences of environmental changes on geomorphic processes first reflect by the small sculptures and designs. Selby (1985) and Bloom (2002) accorded that generally smaller features of landform can be created and destroyed more rapidly than larger ones with little fluctuation of processes. It depends on the variability and efficiency of processes under specific environmental conditions. Two very remarkable micromorphological studies done by Bryant and Davidson (1996) and Davidson and Simpson (2001) deal with some of the important significance of the small features. Encouraged by the earlier works this author also conducted detailed study on small features of coastal beach like rip channels, different types of wave ripples, tidal ripples etc (Dey 2003a). In this work, forms and origin of the micro-features were discussed. In 2004 in IAG symposium (held in Calcutta University) I focused some paleo-coastal micro-features which preserve the early history of shoreline shifting of Bengal basin. The complete work is now in press which also deals with a genetic classification of micro-features (Dey 2007).

Modelling for the assessment changes of small features of coastal area

During the 1999-2000, when I was working on seasonal variation of sea beach by sea level change I decided to concentrate on analysing the impact of sea level fluctuations upon beach profiles. In this time the "Topo-Scaling Factor" model (Dey 2000) was formulated which was as follows:

$$T_S = \frac{W}{w} \tan \beta$$

Where,
T_S=Topo-scealing Factor, W=width of the beach during low tide, w=width of the beach during high tide, β= angle of the beach.

This model proved that it can assess the changes of geometric forms of the beach profiles. Then I became more interested to detect changing nature of sea beach by seasonal variation of coastal processes. Thus another model "Process Efficiency Index" was prepared (Dey 2002) which can be calculated by the following equation:

$$P_E = H\left(\frac{0.246U}{g}\right)\tan\ \beta$$

Where,
P_E=Process efficiency value, H=average wave height, U=wind speed/hour, g=acceleration due to gravity, β= angle of the beach.

Later these two models were used for change detection of the microgeomorphology of rip-channels in different parts of the West Bengal coast (Dey 2003; 2004; 2005b). It was also proved that these models are suitable for risk assessment as they deal with the immediate impact change of sea level or coastal processes on small designs of landform or microgeomorphic features.

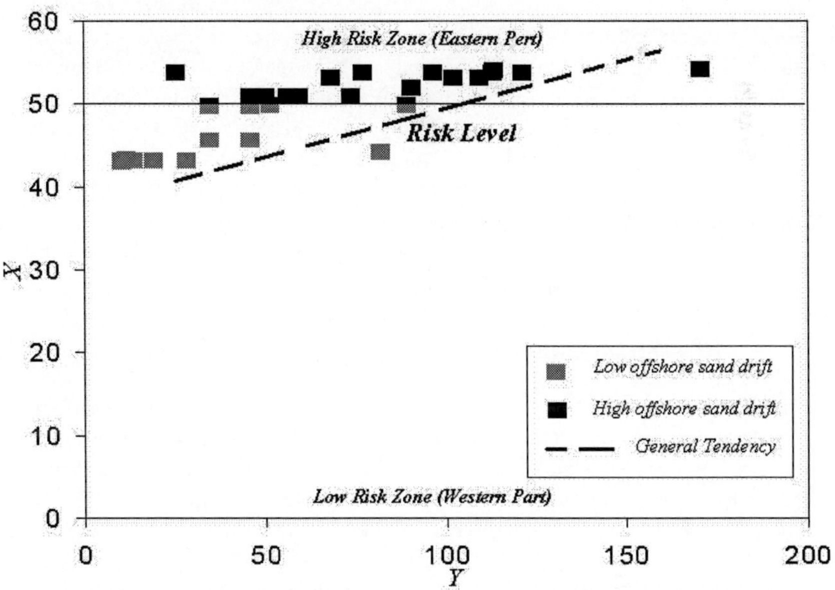

Fig. 8-1: An example of risk analysis by "Process Efficiency Index"
(*Source*: Dey 2003)

Geomorphic Studies in Different Micro-Regional Environments

Conceptual modelling for understanding a new study area

At the end of 2003 I realised that only coastal studies are not enough to establish the concept of "microgeomorphology" and it is very important to study the smallest units of landform developed by different environmental conditions. Fortunately I joined in Govt. Women's College of Agartala, Tripura and just after 9 months I joined in the Department of Geography and Disaster Management of Tripura University. I found that Tripura was a completely new study area for me where only a few scientific researches have been done in geomorphology. So, I decided to take the opportunity to start working in this new field area with my research scholars. My initial works on the geomorphic succession of Tripura shows that during the Tertiary period this part of Eastern Geo-Tectonic Province of Bengal basin was nourished by marine-coastal processes which gradually changed by the south ward shifting of shoreline. Sediment deposition during the Miocene epoch in the Bengal Basin plays a vital role for the southward shifting of the shoreline and formation of this landmass. Tectonic upliftment and climatic changes might play additional roles with regards to the relative fall of sea level. Nandi (1985) accorded that maximum sediment deposition occurred under warmer climatic conditions than recent times. The Pliocene and Quaternary periods are marked by tectonic changes in this area. Tectonic activities are believed to have resumed even in recent times (Nandi 1977; 1985) and have exerted significant influence on landform evolution in this area. It was an attempt to explain the Tertiary-Quaternary geomorphic development of this paleo-coastal area by formulating some conceptual models (Dey 2005c).

Bio-geomorphology: Study in highlands of Tripura

Geomorphologically the highlands of Tripura attracted the present author due to its specific environmental characteristics. During the field work in a natural lake of Agartala, this authors and his research team observed a typical pattern of variations of plant types and density with the change of micro-level slopes on the highland (Dey *et al.* 2006). Plant coverage and plant base area on the bottom of the highlands, slopes and top of the highlands were measured on the basis of their height of the plants also measured systematically. They are measured by the followings (Dey *et al.* 2007):

a) Plant coverage= Number of plants / 25 sq metre
b) Plant base area= Horizontal extension of roots of the plant

Due to uneven and rugged landscape, dense undergrowth (in some parts) and regular landslides on the slopes, it is very difficult to measure wide areas plant coverage analysis. So we decided to measure number of plants on the selected slopes of the highlands in an area of (5m x 5m) 25 sq metre. Besides that existence of different animal species were identified during the field work. Seasonal variations of slopes and runoff channels were measured for the assessment of seasonal control upon tropical geomorphology and the affected places by human intervention were also identified during the field work. A complete influence of micro-relief and microgeomorphology upon the micro-level plant ecology of the highlands in a rural area near Tripura University campus has been recently described in a recent work which is already in press(Dey *et al.* 2009a).

Technical Applications: Some Recent Works

Mapping by digital method

Personally I was interested to develop the technical application in geomorphic studies to touch the international standard. We started work on mapping of the vertical run-off tunnels of an artificial cut profile of a highland area. Data collected through observation have used for analysis, mapping and statistical testing. Field maps and photographs are used for preparing the maps of small features like vertical tunnel or weathering marks. Simple GIS software has been used for digitisation and mapping.

Use of high resolution images for geophysical signature identification and assessment

This author always believes that radiometric data is more useful than multi-spectrum data for detecting geophysical signatures. Early works by Whithe and El Asmar (1999) and Bagli and Soille (2003) presented the technical concepts of grayscale use for geo-physical analysis. Very recently I have worked with an international research group and successfully used grayscale for the detection of shoreline changes of a part of Indonesian coast (Marfai *et al.* 2008). The international acceptance of the methodology encouraged me to concentrate with my research team to apply greyscale analysis for microgeomorphic study (Dey *et al.* 2008a). Another successful application of radiometric data by this author and his team was the measurement of sediment in the water of Rudrasagar Lake of Tripura (Dey *et al.* 2008b). Already we were concentrating to identify tectonic signatures of Tripura and thus we applied many geophysical methods of remote sensing data interpretation. Considering the solid earth a reflectance scale was prepared based on Edinsink's calculation (1990) and tectonic signatures were identified successfully in a recent work. The equation was as follows:

$$R = (d*d*a)/z + (d*d*b*c)/z$$

Where,
"R" = reflectance, "d" = mean earth-sun distance in astronomical units, "z" = cosine of the solar zenith angle, "a" = the intercept, "b" = gain coefficient, and "c" = digital count.

Apart from that we also experimented by BRDF imaging for detecting the fault zones of Tripura, India. The raw image was prepared on the basis of two parameters namely, wave length of light and bidirectional reflection distribution function (BRDF). Thus BRDF models stand on the following two very common algorithms:

$$L = [L_{\lambda_1} \dots L_{\lambda_k}]^\top .$$

Where,
λ = wavelength of light, L = radiance

$$f_r(\mathbf{x}, \omega_i \rightarrow \omega_o) = \frac{L(\mathbf{x} \rightarrow \omega_o)}{L_\mathrm{i}(\mathbf{x} \leftarrow \omega_i)\, \mathrm{d}_\perp \omega_i} .$$

Where,
fr= bidirectional reflection distribution function for a surface point, x= ratio of the reflected radiance, *Li* = incoming radiance, *L*= outgoing radiance, *i*= differential irradiance for an incident direction

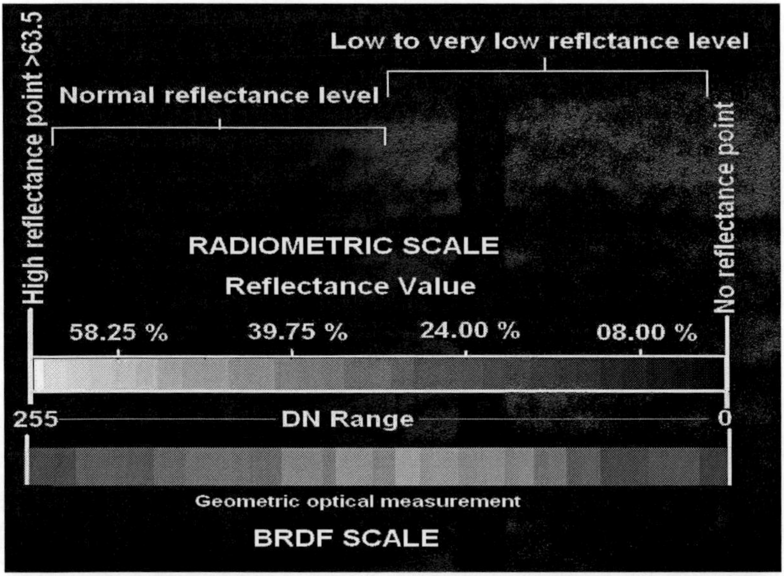

Fig. 8-2: Reflectance scale based on Edinsink's equation and BRDF scale.

In a recent work (Dey *et al.* 2009b) we successfully use the radiometric scale of reflectance and identify three unknown fault line in southern part of Tripura which is a very earthquake prone zone.

Development of digital method for microstructure study

We realised that if we want to touch the international level by working in the field of microgeomorphology we should concentrate on study on microstructure which play very important role for the development of micro-features. Microstructure studies of the sediment are directed towards two basic objectives, firstly to understand the forms and patterns of sediment depositions and secondly to understand the environment and genesis of depositions. This is to understand the evolution of landform by microgeomorphic studies. In the 70s of the last century Brewer (1976) analysed the soil fabric at micro level which have been very important for the growth of systematic micro-fabric studies especially the development of terminologies in micro-fabric study. After that microstructure studies by van der Meer (1987; 1993), van der Meer and Laban (1990), Menzies and Maltman (1992) also dealt with the evidences of genesis and deformation.

An early study of Lachniet *et al.* (1999) shows that microstructure study can also explain the environment of deposition (wet type and dry type). Piazolo *et al.* (2002) discusses the complex impact of microstructure process on microstructure development. They also proposed the macrostructure mapping in their work. With the development of computer software the digital technology has become very popular for micro level studies now-a-days. A very recent work by Mamtani *et al.* (2007) deals with the geometric forms and depositional nature of BIF (Banded Iron Formation) by high resolution microphotography. Microstructure mapping has also been proposed by those authors which permits an estimation of strain rate and temperature condition. Considering the recent trend, we decided to perform an experiment on digital techniques for microstructure analysis of sediment layers. The main encouragement of this study is the observation of different reflectivity of light of different minerals like the glittering micacious and dark coloured clay perticals. In a recent work the research team of the present author applied DCA (Digital Colour Analysis) method for microscopic studies (Dey *et al.* 2009c). This analysis was successfully used to detect for the temporal change of mineral deposition in a part of Bengal basin.

Digital data sets or micro-images from sediment thin layers by microphotography were prepared for analysing (i) siliceous and micaceous fabrication and (ii) pattern of sediment and clay penetration. Since the scanned area consists of mainly sand, mud and micacious materials, the red band within total of R-G-B scheme can be chosen for its better contrast and higher reflectance value to differentiate grains. Thin layers of the sediment samples are used as the reflector or objects of reflection. For DN interpretations and micro-fabrication mapping Red Band (from RGB combination) has been taken.

To understand the sediment layer characteristics and micro-fabric mapping the percentage of sediment coverage is measured by the number of pixels of each type of sediment deposition in the sample. The following mathematical explanation (Dey *et al.* 2009d) is suitable for measuring the area coverage:

$$A = \frac{100 \times P}{P_N}$$

Where,

A= Percentage of area converge; P=Number of pixels in particular class (sediment); P_N= Total number of pixel in images

BRDF (Bi-directional Reflectance Distribution Function) was calculated in our laboratory for mapping the micro-facets of the sediment samples under reflected lights.

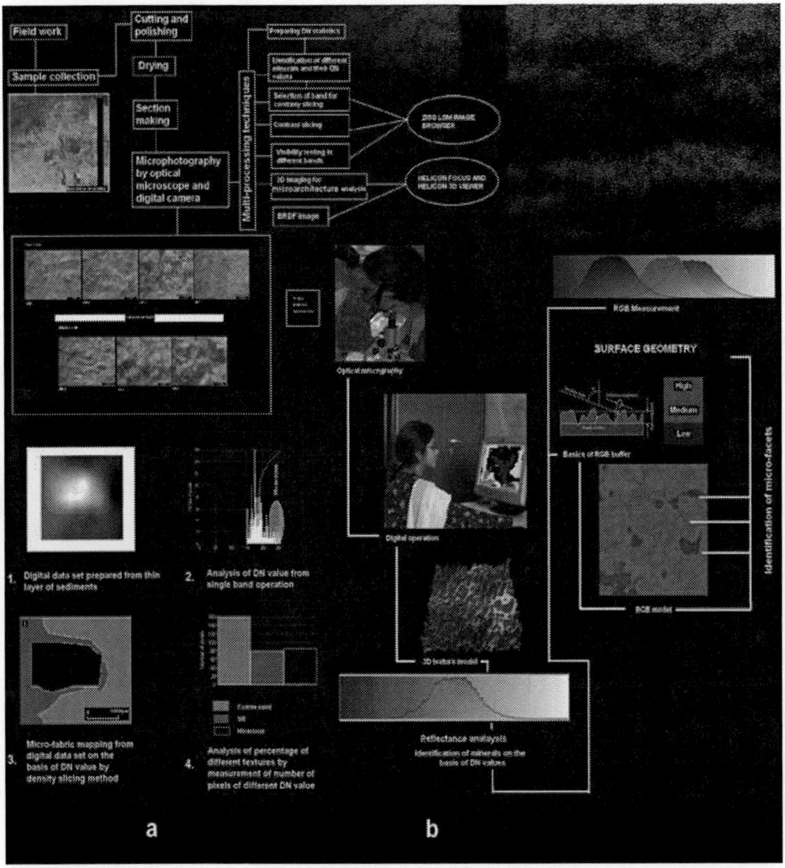

Fig. 8-3:
(a) A laboratory sheet showing the methodological approaches of recent experiment result of micro-fabric mapping and analysis
(b) BRDF micro scale and monochromatic nano scale imaging

Recently nanostructure imaging is also done for understand the physical characteristics of the depositions (Dey *et al.* 2009e) though the initial experiments were not much successful because of bad image productions. To be very honest still this research group is under development and trying to get success in nano scale imaging and its application for landform evolution study. In many cases we applied multi processing technique which is actually an integration of different imaging software and mathematical formula. The targets of imaging experiments are to understand the change of source zone and change of current flow in temporal scales.

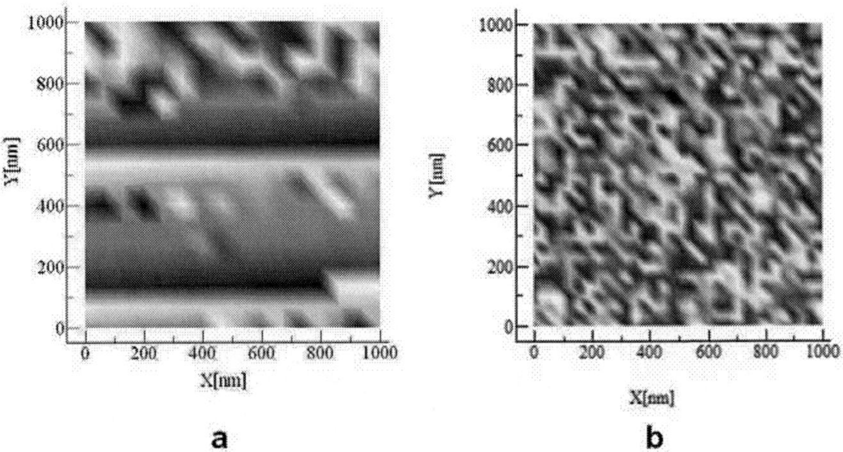

Fig. 8-4: Nanoscale images (a) low quality image and (b) result of satisfactory imaging operation

Planning for the Never-Ending Journey towards the Dream Destination

Theoretical planning

The destination is to understand the language of the nature through the study of the smallest units of landform. It is not very easy task but very encouraging when we think we are going through an unknown path where we have to face many research challenges. Actually the boys and girls of this research team like to take challenges. Since research is a job of regular learning, rectification, updating and modification of thinking it is not possible to make a plan for very long term. Thus we prepared our plan just for upcoming years. Actually presently our target is to identify the micro-geomorphic features, their development and form through geological time. It is also our aim to establish their significance with environmental changes.

Through the in-depth studies of landform it has been proved that the landform is more complex than it looks like. Normally these small features are the result of geomorphic processes. But through our studies it has been observed that the other natural environmental processes like exogenetic (denudation), biological etc also play important role for the development of small sculptures and designs on the surface. Thus it is also our target to establish a hypothesis, e.g. geomorphology consists of many small features which are the result of combined works of different environmental processes within a very short area of the earth's surface under specific environmental condition.

It is also our target to establish that the study of the forms of microgeomorphic features is not only a simple tool for in-depth explanation of landform of any area but also very helpful to assess the geomorphic succession under certain environmental condition, micro-level function of environment and environmental fluctuation.

Quaternary landform research group

Recently Quaternary Landform Research Group has been established in the Department of Geography and Disaster Management, Tripura University. For the development of the group my teacher and senior glacial-geomorphologist Prof. Guru Prasad Chattopadhyay encouraged me and my research students Ms. Chandrani Debbarma and Ms. Prasamita Sarkar took the necessary initiatives. The researches by QLRG focus on last two million years landform evolution in the tropical coastal and paleo-coastal areas. The principal aspects of researches are: Evolution and dynamics of Tropical Coasts, Microgeomorphology and Cenozoic Microstructure, Lacustrine and Tectonic Geomorphology. Geological, geomorphic and ecological evidences of coastal changes are the major interest of study this group. The members of this research team are engaged in studies of Quaternary period depositions especially microstructure and microfabric of the sedimentary rooks in the western and eastern geo-tectonic provinces of Bengal basin. Experiments on petrographic image analysis (PIA) and microstructure mapping are going on by using digital technology. Small units of geomorphology or microgeomorphic features of coastal and tectonic areas are studied by

QLRG as one of the important evidences for the assessment of recent landform changes and their influencing geo-climatic factors.

Rudrasagar Lake of Tripura is studied for last five years by this research group. The QLRG concentrates on the history of landform evolution and present geo-ecological status of this lake of international importance. The geo-tectonic signatures on the geomorphology of Baromura hill, Tripura also attracts the researching members of QLRG. This research group is engaged in the experiments of mathematical modelling and their application for scientific assessment of short term and long term landform dynamics. The group members are also using Remote Sensing and GIS techniques for detecting shoreline changes, changes of wetlands, signatures of tectonism, etc. (Dey *et al.* 2009). A recent interest of this research group is the experiments on analysis of various spectral resolutions for geo-spatial data analysis. Attempts are going on to study the changes of geo-physical characteristics of the earth surface through the spectral and spatial resolution in the Environmental Geomorphology Laboratory and Digital and Micro-Analysis Laboratory of QLRG.

Technical applications

Keeping the recent development of Earth Science at international level we are also planning to take more help of technology and use Microgeomorphology for the assessment of land degradation and hazards. Presently we are planning to develop two experimental works in our digital laboratory:

a) To identify the range of DN values of some specific resolution in different micro-relief conditions. Actually we are working to prepare the detail data sets of different micro level slopes and their environment.
b) To understand the paleo-geomorphic condition by analysing the micro-fabrication of the sediment layers by digital analysis (single band operation of the micro-images by sophisticated micro-image processing software).

The digital micro imaging laboratory has concentrated on using the hardware models prepared by ourselves. We are planning to continue the two main experiments of this laboratory which are as follows:

a) To formulate an acceptable model by which land degradation can be assessed under different slopes of landform

b) To understand the influence of the range of different environmental processes for the development of small designs and sculptures of landforms.

Looking for the Future

Being a just ordinary member of the scientific society I know very well that it is not possible for any honest and dedicated researcher to reach in all-satisfaction room. But we should target the current international level. Through my research experiences I realised that there are few negative points in researches in this country which are playing as barrier. I agree with my colleagues that there are so many problems in a developing country like India, but at least we can try to find out the way to overcome. For example, recent trend of international researches says that without an efficient research team, high quality work is very difficult today. Thus we can start encouraging our students from their PG level by research oriented syllabus. Then we can get at least some good students at M.Phil. and Ph.D. level, who have some idea on research and can play important role in a research team. In our Tripura University, many M.Sc. theses of Geography and Disaster Management are original works and accepted in national and international journals. These M.Sc. students are actually playing the supporting role in our research team and at the same time they are learning the technical skills for their future. Lack of instrumentation and laboratory development is another serious matter in Geography in our country. I am also facing the problem for developing a good digital laboratory and a sedimentology laboratory. I wonder why in the Geography departments of Indian universities sediment analysis laboratory, seismology laboratory, paleo-environment laboratory etc are ignored. I have seen only few Geography departments in Indian universities have well instrumented Geomorphology, Hydrology and Climate / Weather monitoring laboratories. True fact is that Physical Geography is now little deprived in India, which is not good at all for the development of the subject. Laboratory skills should be developed among the budding geographers. Otherwise collection of secondary data and manipulation will be continuing. This is a big problem for Indian researches which is not encouraging the young geographers to conduct experimental works in their laboratories and to search new methods to generate field data by their own. Yes, funding is a problem, but instead of that we can try to develop some modern laboratories in our departments. We can start our process from M.Sc. level. Though I feel it is not enough, there should be a modification of Geography Practical syllabus from UG level like other science subjects.

Recently the UGC is giving us the opportunity to access the international journals through inflib.net. Now it is the time that we should understand the international level of scientific researches. The editorial and reviewing standards of the Indian journals should learn from them that also can be considered as international standard. It is very important to upgrade the national research quality. Finally the official hazards in the universities should be removed that the research teams can work freely.

One thing we should remember that if there are “so many” problems, there must be “so many” ways to solve the problems. Due to paucity of space and focal limits, it has not been possible here to present a comprehensive discussion. Senior Indian geomorphologists have established a strong tradition of scholarly works and many of them are still active. The appreciable point is that the next generation is preparing itself under the careful guidance and encouragement by the senior professional geographers. Some geographers have recently said that the colour of new generation Geomorphology in India is little fading. But, I know personally many young colleagues who are working rigorously and consistently in the fields of coastal geomorphology, glacial geomorphology, hill slope geomorphology, fluvial geomorphology and arid geomorphology. That gives me satisfactory basis to conclude that in future Indian geomorphologists will not only produce better works but also claim prestige, higher status and laurels at international level. So, let us dream and work for the brighter future of landform science in India.

Acknowledgement

The author is grateful to Prof. G.P. Chattopadhyay, who introduced him to the field of Geomorphology. He is also thankful to the members of his research team and students. Finally, the author would like express to his gratefulness to own parents Mr. Prabhas Chandra Dey and Mrs. Minakshi Dey, wife Dr. Jaya Ghosh (Dey) and children who are always the source of energy to take new research challenges.

References

Bloom, A.L. 1998. *Geomorphology-A Systematic Analysis of Late Cenozoic Landform* (Third Indian Edition). New Delhi: Prentice-Hall Pvt. Ltd.

Bryant, R.G. and D.A. Davidson. 1996. The use of image analysis in the micromorphological study of old cultivated soils: An evaluation based on a case study from the island of Papa Stour, Shetland. *Journal of Archaeological Science* 23: 811-822

Chatterjee, S.N. 1972. Some aspects of batch erosion and littoral drift at Digha, West Bengal. *Journal River Research Institute* 3: 1-15.

Davidson, D.A. and I.A Simpson. 2001. Soil micromorphology and archaeology. In *Handbook of Archaeological Sciences,* ed. D. Brothwell and M. Pollard, 167-177. Chichester: John Wiley & Sons,

Dey, S. 1999. Natural environmental control through seasons on morphology of Digha-Shankarpur coastal tract of Medinipur, West Bengal. *Indian Journal of Geography and Environment* 4: 16-23.

—. 2000. Measurement of seasonal changes of beach profile in tropical environment by "Topo-Scaling Factor". *Indian Journal of Geography and Environment* 5: 82-85.

Dey, S., S. Dutta and S.B. Adak. 2002. Holocene sea level change of West Bengal coast. *Indian Geographical Journal* 77 (1): 7-20.

Dey, S. 2002. Impact of sea level change upon landform development along Digha coastal tract and risk assessment by process efficiency process. *Indian Journal of Geomorphology* 7: 111-117.

Dey, S., and R. De. 2002. Evidences of environmental changes at ancient coastal civilisation of Tamralipta. *Indian Journal of Geography and Environment* 7: 80-85.

Dey, S. 2003a. Morphodynamics of tropical Mesotidal Sea beach of Digha. *North-Eastern Geographers* 32 (1 & 2): 70-80.

—. 2003b. Morphodynamics modelling for risk assessment along tropical coast: A microgeomorphic study at and around Digha (Abstract). Paper presented at the 25th Indian Geography Congress, 27th-29th December, in Ghaziabad, Uttar Pradesh.

—. 2004. Microgeomorphic modelling for rip channel study at Digha coastal tract, India: A review. *Journal of Scientific Development and Environmental Research* 3: 22-25.

Dey, S., P. Ghosh and A. Nayak. 2005. The influence of natural environment upon the evolution of sand dunes in tropical environment along Medinipur coastal area. *Indonesian Journal of Geography* 37 (1): 51-68.

Dey, S. 2005a. Natural environmental dynamics of western geo-province of the Bengal basin and its impact upon historical port of Tamralipta (Tamluk), India. *National Geographer* 11 (1 &2): 75-87.

—. 2005b. Microgeomorphology: A simple tool for landform analysis in the tropical coastal area. *Swansea Geographer* (Doearyddwr Abertawe) 40: 35-44.

—. 2005c. Conceptual models for the assessment of tertiary-quaternary geomorphic evolution of paleo-coastal Tripura. *Annals of the National Association of Geographers, India* 25 (1): 73-80.

—. 2006. Holocene sea level change and impact upon geomorphology of Digha-Shankarpur coastal tract, West Bengal. In *Natural Hazards and Disasters Management-Vulnerability and Mitigation*, ed. R.B. Singh, 289-297. Jaipur: Rawat Publications.

Dey, S., S.K. De, K. Debbarma, S. Sahi, C. Debbarma and R. Bhowmik. 2006. Natural urban lake ecology of Agartala, North East India. *Indian Journal of Geography and Environment* 8 (In press).

Dey, S. 2007. Marine-coastal micro-features and their importance in geomorphic assessment: An analytical study along East Medinipur coastal tract, India. In *Issues of Geomorphology and Environment*, 56-61, ed. Sunil K. De, Calcutta: ACB Publications.

Dey, S., D. Choudhuri, C. Debbarma and S.K. De. 2007. Micro-environmental status of a dissected highland (College Tilla) of Agartala City, North-East India. *Indonesian Journal of Geography* 39 (2) (In Press).

Dey, S., C. Debbarma and P. Sarkar. 2008a Microgeomorphic assessment in "Grayscale Environment". *Indian Journal of Landscape system and Ecological Studies* 31 (1): 223-230.

Dey, S., C. Debbarma and P. Sarkar. 2008b. Wetland environment and mapping: a multi-sources spatial data analysis in Rudrasagar Lake, North East India. *Journal of Water and Land-Use Management* 8 (1 & 2): 125-143.

Dey, S., J. Debbarma, R. Dadebbarma, S. Paul and B. Roy. 2009a. Nature of plant diversity on tropical-monsoon highland slopes: a micro-environmental study in Suryamaninagar of Tripura, North-East India. In *Biogeography and Biodiversity*, ed. R.B. Singh, 103-115. Jaipur: Rawat Publications.

Dey, S., P. Sarkar and C. Debbarma. 2009b. Morphological signatures of fault lines in an earthquake prone zone of southern Baromura Hill, North-East India: A multi sources approach for spatial data analysis. *Environmental Earth Sciences*, DOI 10.1007/s12665-009-0033-5.

Dey, S., S. Ghosh, C. Debbarma and P. Sarkar. 2009c. Some observations on regional evidences of Tertiary-Quaternary geo-dynamics in a paleo-coastal part of Bengal basin, India; *Russian Geology and Geophysics* 50 (11) (In press).

Dey, S., S. Ghosh, C. Debbarma and P. Sarkar, M.A. Marfai, and S. Maiti. 2009d. Some observations from radiometric '8 bit' data of sediment thin sections based on alternative petrographic image analysis method. *Journal of Earth System Science* 118 (2):1-11.

Dey, S., C. Debbarma and P. Sarkar. 2009e. Imaging and visualising nanometre scale surface geometry of a crystalline mineral (SiO2) in monochromatic spectra. *Journal of Applied and Natural Sciences* (AANS) 1 (1): 31-35.

Lachniet, M.S., G.J. Larson, J.C. Strasser, E.B. Eveason, and R.B. Alley. 1999. Microstructures of glacigienic sediment-flow deposits, Matanuska, Glacier, Alaska. *Geological Society of America* (Spl. Paper 337): 45-57.

Mamtani, M., A. Mukharji and A.K. Choudhuri. 2007. Microstructure in banded iron formation (Gua Mine, India). *Geological Magazine* 144 (2): 271-287.

Marfai, M. A., H. Almohammad, S. Dey, B. Susanto and L. King. 2007. Coastal dynamic and shoreline mapping: multi-sources spatial data analysis in Semarang Indonesia. *Environmental Monitoring and Assessment*: DOI 10.1007/s10661-007-9929-2.

Merh, S.S. 1987. Quaternary sea level changes: The present status vis-à-vis records along coasts of India. *Indian Journal of Earth Science* 14: 235-251.

Menzies, J., and A.J. Maltman. 1992. Microstructures in diamictons: Evidence of subglacial bed conditions. *Geomorphology* V 6: 27-40.

Nandi, M. 1977. Physiography of Tripura State. *Geographical Review of India* 39 (4): 367-73.

—. 1985. Geographical Account of Tripura. In *An Anthology of Tripura*, ed. J. Gan-Chowdhui, 1-9. New Delhi: Inter-Indian Publication.

Umitsu, M. 1987. Late quaternary sedimentary environment and landform evolution in the Bengal lowland. *Geographical Review of Japan*, Series B 60, 22: 164-178.

CHAPTER NINE

HYDRO-GEOGRAPHY: CONTEMPORARY ISSUES AND FUTURE AGENDA

L. N. SATPATI,
CALCUTTA UNIVERSITY

A Contextual Introduction

As an academic discipline, especially in the higher education system, geography is a relatively new branch of study among the sciences and humanities. Initially in India geography used to be regarded as a school subject and its vertical development started within the teachers' training colleges. Then geography was taught and nurtured by scholars of other disciplines. However, geography education in twentieth century witnessed several important developments that had great influence in shaping modern geography in the country. Some of the important features of development of the discipline in India can be outlined as follows:

Indian geography took a modern path since the Independence of the country along with the restructuring of Indian economy through the Five Year Plans. In the beginning, involvement of foreign geographers, like O. H. K. Spate, and academics of other disciplines like sociologists, economists and statisticians is commendable.

The change in content and approach of geography, from traditional to modern, incorporated the specializations of the pioneering practitioners and makers of the subject in consideration with the national goal of the country. Thus, geography included many themes of social, economic and regional issues along with its physical base. The traditional cartographic techniques gradually took merger with sophisticated statistical and other quantitative techniques for better analysis of the subjects. Building of mathematical/ statistical model became a passion among the modern geographers.

The last quarter of the twentieth century witnessed, quite justifiably, large number of themes of social interest in geography, making it more inter-disciplinary. But, this trend gave the subject more social (or human) orientation than it was before.

The contemporary Indian geography, in the era of globalization, cannot be placed in isolation. So in terms of its themes of investigation and methodological issues it has become multi-dimensional, and sometimes trans-disciplinary in the line of global geography. Now individual problems (i.e. themes or topics) of contemporary relevance have become more important for investigation in a suitable scale of spatial context. This has led to the emergence of many new areas of specializations in geography.

Growing interest in geography among the people and its scope for employment has created almost mass expansion of the discipline. More than one hundred universities and over a thousand colleges in India now teach geography at graduate and post-graduate levels. Geography is also a very popular subject among the candidates appearing for the civil services and other competitive examinations. The generalized nature of the subject has given it a wide breadth (unfortunately not so significant depth) to include almost everything below the sky under the purview of its area of interest.

Development of any special area of interest of the discipline needs to be framed under this broad spectrum of contemporary geography in India. Historically, influence of the older post-graduate departments of geography like that of Aligarh Muslim University, Calcutta University and Allahabad University remain quite long and thereby interest on geomorphology, agriculture, industry, settlements and population studies were much popular among the geographers. But later on new areas of interests were added to the list of specializations and some departments became centres of specialized fields in geography. To quote one UGC report (www.ugc.ac.in), "some subjects in the discipline of geography have moved faster than the others, for example; regional planning, population geography, settlement geography (urban and rural), cartographic techniques (geographic information system/ remote sensing, computer mapping, field-work), and geomorphology. The inputs from disciplines other than geography in terms of data availability, methodology and research design have enabled above branches in geography to grow faster than the others. Besides, new branches of geography appeared with interest shown by some scholars (like administrative geography, electoral geography, geography of central places, gender geography, geography of social well-being, environmental management, medical geography, population geography, geography of social groups, human ecology), as offshoots of several branches in geography".

Why another Specialization Called 'Hydro-Geography'?

Long ago UNESCO organized a seminar on "The Teaching of Geography as a Means of Developing International Understanding" held at Montreal, during summer of 1950. It was then opined that, "geography deals mainly with the static and relatively static, the permanent and semi-permanent conditions of the earth… (it) is not concerned with all the features in a landscape. It is rather the selective study of the humanly significant relationships which exist between the cultural and physical patterns and which give each region its distinctive personality" (Scarfe 1954, 5-6). The statement is quite relevant in these days too, and water resources can be attributed as one such selective study pertaining to any geographic region, be it administrative, natural or quasi-natural.

Study of water resources is not new in geography. This is because among the important elements on the earth's surface water is probably the most significant one, in terms of shaping and reshaping the landforms, and the very existence and sustenance of mankind. The role of water has been duly given importance both in physical geography, as in geomorphology, climatology, biogeography etc. and in human geography, as in commercial geography, agricultural geography etc. UGC appointed Curriculum Development Committee (CDC) in Geography also has enlisted 'geography of water resources' as one of the special areas to be taught at college and university level. But baring a few exceptions (e.g. Department of Geography, DSE, Delhi University) water has not got comprehensive weightage in the syllabi of geography.

A complete deliberation on geographical issues of water resources should seek appropriate terminology, and it will not be improper to study this important area as 'hydro-geography' in place of 'geography of water resources'. The term 'hydro-geography' is neither new, nor uncommon, as in many countries, including India; spatial issues of water resources are studied under this broad heading. The present author has some reservations regarding the 'geography of water resources', as if this is to study only the distributional aspect of water as a resource. Whereas the term 'hydro-geography' seems to be more comprehensive and may be used to encompass the distribution of water in a spatio-temporal frame and also various other aspects related to water- its quality, demand and supply, importance as an element in the earth-atmospheric systems, management of its extremes, etc.

It is relevant to point out here that many of the geographers' basic understandings pertaining to water resources, be it surface and/ or underground, have their roots in other disciplines, mainly like water resources engineering, hydrology, fluvial geomorphology, atmospheric sciences, economics, etc. Although a complete understanding of real world problems require inter-disciplinary outlook, in practice the role of geographers is almost negligible in this regard. Therefore, the scholars of other disciplines topically as well as methodologically influence whatever of water resources is taught in geography. The importance of this aspect lies in our view regarding water resources. The holistic approach is yet to be popular in geography. Works of geographers in the field of water resources are rather sporadic and scattered, but often very brilliant in exposition of water-related spatial problems, like flood, drought, water pollution, social impact of water scarcity, water and gender discrimination etc. If compiled together and framed in a systematic manner these works may lead to proper development of this emerging branch of geographical knowledge. This article is a modest effort to highlight issues related to the significance of study of hydro-geography in respect to contemporary water scenario of the world in general and India in particular.

Water Resources: Importance and Current Status

Water is certainly the most important of all environmental resources that play vital role in maintaining various life support systems. To mankind water is an indispensable item and without it we cannot think of our existence. But in many countries including India the scarcity of water is being felt increasingly. It is so experienced in some places of the world that water has already become an agenda in the political discussions of different levels. Sometimes it is also being warned that if there be another world war that will be fought for water only. The critical situation thus reminds us to quote John F. Kennedy, a former U S president, "Anyone who solves the problem of water deserves not one Nobel Prize but two-one for Science and the other for Peace".

Water is in fact a wonderful gift of nature to mankind. No other resource, *per se*, on the earth's surface can be considered as important as water. Through geo-bio-chemical processes it provides the dynamic inter-linkage among all the spheres influencing life processes particularly at local levels. Water is, therefore, possibly the most important vector of the earth's environmental system (Chaturvedi 1987). The traditional relation between water and man is quite intimate since time immemorial. Almost all human activities are directly or indirectly influenced by the availability as well as the quality of water. Different water sources, whether surface or underground, play crucial role in the life and living of the people of the area concerned.

It is water that has made life possible in the earth, and again it provides crucial support systems for sustenance of life. Water plays key role in transporting nutrients to plants to make primary production i.e. food for themselves and also for the animal kingdom. So every drop of water should be considered precious, particularly when it is short of supply while its demand is on the rise (Aswathnarayan 2001).

Anthropocentric Use of Water

Water is needed in almost every sphere of human activity. It is required for direct consumption, or indirectly for washing, cleaning, cooling, transportation or even waste disposal. Important sectors of human activities that require water can be grouped as follows: irrigation to crops; generation of thermal power; industrial use, e.g. cleaning, cooling, mixing etc.; domestic requirements for drinking, washing, cooking etc.; livestock rearing; and generation of hydro-electric power, pisciculture, navigation, recreational purposes etc. Therefore, in order to ensure the availability of sufficient quantity of good quality water, it becomes almost imperative in a modern society, to plan and execute suitable water management systems that may provide potable water to the various sectors of the community in accordance with their day-to-day requirements. Again, planning of water resource projects fundamentally depends on the availability of water in a river basin, where both planners of water resource development, and environmentalists with a goal of water conservation, share common concerns. While planner's intentions are to put all available water resources to productive uses, the environmentalists' efforts are directed to maintain the water supply on sustainable basis (Sharma 1985, 81-82, 96-98; Southorn 1997).

A rational system of water management should combine both of these two approaches within a single framework. Any attempt of water resource management should actually imply the management of the concerned watershed/ drainage basin including conservation of its soil, mitigation of floods, enhancement of groundwater, maintenance of water quality and other related environmental parameters. Most of the ecosystems heavily rely on the supply of adequate usable water, but human dependence as well as interference to water resources has reached to such a level that it is better in using the term 'anthropocentric use' rather than 'anthropogenic use' of water.

An Overview of Water Resources of the World

The planet earth has a huge quantity of water, almost 1500 million cubic kilometres, but most of it (about 94 percent) contains too much of salt. Total amount of available fresh water on earth is only about 84.4 million cubic kilometres (Penman 1970, 99). Although about three-quarter of the earth's surface is covered with water it is neither ubiquitous nor abundant. Only a small fraction, that is 0.52 percent of water on the surface of the earth, and 9.86 percent of groundwater, is actually exploitable (Baugartner and Reichel 1975). The survival of mankind depends on this small amount of total water resource, which is to be used not only for domestic purposes but also for crop production, livestock rearing, industrial activities, power generation, and recreation purposes. Furthermore, this very amount of about 6 million cubic kilometres of usable fresh water varies with seasons, and is extremely unequal in its spatial distribution. It has been estimated that about one-hundredth of one percent, i.e. 139,211 cu. km., of the world's water supply is stored in lakes, rivers, soils and in the atmosphere (Shiklomanov 1996). This small amount is vitally important and is indeed essential to all life forms on the earth.

Although water is a renewable resource and consequently inexhaustible in a sense, the supply of water for human use in any one location may be limited because of low rainfall, prior consumptive use of available water, mismatch between number of population and available water resources, and pollution of water through wastes dumped by the people of that area (Purdon 1980, 97-98). So the usable amount of water is becoming increasingly reduced with the rapid growth of world population and degradation of water quality by various types of pollutions and contaminations.

Therefore, like all other natural resources, global fresh water resources also have their own limitation. There is a final limit up to which mankind can draw water available in various deposits on the earth, without damaging the natural resource base or without causing any adverse and irreparable change in the concerned ecosystems.

Global Water Use and Problems of Concern

It has been estimated that world water consumption has increased by six times during the period 1900-95, more than double the growth rate of world population (World Resources 1998). As per the United Nations assessment, the global water situation will worse with rise of population and consequent growing demand for food over the decades ahead.

The rural areas i.e. the producers of agricultural goods still use maximum water, about 75 percent of total water consumption of the world. In many of the countries having agriculture as the base of national economy, demand for water is even more than 90 percent in the rural areas. But the rise of demand for water and deterioration of quality of available water will actually diminish the water capacity of these areas. The problem of water resources in rural areas is very sensitive as the supply of water here principally depends on natural processes and with the change in socio-political and economic systems the pattern and mode of water utilization has also undergone tremendous change sometimes without any consideration of the environmental issues. Over utilization of water, degradation of water quality, damage to ecosystems, water disputes etc. are some of the various problems of rural water resources that must be taken care of in planning for rural water resources. In India, where agriculture is still the most important activity and about three-fourth of the countries population live in the villages, sustainable management of rural water resources is crucial for the development of the country.

A recent study on 'Water Vision – 2025 for the South-Asia Region' (as reported by Sue Williams of UNESCO to *Science and Culture,* 2000: *66 (5-6)*, ISNA, Kolkata) including Bangladesh, India, Nepal and Pakistan states that the region, which already suffers food shortage, will move headlong towards a food crisis, as its water future is hardly sustainable. The study has also pointed out that Pakistan, for example, will be short by 28 million tones of different food crops by 2025 owing to shortage of agricultural land and more importantly water crisis, since the country's groundwater resource is at the brink of exhaustion. By the same time, in India, water availability will have to be raised from around 520 billion cubic metres in 1997 to 1000 billion cubic metres.

In many developing countries over-use of both surface and ground water resources is leading to scarcity of water and the deterioration of its quality. In many coastal areas, polluted surface water is causing local people to overdraw the unconfined aquifers, which in turn cause salinization, particularly of the shallow ground water. Water scarcity has been exacerbated and human health gravely damaged by accelerating contamination of usable water supplies, especially in rapidly urbanizing areas. Recent Arsenic pollution of groundwater in many parts of Bangladesh and West Bengal Plains has been related to over-exploitation of groundwater for irrigation. Degradation of water quality has also been reported from the developed world. For example, in Europe, more than 80 percent of groundwater lying beneath agricultural land exceeds the European Union's guidelines for nitrate concentration levels in drinking water, possibly due to over-drawl of groundwater, and excessive use of pesticides and fertilizers (The World Bank 1990; Stanners and Bourdeau 1995).

So, the management of water resources is very complex, as the problems of allocating water among the competing users are becoming increasingly difficult, but of crucial importance as supplies diminish when the demand for water is steadily on the rise. The problems are becoming more serious with the rapid growth of population. A rational management of water is the key to mitigating water scarcities of the future and to avoid further damage to aquatic ecosystems.

Water of India: Issues and Concerns

India's water wealth and its present scenario

India has already crossed the mark of one billion persons of her total population and out of this the urban population is around 400 million. This entails a large responsibility on the governments and civic bodies to provide safe drinking water to all of their communities, as also to plan for agricultural, industrial and other water needs.

India has a total geographical area of about 3.3 million square kilometres. The country experiences extremes of climate both over time and space. The annual average rainfall over the country is of the order of 1170 millimetres, i.e. a total fall of nearly 400 cubic kilometres. The average flow in the river systems of the country has been estimated as 1880 cubic kilometres (Rao 1979). But over 90 percent of the annual runoff in the Peninsular Rivers and over 80 percent of the annual runoff in the Himalayan Rivers occur during the four months of the rainy season i.e. June to September. This particular climatic condition is a major constraint in exploiting available surface water resource in India. Besides, there are other problems like lack of appropriate technology, proper coordination among the states (Chandra 1993) and other infrastructure facilities. A recent estimate made by the Central Water Commission (CWC, *refer Annual Report of the Ministry of Water Resources 2004*) indicate that the water resources utilizable through surface structures are about 690 cubic kilometres (i.e. only 36 percent of the total). Quantum of water that can be extracted economically from the ground water aquifers every year is generally recognized as ground water potential. The Central Ground Water Board (CGWB) has made scientific assessment of the quantum of replenishable ground water in the country, which has been estimated to be about 450 cubic kilometres.

The main user of India's water wealth is the agriculture sector on which most of the country's population especially the rural population is still heavily depending. Water supply as irrigation to crops has played a dominant role in the development of Indian agriculture and has made the country self-sufficient in food grain production.

Water resource projects in India have been established not only for irrigation, but also for generating power, providing flood control, meeting the municipal and industrial water requirements, navigation and water based recreational facilities. The installed capacity of hydropower in the country has gone up by thirty folds during the last fifty years. India has mere 4 percent of the world's freshwater resources, while in terms of population her share is 16 percent of the world's total. The demand for water for various purposes, as in the year 2000, has been estimated by the Central Water Commission as 750 billion cubic metres, of which 630 billion cubic metres is for irrigation to crops, 33 billion cubic metres for domestic uses, 30 billion cubic metres for industrial purposes, 27 billion cubic metres for energy and the rest 30 billion cubic metres for other uses.

Since Independence, India's water policy has mainly concentrated on highly visible large dams, reservoirs and canal systems, but it has ignored minor water works such as tanks, dug wells and tube wells etc. The so called multi-purpose dams have failed measurably to arrest devastating annual cycles of floods and droughts, but the bureaucrats still swear by the huge and expensive irrigation systems. Moreover, information about India's water resources are not only inadequate, in many cases these are inaccurate. Till 1990s there was practically no systematic record of India's water resources. According to one observation of Dutt and Sundaram (1991), "Unbelievable it may seem, till now we have no arrangements in this country to compile and publish, on an annual basis, comprehensive data regarding various aspects of water which are important for policy analysis and programme formulation and for monitoring the efficiency of use of our scarce water resources". During the last two decades, however, there has been considerable improvement regarding assessment of water potential of different parts of the country.

Sources and use of water in rural areas of India

Water may be obtained from surface flows and reservoirs as well as from underground sources. Among the surface water sources of rural areas wetlands of various size, shape and nomenclature are there, e.g. ponds, *dighi*, lakes, *beels*, *jheels*, rivers, *jores* etc. The underground sources may be of either confined or unconfined geological formations that lie at different depths at various locations. Springs and artesian wells can largely contribute in areas where the availability of water is relatively scarce either seasonally or annually. However, the use of water from any of these sources depends on availability of alternative sources, quantity and quality of utilizable available water and also few other socio-economic considerations. In a rural environment the main use of water is in agriculture and agriculture related activities like fishery, poultry and animal rearing. Other essential uses of water are direct consumption, and indirectly for washing, cleaning, cooling, fire-fighting, waste disposal and transportation. Some water is also used in mining and industrial activities.

India has an estimated utilizable annual water resource potential of 1122 cubic kilometre (km^3) out of which 690 km^3 is from conventional surface storage or diversions and the rest 432 km^3 is the replenishable groundwater. The national average of per capita utilizable water was 1869 cubic metre (m^3) in 2001, which was almost one-third of what was in 1951 (Central Water Commission, 2002). As the trend shows India is heading towards water stress (1700 m^3 per capita per annum) and in many parts of the country it is even less than 1000 m^3 per capita per annum. However, if the quality of water is taken as one of the parameters many so-called water abundant areas of the country may be regarded as water famine (500 m^3 per capita per annum) areas. For example, in many parts of the Gangetic West Bengal owing to the menace of Arsenic, utilizable water has sharply declined (ref. www.soesju.org). Similarly in many other areas of the country contamination of water by toxic elements/minerals and pathogens have caused water unfit for consumption and multiple uses. In rural India bacteriological and viral presence in drinking water are causing lots of health problems.

According to the BIS standard in terms of quantity of domestic supply of water, in rural areas of India it should be at least 40 litres per capita per day. In desert areas or in a drought condition this should be 70 litres per capita per day of which 30 litres per capita per day is meant for the cattle population. But it has been remained true for the last decades that this much of water, irrespective of its quality, is actually not available to the people of many rural areas of the country. In many habitations of Gujarat, Rajasthan and Tamil Nadu, and few areas of many other states including the Meghalaya the erratic seasonal distribution of rainfall compounded with lack of appropriate water management has resulted in severe water crisis at times. Although the governments- both at the centre and the states claim that ninety-nine percent of the rural habitations have access to drinking water facility, till June 2003, the total number of villages not at all covered under any sort of safe drinking water was 12299 and partially covered was 99477. In many villages the people, mostly women, are still compel to travel long distances for procurement of domestic water.

Government initiatives in rural water of India

The government of India and the state governments have undertaken many schemes since the Independence to alleviate the problems of the village people. But many of the problems are still alive and new problems have come out as the consciousness and aspirations of the people are also increasing with the spread of information, communication and technology. The problem of water in rural India has been tried for sort out through planning and implementation of various programmes from time to time, for example, irrigation to agricultural crops, rural development, health and sanitation, drinking water supply etc.

The irrigation potential has increased by about four times during the last fifty years. But the net sown area has increased only marginally, by about six percent, although the gross cropped area has increased by almost seventy percent, and area under High Yielding Crops by one hundred times during the same period. From this it is clear that there has been huge disparity in extension of irrigation to various parts of the country, and only some areas got the benefit of assured supply of water while other vast areas remained rain-fed. The problem associated with over irrigation are also many, but without assured supply of water to the crops through proper irrigation the development of rural communities cannot be expected.

Work on providing safe drinking water to the rural areas started in a commendable way with the Rajiv Gandhi National Drinking Water Mission, which was followed by *Swajaldhara* scheme. Under the later programme there has been an effort to institutionalize the community based rural water supply systems with people's participation. This along with newly started Total Sanitation Campaign has targeted as community-led and people centric (*India 2006*, a Ministry of Information and Broadcasting, Government of India publication) programme. But in many occasions the schemes utterly have failed due to many reasons e.g. apathy of the people in accepting the system of toiletry, as the common people are habituated in defecating on the open fields.

The present UPA government at the Centre has put emphasis, in principle, on agriculture, water, education (mainly primary), healthcare, employment, urban renewal and infrastructure. This has obvious consequence on development of the rural areas, since problems of agriculture were one of the major agenda in the political sphere of the last Parliamentary general election of India.

Mention may be made of another gigantic project, called the River Linking Project, which created much controversy and has become a part of the political agenda. Without going into the details of its pros and cons it may be pointed out here that the project was certainly not meant to perpetual solution of water crisis in some parts of India, where the crisis is more man-made than natural. Intensive cultivation of high value crops like cotton and rice, which require more water, improper farming practices, and mismanagement of soil and water are said to be the root cause of water crisis and resultant ill consequences in these parts of the country.

Urban water supply in India

In urban areas the principal requirement of water is for the purposes of drinking, cooking, bathing, toilet, washing of vessels and clothes, other cleanings, and gardening. In most of these sectors the quality of water needed is of potable quality. The supply should be adequate to mitigate the demand of the urban residents that comprise about forty percent of the total population of India. The urban supply of water is usually through pipelines, and the leakage and mismanagement of them attribute to a loss of more than thirty five percent of potable water, on the average. There is also the threat of contamination of water through the leakages particularly during when pressure in the pipe is less that pressure of contaminated water outside. This poses a serious health hazard in urban areas especially in the slums and settlements of poor people who solely depend on the municipal supply and use water directly for consumption without any further purification of it.

A study conducted by Tata Institute of Social Sciences (Shaban 2007) show that the availability of water in the following cities are far less than the World health Organization (WHO) prescribed (150 litres per capita per day) limit of water requirements: Delhi (78), Mumbai (90), Kolkata (116), Hyderabad (96), Kanpur (77), Ahmedabad (95) and Madurai (88). Among these Kolkata still has better position in terms of volume of supply, but studies have shown that a large part of the city depending on tube wells are experiencing the problem of arsenic, resulting in the closure of the tube wells, and extension of municipal supply to these areas. The escalation of cost of urban water supply has insisted the Kolkata Municipal Corporation (KMC) to consider re-imposition of water tax on the assesses. Pollution and contamination of municipal water sources i.e. the rivers, e.g. the Ganges, the Jamuna etc. is such that purification of the water for municipal supply requires involvement of sophisticated infrastructure and huge investment.

Policies and politics of water resource management in India

India has a long history of water management practices for the last few thousand years. The ancient Indians, especially the social thinkers were very much concerned about the proper upkeep of the country's aquatic resources so as to help conserve moisture, maintain water quality and make equitable distribution of water to all concerned (DasGupta 1998, 118-124). At present, the Ministry of Water Resources lays down policies and programmes for development and regulation of the country's water resources.

Recognizing water as a precious national asset, the National Water Policy (NWP) adopted in 1987 embodied Government of India to resolve that national perspectives would govern planning and development of water resources. The policy has recognized drainage basin as the basic unit of planning for water resources development and calls for appropriate means to optimize utilization of these resources, not only for the benefit of the people living in the basin(s) concerned, but also to transfer surplus water to meet the requirements of water deficient areas. But, right from the adoption of the NWP, a number of problems and challenges have emerged in the development and management of the water resources sector. Accordingly, the National Water Board (NWB) in its meeting on October 29, 1998 reviewed the same policy, finalized and updated NWP- 1998, and after considering proposed changes the National Water Resources Council has adopted it on the 1st day of April, 2002. However, many issues of the policy are highly debatable and environmentalists are still arguing against it. The issue of gigantic River Linking Programme, as has already been mentioned, in its form as projected by the earlier government at the Centre, has got stiff resistance from various corners. The issue, although dormant for the present moment, is said to be highly political than scientific (Bandyopadyay and Praveen 2004, 8-38) and can be connected to the cultivation of market-oriented crops in the southern states of the country.

Pricing of water is another sensitive issue in Indian context. But a moderate taxation will not only restrict overdraft of ground water, especially when it is used for irrigating crops, but will reduce its wastage (Planning Commission 1992). However, the rate of tax and the modalities of payment must be reasonably settled by the concerned authorities in consultation with the end-users. While fixing the tax rates the amount of water should not be considered alone but the purpose and place of its use along with the socio-economic status of the users should also be taken into account.

In Indian perspective conservation of rainwater and checking soil erosion are essential conditions for the sustainability of dry land agriculture in particular (Sharma 1991) and the country's agricultural economy in general. Integrated watershed development programme (IEDP) is one such policy instrument for achieving the goal (Shah 1998: A66, A 70). The main ideas of the programme are- (a) simultaneous development of land, water and bio- mass resources in the light of the symbiotic relationship among them; (b) integrated farm management; (c) meeting food, fodder and fuel requirements of human and livestock population that depend on these resources;(d) ensuring environmental quality along with sustainable economic viability by promoting low-cost technologies; (e) improving land productivity by promoting better agricultural practices and inputs; (f) releasing population pressure on land by creating non-farm employment; and (g) development of local institutions for future management through participatory approach.

Restoration of old irrigation systems based on surface structures like tanks, canals etc. (Palanisami, 1990) should also be emphasized not only to catch the rainwater over the catchments, but also to enhance the rate of recharge of ground water to a considerable extent. This is another area of water resource management where the application of age-old traditional knowledge may work. Besides, alternative agricultural practices, for example, crops requiring less water but having more commercial value like medicinal plants, bio-fuels etc. may be encouraged for cultivation. Very recently there has been started an ambitious propagation of the idea of rainwater harvesting both in rural and urban areas of the country. Sanctioning of building plans is now becoming an integral part of rooftop (rain) water harvesting in many municipalities of India.

Hydro-Geography in India: Planning for Future Agenda

The goals of sustainable water management

Sustainable development of water resource depends on sustainability of the natural systems, and for that human intervention in the natural hydrological system should be rational and balanced. As every locality has its own water cycle with certain input and output as well as storages at various points of the cycle, estimation of availability of water- both surface and sub-surface should be made as accurate as possible. Estimation of assured rainfall for different periods of the year should also be done and the selection of land use and water use should be made accordingly. Democratic participation of local community in water development/ use processes should be ensured. Water budgets of every local level i.e. of micro-watershed or revenue village level may be integrated to build a huge database for preparation of water budgets at state level and also at the national level.

Facets of water development in rural India

Until now in most of the rural areas in India water is considered as a free good, a gift of the nature. The price of water is determined in terms of its cost of supply for different purposes. The ownership of water rests on the provider of water, irrespective of the source. In most of the occasions the provider of rural water, mainly for irrigation to crops, obtain this valuable resource for free of cost- either from a surface source or from underground reserves. The supplier of irrigation water thus in many cases has nothing to pay for obtaining water, and the profit is made only through investing on the equipments and fuels for extracting water. So valuation of water and other related economic exercises like ownership, taxation, equity etc. are very sensitive and conflicting in the rural use of water in the country. But without proper valuation of water as an economic asset its importance and optimum utilization cannot be assured. However, sole business objective of water and its commercialization may restrict the access of this vital resource to the rural poor. Therefore different stakeholders at various levels should act responsibly for good governance of this valuable resource.

Another vital aspect is the maintenance of proper quality of water and its justifiable allocation for the sustenance of the ecosystems. Degradation of water quality actually diminishes its quantity. Whole-some water is capable of multiple uses, but its degradation may restrict it in using some areas like drinking, bathing, washing and even in fisheries and agriculture. In some occasions the reserves, especially the underground ones may dry up permanently. So a judicious extraction of water from such sources should be planned.

In India the growth of population is very rapid and the increase of the absolute number of human population along with cattle population is very high. So the amount of food and fodder that need to be produced is also huge, which necessarily demands more water. Many crops, like rice, require much water and such crops are cultivated in India not only to meet the domestic market but also for export. Therefore, altogether the demand for water in Indian agriculture has increased manifolds along with the increase of demand for water in other sectors like industries, domestic etc.

Holistic development of rural water resources–concepts and procedures

Although at global scale the hydrological cycle is a closed system, in regional or at local scale it is an open system (Briggs *et al.* 1997, 234-35). This regional or local hydrological system operates through surface and/or sub-surface drainage basins, including sub-basins, of various shape, size, land cover/ land use etc. Each drainage basin is different from the other, even from the adjacent ones, in terms of its morphological, fluvial, and other characters. So any kind of intervention in the drainage basins actually affects the character of the concerned hydrological system and in turn the system behaves in a different way that may not be acceptable to the people of that area.

It should be remembered that before any kind of human interference in any area on the surface of the earth the things are balanced in a natural way, and the land surface might be regarded as a natural landscape. But, introduction of human activities substantially changes the natural systems; and, the area becomes an intervened (very often called "cultural") landscape. The natural landscape takes many years to be of an appropriate and mutually adjusted system. But human actions are more rapid and take only a few years to change the whole thing definitely not in adjustment with the nature, and thus some imbalances appear, which the natural processes always try to sort out in its own way. So whenever man intervenes in the hydrological system, as man finds it convenient, some changes certainly occur in the system and the hydrological cycle tries to readjust in the changed situation.

Therefore the holistic approach of water development may be actually placed in the ecosystem framework that includes and recognizes the importance of every living as well as non-living things of an area. This is different from Integrated or Participatory Water Resources Management (Calder 1999, 151-165), but it definitely includes some of the principles of these approaches too. The basic difference is that in these two approaches the supremacy of man over nature is taken for granted and man has been advised to take certain actions with his technology and other skills for his betterment; whereas in the holistic approach man is advised to see himself as part of the nature and he should restrict himself to take certain actions cautiously and wisely or through experience for the sake of sustenance of natural systems and obviously of mankind, even if some of his immediate achievements seem more beneficial. The Integrated Water Resources Management as adopted in the U N Conference on the Environment and Development in 1992 has nevertheless mentioned that water is an integral part of the ecosystem, a natural resource and social and economic good. So the holistic approach seeks to deal with the concerned ecosystem in totality, obviously as perceived by man at the time of interactions.

Holistic development of water resources in rural areas should start from a small village and it should be integrated with the national goals and missions. However, this approach needs to involve the people at the end user or grass root level. The role of women and landless poor cannot be ignored at all, as they are the key people in conservation of the natural resources and the natural systems as may be substantiated through various incidents that happened in different parts of India. This again necessitates social mobilization, integrated with the socio-political goal of the country as well as the geographical need of a particular area (Khan 2004, 48). In the holistic management system people should look beyond the technical skills, and therefore, consideration of eco-friendly options should always be adhered to. A full knowledge of the ambient environment, it's functioning and a proper appraisal of it is necessary for efficient allocation of water resources for various purposes. An efficient management practice is possible with proper co-ordination among the collectors/ providers of water and its users. But for this a scientific database of various aspects of water resources should be created at village level and that should be linked to the perspectives of respective regional levels having similar environmental conditions.

Transfer of knowledge and technology from the institutions of higher learning, research laboratories and industrial houses to the village people for building up village level water management systems is one of the primary requirements in holistic water management. This is especially required to explore the technical opportunities by the rural people. However, the technicians and researchers may find themselves enriched in the exercise as this is a two way process. Ultimate objective of the whole exercise would be to set up water laboratories with sufficient spatio-temporal data including maps by the rural people themselves as Non-governmental Organizations or otherwise for holistic rural water management. The success of the approach obviously depends on proper education and its integration with sanitation and health, generation of employment opportunities and provision of suitable markets for rural (agricultural) produces.

Role of geographers in management of water resources and water-problems

One of the basic foci of geography is to understand humanly significant relationships between physical and cultural patterns of a region. Problems related to water includes understanding of both physical and human behaviour in a spatio-temporal context. It is therefore the geographers who can take the lead role in this regard, as geography has an edge over other disciplines in addressing the problems concerning man-nature interactions.

The problem of water resource management cannot be solved through technical solutions by either supplying more water or by reducing the level of pollution/ contamination of water only. Human need of water should also be judged according to the socio-economic goal of a country, and if necessary such needs should be regulated for the greater benefit of the society of the present, and also for the future generation. The regulation can be made through wise use of water reducing the quantity of intake, restriction on misuse or wastage, and recycling of it wherever possible. Alternative land use practices can be effectively adopted, e.g. cultivation of crops requiring less water can be encouraged in place of water loving plants. This, however, may necessitate change in food habit. But to tide over future crisis of water and to avoid possible conflict on ownership of this precious resource, such measures may be taken at the earliest. Geographers can chalk out the plan for future land use patterns for different geographic regions in commensurate with the availability of water and its demand. The holistic approach in geography can incorporate various parameters of water related problems that a nation may face in future for a well-directed solution.

But the geographers are, in many occasions, handicapped due to paucity of reliable data/ information. Since geographical problems are mostly field oriented, generation of primary level data on the field should be made continuously and densely in a space-time frame. This means there should be adequate database at micro level for a relatively long period of time, for better understanding of the problems. The entire exercise requires huge investment in the form of money and skilled labour force. But this can be reduced substantially if people are motivated to voluntarily participate in the process.

Although in some cases modern remote sensing techniques have provided ample opportunities in getting reliable high resolution repetitive spatial data for micro level land use study, there are not enough non-spatial data, e.g. data on quality of water, discharge data of a micro watershed etc. for proper planning at this level, where man-nature interaction is direct. It is thus utmost necessary to build databanks in different centres of geography, and a mutual co-operation among these centres are required to exchange not only data/ information, but the ideas that may enrich all through dissemination of contemporary advancement of geographic knowledge.

The same principle is also applicable in case of hydro-geographic study, in which generation, storing, updating and dissemination of data is necessary to build up a strong conceptually developed network of water management personnel in the country and abroad. This is because the problems of water can be found at all levels- micro, macro, meso, national, regional, international, etc. and these problems cannot be solved without a co-coordinated effort at these levels. Various examples can be cited in this regard, e.g. water dispute between Karnataka and Tamil Nadu, problems of water sharing between India and Bangladesh, problems of water crisis in West Asia and Africa, problems of flood, erosion and siltation in the Ganga-Brahmaputra-Meghna (GBM) Basin, etc. Hydro-geographic study based on scientific attitude and adequate database free from any error or prejudice should include all these aspects to solve the water related problems for a better future of mankind.

Suggested Curriculum for Development of Hydro-geography

It has been already pointed out that any discussion on water resources in geography should be holistic, keeping in mind the more realistic man-environment approach, in which man is usually placed at the centre-stage. This does not mean that other components of the environment should be taken for granted, as man-environment relationship is a two-way interactive process in which there are enormous opportunities to get mutual benefit. Whenever the process is forced to operate in just one direction favouring man, it has ultimately proved disastrous. Our education system, particularly those aspects, which deal with such investigations, should encourage the people to live in harmony with other components of the nature that have very close association with day to day activities of man. Our instruments of teaching-learning process should be so designed that they take into account the natural extremes as an integral part of our life and the adoptive measures should be taken accordingly.

Keeping in view of these principles, a tentative outline of curriculum for the study of 'hydro-geography' is suggested for further development in the Indian context, as follows:

Unit-I
Geographers' approach to water resources; Global water crisis; Water in the hydrological cycle and human intervention in it; Hydro-hazards, problems of drought, flood, water-logging, erosion, siltation etc. and related issues

Unit-II
Controlling factors of basin hydrology; Issues related to Inter-basin transfer of water; Wetlands and their problems; Irrigation in relation to agriculture of India; Virtual water and related issues

Unit-III
Water quality parameters; Pollution and degradation of water resources; Population growth and sustainable use of water in different sectors; Right to water as a basic need; Water movements; Water ethics and values

Unit-IV
Water pollution and health in reference to arsenic, fluoride, and contamination of water by pathogens, pesticides & chemicals; Socio-economic impacts of water pollution; Water and gender

Unit-V
Water resources management; International perspectives of water management; Water pollution acts and regulations in India; Water Policy 2002; Rural and urban water supply; Rain harvesting, Climate change and water

Concluding Remarks

In India the nature of sources and availability of water is diverse according to the physical environment of the areas concerned. The traditional extraction and use of water have been governed by this spatial peculiarity. But with the rise of demand for water in almost every areas of the country, many problems have emerged in regard to extraction and supply of water from various sources. Therefore, people should undertake proper management of this precious resource. A holistic management practice is the only option of proper utilization of water as it considers almost every aspect of water management that too in harmony with the nature. But to achieve this, proper education, governmental and institutional support especially in the beginning, and co-ordination among the stakeholders are necessary. Research and teaching of hydro-geography will definitely encourage our new-geographers to disseminate various ideas to the people of India leading to proper management of water resources through wise use and minimum misuse of this precious gift of the nature.

Considering the Indian situation hydro-geographical studies should include the problems as outlined earlier, and in terms of quality of information and methodological issues these should be at par with other disciplines studying water resources in the country.

References

Aswathnarayan, U. 2001. *Water Resources Management and Environment.* Leiden (The Netherlands): A. A. Balkema.

Bandyopadhayay, J. and S. Praveen 2004. Interlinking of rivers in India: assessing the justification. In Proceedings of the All India Seminar on Interlinking of Rivers in India: Myth and Reality, 8-38. Environmental Engineering Division, Inst. of Engineers (India), Kolkata: W. B. State Centre.

Baugartner, A. and E. Reichel. 1975. *The World Water Balance.* Amsterdam: Elsevier Scientific Pub. Co.

Briggs, D. *et al.* 1997. *Fundamentals of the Physical Environment.* London: Routledge.

Calder, I. R. 1999. *The Blue Revolution, Land use and Integrated Water Resources Management.* London: Earthscan Publications Limited.

Water and Related Statistics, 2002. New Delhi: Central Water Commission.

Chandra, S. 1993. Sustaining environment in water resources projects. In, *Environmental Impacts of Water Resources Development,* ed. R. S. Goel, 346-347. New Delhi: Tata McGraw-Hill Pub.

Chaturvedi, M.C. 1987. *Water Resources Systems Planning and Management.* New Delhi: Tata McGraw Hill Publishing Company.

Das Gupta, S.P. 1998. Environmental Science in Ancient India. *Geographical Review of India* 60(2):118-124.

Dutt, R. and K. P. M. Sundaram. 1991. *Indian Economy.* New Delhi: S. Chand & Company Ltd.

Khan, D.G.A. 2004. Social mobilization and rural empowerment. *Kurukshetra* 48 (September).

Palanisami, K. 1990. Tank Irrigation in South India, *Journal of Water Resources* (ESCAP, Natural Resources Division, Bangkok) 167: 50-55.

Penmann, H.L. 1970. The water cycle. *Scientific American* 223 (3): 99.

Planning Commission of India .1992. *Expert Committee Report in Pricing of Irrigation Water*. New Delhi.

Purdon, P. W. *et al.* 1980. *Managing the Environment*. Ohio: Merrill Pub. Co.

Rao, K. L. 1979. *India's Water Wealth*. New Delhi: Orient Longman Ltd.

Rosegrant, M. W. 1997. *Water Resources in the Twenty-First Century: Challenges and* Implications for Action. Washington D C: International Food Policy Research Institute.

Scarfe, N.V. 1954. A Handbook of Suggestions on the Teaching of Geography towards World Understanding-X. Paris: UNESCO.

Serageldin, I. 1995. *Towards Sustainable Management of Water Resources*. Washington, D C: The World Bank.

Shaban, A. 2007. Water poverty in urban India: a study of seven major cities. Paper presented in international seminar on Poverty and Development in South Asia, January 11-13, in Dept. of Geography, Calcutta University.

Shah, A. 1998. Watershed development programmes in India: emerging issues for environment and development perspectives. *Economic and Political Weekly* 33 (26): A66 & A70.

Sharma, B.L. 1991. *Dry Land Farming: Perspectives and Prospects*. New Delhi: Daya Publishing House.

Shiklomanov, I.A. 1996. *Assessment of Water Resources and Water Availability in the World,* Background Report to the Comprehensive Freshwater Assessment. St. Petersburg: State Hydrological Instt.

Stanners, D. and P. Bourdeau. 1995. *Europe's Environment, the Debris Assessment*. Copenhagen: Europe's Environment Agency.

University Grants Commission (UGC). 2007. *History of Development of Geography in India* (online). Available at http://www.ugc.ac.in/policy/geo/preface_g.pdf

CHAPTER TEN

NATURAL HAZARDS AND DISASTER MANAGEMENT: EXPLORING NEW AREAS OF RESEARCH AND CHALLENGE IN GEOGRAPHY

N. C. JANA,
UNIVERSITY OF BURDWAN

Conceptual Background

Hazard and disaster are closely related. "…..A hazard is a natural event while the disaster is its consequence. A hazard is a perceived natural event which threatens both life and property…….. A disaster is the realization of this hazard….." (Whittow 1980).

Disasters have been mankind's constant companion since long historical past. Generations after generations, people have been suffering from the consequences of disasters and accordingly recovered from the same, and life continued as usual. With the passage of time, the scenario has changed quite a bit. Of course, there has not been much reduction in the traditional disaster threat. But we have learned a lot to cope with these problems to a certain extent. Whatever may the effects of disasters, the largest sufferers are the less developed countries and economically weaker sections of the society.

In fact, man-made disasters' threats have grown more severe. Increasing population has forced people to live in disaster prone areas. This is a growing phenomenon particularly in developing countries, for example, encroachment of human settlement in the flood prone areas of major river systems as well as on low islands, which increase the susceptibility to flood disaster.

Disaster has changed its face in the modern world, particularly since World War II. Increased social violence has drastically affected many nations and communities. Incidents of hijacking, terrorism, civil unrest and conflict with conventional arms have inflicted heavy burdens on the governments and societies whose existence is already unstable because of poor economic and social conditions. In addition, the threats from hazardous materials and nuclear accidents pose another modern problem for disaster management. The threat from possible nuclear war is almost beyond comprehension and in which even those countries can suffer a lot who are not directly involved in nuclear conflicts.

It may be pointed out that most of the world's worst disasters tend to occur in the countries between the Tropic of Cancer and the Tropic of Capricorn, which are coincidentally the poorer countries/nations of the world. Therefore, disaster can be a strong aggravating factor in making the differences between wealthy and poor nations. Whatever may be the status of the nations, it underlines the need for all countries to develop and maintain an effective disaster maintenance capability appropriate to their requirements. It also underlines the necessity for a coordinated international action in order to strengthen all aspects of disaster management.

The South Asian region represents diverse agro-climatic regions, which subject to diverse natural disasters. Long coastal regions are prone to cyclones, arid and semi arid regions to persistent droughts, the Himalayan mountain terrain and parts of the continental crust to earth quakes and land slides and near perennial rivers to periodic floods. The countries in this region are densely populated and represent low income economies. Recurrent natural disasters inflict setbacks to their efforts in development and aggravate poverty conditions.

India is a country of unity in diversity exhibiting manifold physical features and cultural panorama. The Indian subcontinent is highly prone to a number of natural disasters. Drought, Floods, Cyclones and Earthquakes are major natural disasters in the country, though Landslides, Avalanche and Bushfire also occur in most of the states of Himalayan region. Out of 32 total States/Union Territories, 22 are quite prone to disasters. Due to unique and widely varying geographical and geological conditions of the country, virtually all types of natural disasters take place with ranging intensities and in different regions (UN World Disasters Report 1994). The major natural disasters not only bring untold misery to the people, irreparable damage to property and environment but also affect adversely the economic system as a whole. Every year crores of rupees are spent by the government in mitigating the effects of these disasters which could otherwise have been utilized for development activities.

The most significant global effort made in recent times to mitigate the effects of disasters is the launching of the IDNDR programme by the

United Nations. Recognizing the rapidly rising world wide toll on human and economic losses due to natural disasters, the UN General Assembly in 1989 made a decision to launch a far reaching global programme to save human lives and reduce the impact of natural disasters. With this in mind, the decade 1990-2000 was declared *International Decade for Natural Disaster Reduction* (IDNDR). The objective of the IDNDR is to reduce through concerted international action, especially in the developing countries, the loss of life, property damage and social and economic disruption caused by natural disasters such as earthquakes, windstorms, tsunamis, floods, landslides, volcanic eruptions, wildfires, grasshopper and locust infestations, drought and desertification and other calamities of natural origin.

In the present paper, an attempt has been made to study the contribution of Indian geographers in the field of hazards and disasters, scope of geography in dealing with hazards and disaster aspects, place of geographers in decision-making bodies for disaster management, future line of direction and or the role of young geographers.

Hazard and Disaster Studies: Contribution of Indian Geographers

Geographers' contribution in the field of hazards and disasters are insignificant. Few contributions have come from the physical geographers or physical/geomorphological hazards have been dealt in geography. Social aspects of hazards and disasters have not been analyzed in depth by the geographers, which need to be explored. While delivering the presidential address in the 15th Indian Institute of Geomorphologists (IGI), S. C. Mukhopadhyay (2003) analysed the causes and management measures of natural hazards and disasters in Indian context.

Flood

H. S. Mangat (1994) reviewed the situation of July 1993 floods in Punjab particularly experienced by Patiala city and its environments. He expressed concerns that even after taming almost all the rivers and adopting various flood control measures a highly developed state like Punjab is subject to frequent floods. The devastations associated with floods are directly related with their level and duration instead of the extent of area under floods, which are further associated with rainfall characteristics and man made obstructions not only at local level but at regional scale. Gupta and Sharma (1998) studied the flood situation in

Raipur District of Madhya Pradesh in 1994 and assessed its causes and consequences. In the article entitled 'Floods: An Ecological Hazard of the Semi-Natural Drainage Ecosystem of the Inter State Chandigarh Region, India', Singh (2000) has indicated three factors responsible for the flood hazards in the inter-state region such as continuous deforestation and grazing in the Siwalik hills resulting in increased run-off, change in longitudinal profiles of ephemeral streams due to silt deposition giving a planar to convex profiles to stream beds, and eutrophication of streams due to sewage inflows from Chandigarh and its satellite towns. Mukhopadhyay and Scheidegger (2000) have made a study of the structure of the discharge versus frequency relationship for the Tista River in the Eastern States of India and Bangladesh. The authors concluded that the flood-frequency curve for the Tista River can be explained in terms of a natural process representing the evolution of a quasi-stationary self-ordered states occupy in space and time at the limits of the range of flood-magnitudes investigated. Mahanta and Bora (2002) have made an impact assessment of geomorphic hazards especially floods on the Kaziranga National Park of Assam in India and attempted to evolve a scheme of flood and erosion control measures needed for environmentally sound planning and development of the park. Mukhopadhyay (2002) studied the geomorphology and the various causes of floods in the lower Brahmaputra basin and adjacent areas covering primarily the North-Eastern states of Assam, Sikkim, and Arunachal Pradesh, etc. Jana (2003) analyzed the flood situation in India after fifties and highlighted on the ground reality of policy measures adopted at the government level from time to time. The article entitled 'Major floods of the world in 1990s: Lessons for flood management in India' by Sivaswami and Jana (2005) deals with the major floods of the world during 1990s, their magnitude, damages and subsequent management measures; which can provide lessons for flood management in India. Mukhopadhyay (2005) has done an intensive work on the impact of flood hazard on bank dwellers along Mayurakshi River of eastern India. Deb Sarkar (2005) has evaluated the role of river communication on lessening flood intensity in Deltaic West Bengal, India. In his paper entitled 'Taming the river: Concerns and policies in managing the floods', Sreekesh (2009) attempts to review the causes of flood, flood situation in the country, area affected, damages caused and the interventions made by the government during the last half a century.

There is a common consensus that over-spilling of river channels due to heavy downpours in the rainy months cause flood and inundation in our country. Most of the studies are confined to the analysis of the state of deforestation and siltation of reservoirs in the upper catchment as well as the encroachment into the floodplains and the assessment of flood situation in the downstream segment. But sometimes floods are occurred locally for the stagnation of rainwater due to increasing rural road networks. On the other extreme, floods are common phenomena in the large metropolis during monsoon season as a result of drainage congestion due to unplanned urban sprawling, which obstructs the natural flow of drain waters. Geographers should orient their vision of research to these aspects of flood rather than confined to the analysis of flood phenomena in the drainage basin only.

Drought

Mishra (1992) analyzed annual rainfall data of 19 stations spread almost all over West Bengal for the period of 90 years from 1901-1991 and plotted chronologically the occurrence of moderate and severe drought of each station towards developing any criterion to identify drought of her own. From the analysis it is evident that none of the criteria is fully suitable to identify the actual drought-stricken areas of the State where agriculture and other normal activities of the inhabitants suffer due to negative departure of rainfall from the normal. It is, therefore, essential to fix up a scientifically acceptable criterion for identification of drought over different agro-climatic regions of the State considering all relevant parameters like the principal crop and its water need for each stage, actual rainfall, evaporation, soil moisture status and the length of the period which the crop is capable to sustain water deficiency. Degree of availability of irrigation and its extent as well as adoption of appropriate agronomic practices to combat drought should be considered at a later stage for the purpose of drought management. The problems of desertification have been examined through quantitative comparisons of economic parameters like land use, water availability and crop yield performances between desert areas and bordering regions (Sen, 1992). All available evidences suggest that the Rajasthan Desert, has not, and is not, advancing towards the east but the fresh sand deposition is still going on and towards the west within the desert. The inherent vulnerability of the land and intense biotic pressure have combined to set in a process of deterioration within the desert. Butola and Jana (2003) have made an in-depth analysis of the worst situation of Rajasthan during consecutive droughts between 1999 and 2003. Banukumar, Rajamanickam and Aruchamy (2005) have attempted to calibrate all the hydrological parameters viz. situation of ground water table, its oscillation and surface drainage density to evaluate the drought prone area in terms of hydro-geomorphological condition. From the present study, it is identified that the severance, periodicity, extent of hydrological drought are varied according to the drought-wet year occurrences, seasonal variation of rainfall within a year and also hydro-geomorphological condition of the area. In a study the drought and flood situation in Rajasthan, India have been analyzed (Kalwar, Sharma and Kalwar 2005). In this analysis, district and area-wise position of drought/flood events have identified. The frequency analysis of drought/flood events of Rajasthan during past 45 years has however revealed that 89 per cent area of Rajasthan is categorized as drought area and 79 per cent area can be classified as moderate flood area as per definition prescribed by the Irrigation Commission, IMD, and other allied organizations. The paper entitled 'Natural disasters in India: A case study of droughts in India and its management strategies' by Bupinder Zutshi (2009) attempts to examine the natural disasters confronting India with special reference to droughts. It also highlights the management strategies required to reduce the impact of droughts especially with the help of developing rainwater harvesting structures and other IT based software. It also highlights good practice initiatives adopted by Civil Society Organizations, which have resulted in effective rainwater harvesting and mitigating the impact of droughts.

We do have a general conception of occurring meteorological drought in western Rajasthan, but hydrological and or agricultural droughts occur in some parts of Gujarat, Andhra Pradesh, Orissa, Haryana, Maharashtra, Chhatishgarh, Jharkhand, Madhya Pradesh, Himachal Pradesh, Karnataka, Kerala, Tamil Nadu, and Uttar Pradesh. Whatever may be the nature of drought, the drought-affected areas are gradually increasing all over the country due to marked depletion of surface water, over-extraction of underground water, inadequate soil moisture by human activities especially for high intensity of agriculture and construction works. Geographers can take a challenge in exploring this emerging field of research as the Indian agriculture, backbone of national economy, is more or less suffering from the unavailability of water and acute stress on soil moisture due to the vagaries of monsoon and uncertainty in rainfall.

Earthquake and landslide

Basu (1999) attempted to analyze the causes and consequences of soil erosion and landslides in Darjeeling town. Karlekar (2002) provides an overview of the tendency of slope failure and landslides and its effect on the villages on Vengurla Coast. The specific effects along the said coast are significant especially as short-term consequences dictated by morphological and morphogenetic factors. Mandal (2006-07) has extensively worked on the impact of the occurrences of natural disasters in Rudraprayag Terrain of Uttarakhand, India with special emphasis on earthquakes and landslides and their scientific and systematic methods of mitigation.

The in-depth analysis of earthquakes is the exclusive research area of geologists; it normally does not fall under the purview of research by the geographers except the assessment of its magnitude of consequences only. This field of research necessitates the involvement of geographers with science background for its effective investigation. So far as the study of landslides is concerned, a very few contribution has been recorded. It may be pointed out in this context that in India there is a dearth of expert geographers engaged in the research of highland areas in general and landslides in particular, which needs further exploration in this field for the sustainable development of mountains.

Tsunami

Mukhopadhyay (2005a) studied the problems of recent tsunamis and consequent impacts on the coastal geomorphology of India. He also commented that during the recent tsunami some earthquakes cause tsunami while others do not (Mukhopadhyay 2005b). Whereas Joshi (2005) has described in details the causes and consequences of 2004 Tsunami along the Asian Coasts. In another study (Vasudevan, Rajamanickam, Arumugam and Sakthivel 2006-07) have made a geomorphological assessment of the Cuddalore Beach of Central Tamil Nadu in India, which exhibits that the giant tsunami waves of 26th December 2004 has expunged and significantly autograph over the ordinary wave generated previous inscriptions. Jana (2008) has made a detailed study of tsunami has given an impact assessment of 2004 Tsunami occurred in India and its mitigation strategies.

The devastation of 2004 Tsunami has necessitated a comprehensive research to protect the coastal lives and coastal ecosystems against future disasters. It is a matter of hope that quite a few geomorphologists have done commendable work in this regard. Now, it is time for the young geographers to come forward and explore this emerging field of research.

Tropical cyclone and coastal hazards

Paul's (1996) paper deals with natural hazards such as tropical cyclones, coastal flooding, coastal erosion and different sand hazards and their impact on mankind and environment of coastal districts. The identification of hazards has been made in terms of space, time, rate, trend, impact and risks. South 24-Parganas District in West Bengal is most vulnerable to cyclones, flooding, erosion, and sand hazards. Concerns have been expressed over the weather hazards or disastrous weather events in West Bengal, which not only cause huge loss of life and property of the people of the state but stands on the way of its economic development (Mishra 1999). Gadgil (2000) has dealt in detail the Orissa Super Cyclone of October 1999 and assessed the extent of damage and destruction. Based on their studies and comparisons, scholars emphasize upon the tracking and forecasting, monitoring and warning system and precautionary measures before and during the occurrence of cyclones in India (Jana 2006, 2009).

Meteorologists have systematic knowledge and expertise in explaining the mechanism, direction of movement, velocity and time of occurrences of tropical cyclones in the coastal regions. It is a good sign that some geographers with specialisation in climatology are pursuing fruitful research in this field, which urges for the involvement of more geographers towards the management of vulnerable coastal zones.

Land degradation

Savindra Singh and his direct students have contributed profusely to land degradation studies. Singh and Agnihotri (1987) have worked on the nature and magnitude of rill and gully erosion in Teonthar tahsil, Madhya Pradesh. Similarly, an intensive work on ravine erosion in the Deoghat area of Allahabad District, U. P. by Singh, Dubey and Singh (1991) is noteworthy here. Singh and Dubey (1996) have critically analyzed the genesis, micro-geometry, morpho-cycles and integrated management of gully watersheds in the Belan-Seoti confluence of Deoghar-Allahabad in U. P. Attempts have also been made to determine the nature, extent and distribution of gully networks particularly in association with climatic, pedogenic, geomorphic and anthropogenic aspects and to define the implication of gully development in terms of the extent of gullies at work in to usefully potential and even cultivated/forest land. In addition of this, the classification of the gully affected land according to the nature of gully erosion has been made for devising remedial measures of the study area (cf. Jha and Kapat 2003). Kuntala Lahiri-Dutt and Prasanta K. Jana (2003) have discussed some crucial issues related to land degradation in the Raniganj coal belt such as decline of agriculture and forestry, dereliction of land due to coal mining. Recently, modern information technology has been put to good use in examining land degradation. For example, Jha and

Gupta (2003) have studied Darakeswar River Basin (West Bengal) using remote sensing and GIS. Sutapa Mukhopadhyay (2003) has tried to establish a correlation between river bank erosion and land degradation in the River Ganga whereas Mukherjee and Mukhopadhyay (2003) have studied the role of flood in land degradation in the Lower Ajoy River Basin, West Bengal.

We are going through a time when there is an urgent need for food and nutritional security on one hand and a strong demand for infrastructural development and urbanisation for the betterment of human life on the other, which requires a large quantum of land resources. As a result, increasing tendency of land conversion is being observed that has given rise to growing socio-political conflicts in different parts of India, as elsewhere in the developing part of the world where population pressure is considerably high. Keeping in view with the above situation and the finite land resources of our country, a comprehensive research on land management strategies for preventing land degradation as well as augmenting optimum land utilisation is essential.

Disaster preparedness

An emerging area in disaster studies, which has appealed the policy makers and administrators more, is disaster preparedness. Dutta (2009) has tried to analyze and evaluate the shifting emphasis on disaster preparedness than relief and rehabilitation with special reference to Burdwan District Disaster Management and Response Plan. It is encouraging to note that a young scholar in geography has felt the relevance of disaster preparedness in the line of thinking of both the international and national organisations involved in disaster management.

Scope of Geography in Hazards and Disaster Management

The discipline of Geography is a spatial science, which deals with variations and differentiation on space, whatever may be the aspect, be it climate change, deforestation, pollution or differential hazards and disasters. Thus, it can efficiently handle spatial variations in terms of environmental and socio-economic impact assessment of hazards and disasters that can be of great help in development planning and management. A good example of it is the paper by Singh (2007) which looks at the disaster in holistic manner at micro level. Cartographic techniques are another area, better to say, the most important and exclusive area of geography, which help in the vulnerability analysis and mapping of hazards and disasters in a region. In addition, modern methods and tools especially remote sensing and GIS can play a vital role in the proper assessment of environmental impacts and risks of hazards and disasters.

Place of Geographers in Decision-Making Bodies for Disaster Management

At the Central Government level, there are organisations or bodies like National Disaster Management Authority (NDMA) under the Chairmanship of Prime Minister, National Institute of Disaster Management (NIDM) under Ministry of Home Affairs, several inter-ministerial groups and task forces looking after relief and rehabilitation during and after the occurrences of disasters in coordination with the Prime Minister's Office (PMO), expert committee of the Planning Commission in the assessment of damage and destruction after the disaster and budgetary allotment in the post-disaster rebuilding and reconstruction. At the state government level also, there are a number of bodies under the Chairmanship of Chief Minister dealing with pre- and post-disaster crisis situation. A good example of it is Orissa State Disaster Management Authority. In West Bengal, Orissa and Gujarat, there is a separate Ministry for Disaster Management.

It is quite unfortunate that the geographers' expertise has not been utilised so far and they usually do not find any place in the decision-making bodies engaged in dealing with aspects of disaster management at both the national and state level. This presents a challenge for the present young and future generation of geographers to develop expertise and benefit from this emerging area with greater potential.

Future Agenda and the Role of Young Geographers

From the above brief discussion, on geographers' contribution to the field of natural hazards and disasters, it is quite clear that there has been a little contribution from the geographers. Most of the contribution has been confined to the physico-geographic aspects of hazards and disasters and it is still the experimental field of geologists and to some extent geomorphologists. There are only a few contributions from geographers as such to the socio-cultural aspects of hazards and disasters (such as health, trauma and social-psyche of affected people, corruptions involved during relief and reconstruction etc.), which is not less significant than physical aspects. It may further be pointed out that all the contributions in this field are made by the senior geographers and young geographers' presence is least in relative terms. So, this situation necessitates an arduous task by the young geographers to study the regional variations in the magnitude of hazards and disasters due to the physical diversities, climate change as well as degree and methods of resource exploitation. Obviously, such studies can open up new vistas and will lead to exploration of new areas of research like how natural hazards and disasters impact the people living in the vulnerable zones of mountains, coasts and riverine plains in general and women and children as well as elderly and downtrodden in particular. In this way, geographers can directly contribute to society and geography can prove itself socially responsible—an area in which it is allegedly poor! All these aspects are considered to be the essential components of planning for balanced regional development in the present juncture of crisis and uncertainty. In this regard, preparation of 'Risk Zones Mapping' by the geographers is an utmost need. Besides, in the present day geographical research, whether physical or socio-economic or contemporary developmental issues, the components of hazards and disasters should be incorporated and accordingly be debated nationwide. Much has yet to be done in geography keeping in view with different dimensions of natural hazards and disasters in time-space continuum. Geographers, particularly young ones, should take a step forward to explore this emerging field. The day is not far when we will see a separate specialized field in geography i.e. *Geography of Natural Hazards and Disasters*.

Concluding Remarks

Going through the existing literatures on disaster studies, it has been observed that in India the geographers started to contribute in this very field since mid-nineties of the last century. In that sense it is one of the most recent fields of research in which the geographers can explore further and contribute a lot. It may, however, be pointed out that no place in the world is free from hazards and disasters. Due to climate change new areas are being added under the grip of natural hazards and disasters. Certainly any developmental framework should take into account 'hazards and disasters' as the essential component. It is a good sign that *Disaster* is being considered important during 11th Plan Period and it is an important component in each and every Ministry of the Government of India.

Finally, it may be concluded that in tackling the natural hazards and disasters with full courage and strength, geographers' expertise should count. We cannot wait but to incorporate the science of natural disasters management within the framework of development and educational curricula of Geography so that the geographers can effectively contribute in this emerging field and find a respectable position in the decision-making bodies engaged in hazards and disasters management.

References

Banukumar, K., G.V. Rajamanickam and S. Aruchamy. 2005. Study of drought prone areas in Pudukkottai taluk, Tamil Nadu: A hydro-geomorphological approach. *Indian Journal of Geomorphology* 10 (1 & 2): 23-36.

Basu, S.R. 1999. The causes and consequences of soil erosion and landslides in Darjeeling town. In *Natural Disasters*: *Some Issues and Challenges*, 44-63. Santiniketan: Natural Disasters Management Cell, Visva-Bharati.

Butola, B. S. and N.C. Jana. 2003. Droughts of doom: the story of Rajasthan. *Geography and You* 3 (5 & 6): 39-44.

Deb Sarkar, Manisha. 2005. River communication: some alternative measures to lessen flood intensity in deltaic West Bengal–an overview. In *River Flood*: *A Socio-Technical Approach*, ed. M.B. Rahim, Malay Mukhopadhyay and Debashis Sarkar, 232-243. Santiniketan: Natural Disasters Management Cell, Visva-Bharati with ACB Publications (Kolkata).

Dutta, Shantanu. 2009. A shift of paradigm from relief and rehabilitation to disaster preparedness: an exploratory analysis of district disaster management and response plan, Burdwan in West Bengal. In *Disaster Management and Sustainable Development*: *Emerging Issues and Concerns*, ed. Rajesh Anand, N.C. Jana and Sudhir Singh, 107-124. New Delhi: Pentagon Press.

Gadgil, A.S. 2000. Some aspects of Orissa super cyclone of October 1999. *Transactions of the Institute of Indian Geographers* 22 (2): 77-84.

Gupta, M.P. and Sarla Sharma. 1998. Flood situations in Raipur District-Madhya Pradesh, India. *Transactions*, *Institute of Indian Geographers* 20 (1):1-10.

Jana, N.C. 2003. That muddy red line (Floods in India). *Geography and You* 3 (7 & 8): 40-45.

—. 2006. Tropical cyclones. *Yojana* 50 (May): 32-35.

—. 2008. *Tsunami in India*: *impact assessment and mitigation strategies*. Kolkata: Prayas.

—. 2009. Human Dimensions of Tropical Cyclones: Indian Scenario. In *Disaster Management and Sustainable Development: Emerging Issues and Concerns*, ed. Rajesh Anand, N.C. Jana and Sudhir Singh, 46-77. New Delhi: Pentagon Press.

Jha, V.C. and K. Gupta. 2003. Land degradation in tropical lands: A case study. In *Land Degradation and Desertification*, ed. V.C. Jha, 279-290. Jaipur and New Delhi: Rawat Publications.

Jha, V.C. and Saswati Kapat. 2003. Gully erosion and its implications on land use: A case study of Dumka block, Dumka district, Jharkhand. In *Land Degradation and Desertification*, ed. V.C. Jha, 156-178. Jaipur and New Delhi: Rawat Publications.

Joshi, Veena U. 2005. Tsunami: the killer wave that struck the Asian coasts on 26th December 2004. *Transactions, Institute of Indian Geographers* 27 (1): 1-18.

Kalwar, S. C., H.V. Sharma, and Seema Kalwar. 2005. Drought and floods in Rajasthan. *Annals of the National Association of Geographers, India* 25 (1): 81-93.

Karlekar, Srikant 2002. Slope failure and landslides on Vengurla Coast, Maharashtra. *Indian Journal of Geomorphology* 7 (1 & 2):135-140.

Lahiri-Dutt, Kuntala and Prasanta K. Jana. 2003. Land degradation issues in the coal mining region of Raniganj. In *Land Degradation and Desertification*, ed. V.C. Jha, 178-190. Jaipur and New Delhi: Rawat Publications.

Mahanta, P. and A.K. Bora. 2002. Geomorphic hazards and their impact on the Kaziranga National Park of Assam, India. *Indian Journal of Geomorphology* 7 (1 & 2): 57-64.

Mandal, Bedhas U. 2006-07. Disaster mitigation in Rudraprayag District, Uttarakhand. *Indian Journal of Geomorphology* 11 & 12 (1 & 2): 189-196.

Mangat, H.S. 1994. Patiala floods: Where lies the solution? *Transactions of the Institute of Indian Geographers* 16 (1): 29-33.

Mishra, S. 1992. Drought: a case study in West Bengal (1901-1990). *Geographical Review of India* 54 (4): 71-80.

—. 1999. Weather and climatic hazards in West Bengal. In *Natural Disasters: Some Issues and Challenges,* ed. M.B. Rahim, 24-43. Santiniketan: Natural Disasters Management Cell, Visva-Bharati.

Mukherjee, Mallanath and, Malay Mukhopadhyay. 2003. Flood and land degradation: a study in Lower Ajoy River Basin. In *Land Degradation and Desertification*, ed. V.C. Jha, 374-386. Jaipur and New Delhi: Rawat Publications.

Mukhopadhyay, Malay 2005. Impact of flood hazard on bank dwellers: A perception study along Mayurakshi River of Eastern India. In *River Flood: A Socio-Technical Approach*, ed. M.B. Rahim, Malay

Mukhopadhyay and Debashis Sarkar, 66-100. Santiniketan: Natural Disasters Management Cell, Visva-Bharati with ACB Publications (Kolkata).

Mukhopadhyay, S.C. and A. E. Scheidegger. 2000. Fractality of Tista River flood statistics. *Indian Journal of Geomorphology* 5 (1 & 2): 111-114.

Mukhopadhyay, S.C. 2002. Geomorphology and natural hazards in the Lower Brahmaputra Basin with special reference to floods. *Indian Journal of Geomorphology* 7 (1 & 2): 73-80.

—. 2005a. Recent tsunamis and the coastal geomorphology of India. *Indian Journal of Geomorphology*. 10 (1 & 2): 105-112.

—. 2005b. Tsunami: A new addition to our natural hazard list. *Indian Journal of Landscape Systems and Ecological Studies* 28 (1): 241-250.

—. 2006-2007. Natural hazards and disaster management. *Indian Journal of Geomorphology* 11 & 12 (1 & 2): 139-48.

Mukhopadhyay, Sutapa 2003. River bank erosion and land degradation: A study on river Ganga. In *Land Degradation and Desertification*, ed. V.C. Jha, 366-373. Jaipur and New Delhi: Rawat Publications.

Paul, Ashis K. 1996. Identification of coastal hazards in West Bengal and parts of Orissa. *Indian Journal of Geomorphology* 1 (1): 1-30.

Sen, A.K. 1992. Desertification hazards and environmental degradation in Rajasthan deserts. *Geographical Review of India* 54 (2): 38-46.

Singh, Ravi S. 2007. River, Disaster, and Development: Some Overviews. In *Recent Development of Disaster Management: An Indian Perspective*, ed. Amiya Kumar Ghosh and Sabyasachi Raychaudhuri, 171-180. Kolkata: Progressive Publishers.

Singh, Savindra and S.P. Agnihotri. 1987. Rill and gully erosion in the subhumid tropical riverine environment of Teonthar Tahsil, M. P., India. *Geographiska Annaler* 69 A: 227-236.

Singh, Savindra and A. Dubey. 1996. Gully Erosion and Management: Genesis, Micro-geometry, Morpho-cycles, and Integrated Management of Gully Watersheds in the Belan-Seoti Confluence Area of Deoghat-Allahabad, Report on DST-funded Research Project.

Singh, Savindra, A. Dubey and M.S. Singh. 1991. Ravination in the subhumid tropical environment of India: A case study of Deoghat Area, Allahabad District, U. P. *Indian Journal of Landscape Systems and Ecological Studies* 14 (2): 1-15.

Singh, Yadvinder 2000. Floods: an ecological hazard of the semi-natural drainage ecosystem of the inter state Chandigarh Region, India. *Transactions of the Institute of Indian Geographers* 22 (2): 85-97.

Sivaswami, K.S. and N.C. Jana. 2005. Major floods of the world in 1990s: Lessons for flood management in India. In *River Flood: A Socio-Technical Approach*, ed. M.B. Rahim, Malay Mukhopadhyay and Debashis Sarkar, 33-50. Santiniketan: Natural Disasters Management Cell, Visva-Bharati with ACB Publications (Kolkata).

Sreekesh, S. 2009.Taming the river: concerns and policies in managing the floods. In *Disaster Management and Sustainable Development: Emerging Issues and Concerns*, ed. Rajesh Anand, N. C. Jana and Sudhir Singh, 28-45. New Delhi: Pentagon Press.

Vasudevan, S., G.V. Rajamanickam, M. Arumugam, and R. Sakthivel 2006-07. Devastating tsunami strikes Cuddalore beach of central Tamil Nadu coast on 26th December 2004, East Coast of India. *Indian Journal of Geomorphology* 11 & 12 (1 & 2): 123-126.

Whittow, John B. 1980. *Disasters: The Anatomy of Environmental Hazards*. London: Penguin Books.

Zutshi, Bupinder 2009. Natural disasters in India: A case study of droughts in India and its management strategies. In *Disaster Management and Sustainable Development: Emerging Issues and Concerns*, ed. Rajesh Anand, N.C. Jana and Sudhir Singh, 7-27. New Delhi: Pentagon Press.

PART IV:

RETRIEVING HUMAN GEOGRAPHY

The urge for destruction is also a creative urge!
—Michael Bakunin (1814-1876) in *Jahrbuch für Wissenschaft und Kunst* (1842)

CHAPTER ELEVEN

SITUATING WOMEN IN GEOGRAPHY

TARUNA BANSAL,
JAMIA MILLIA ISLAMIA

Introduction

Geography as a discipline is undergoing the process of re-writing and the most noticeable change in the nature of the subject has been the introduction of women and gender studies. In other words, geography is being rewritten with women in the focus now and this process has produced an assortment of literature with diverse theoretical perspectives. (Mackenzie 1984) This marks the development of a substantial field in geography which is commonly referred as the "Geography of Gender" or "Feminist Geography". The geography of gender nurtured itself on the basis of feminist thoughts which grew from ideas generated by the women's liberation movement during the 1970s and at the same time had radical scholarly traditions (Peet 1998).

From its inception, a defining characteristic of feminist geography has been its ability of intellectual cross-fertilization and multidisciplinarity (Nelson and Seager 2005). And, to appreciate this theoretical capability of feminist research in geography, one has to understand the two underlying themes on which the entire feminist scholarship is based. Firstly, feminist research is not just about women rather it is about how women form an essential part of the human-environment relations. Secondly, it is not an empirical study, rather a process of methodological and conceptual modification of addressing the issues of women in geography (Hayford 1974; Mackenzie 1984; Hanson 1992; Rose 1993; Massey 1994; McDowell 1997; Oberhauser 2000). In the words of McDowell and Sharp (1997), "The major goal of feminist geography is to investigate, make visible and challenge the relationships between gender divisions and spatial divisions, to uncover their mutual constitution and problematise their apparent naturalness" (Dias and Blecha 2007).

The present essay examines the ways in which the spaces for women have been constructed in geography keeping the above background in view. The first part of the article deals with the review of feminist approaches in geography. The second section deals with the academic discipline of geography and how women are placed in this discipline with special reference to India. The third section deals with the prominent areas which deal with the issue of women in geography. Here, the author brings in the various topics of debates related to women's work, sexual division of labour, women's position and status in the labour market and the increasing inequality between the two sexes. The final section deals with the methodological issues and how geography can be made more humanistic by not excluding the other half of the human population in the broader field of human geography.

Feminist Approaches in Geography

Like all the fields of geography, feminist geography has developed in a social and political context that influenced the issues selected for and analysis and the methods adopted to investigate them. (Bowlby, Lewis, Mcdowell and Foord 1989) The sub-discipline encompasses two areas of discourse—geography and feminism—which further include diversified approaches. But when geography embraced feminist scholarship, it provided the later with its approaches to study the concerned subject matter. Hence emerged the geography of gender embedded within the conventional geographic paradigms. This was dominated by perspectives that draw on feminist politics and theories to explore how gender relations and geographies are mutually structured and transformed (Pratt 2000; Dias and Blecha 2007)). In simpler terms, it is about the way which gender is constituted and how this relates to the constitution of the environment (Mackenzie 1989). It therefore looks into the geographical debate about women's position, the social construction of gender identities and changes in gender divisions over space and time.

For further understanding of the concerns of feminist geography one has to undergo a historical survey of the area starting from its inception in 1970s. During the mid-seventies period of time the geographers engaged in this field were trying to create a "*geography of women*". The main purpose here was to document systematically how women were disadvantaged in many spheres of life. This added women to the field of geography both as producers of knowledge as well as subjects of analysis. The earliest feminist geographical works focused on mapping the spatial constraints women face (Davies and Fowler 1971; Hayford 1974; Tivers

1977; Ardener 1981; Monk and Hanson 1982). The studies were strongly influenced by radical and welfare approaches.

The 1980s laid the foundation for many subfields and interests as new theoretical constructs were introduced to study the role of gender as an instrumental force and as a category of explanation in geographical processes. Now the studies were focused on the geographical inquiry of gendered spatial divisions between public and private, particularly in the sphere of work and urban processes (Zelinsky 1982; Christopheson 1983; Rossini 1983; Massey 1984; Monk 1984; Mackenzie 1986; Nelson 1986; Pratt and Hanson 1988). It was during this decade only that the first cooperative study was published as *Geography and Women* in 1984 by the Women and Geography Study Group of the Institute of British Geographers. This was a sustained effort at outlining the critique of human geography and an alternative feminist approach. For the point of view of study group, feminist geography looks at how socially created gender structures form and transform space in a project dedicated to ending gender inequalities through social change (Peet 1998, 270). They chose four streams of feminist approaches which were to be included in mainstream geography. These were (i) feminist urban geography (ii) women's increased participation in the public sphere (iii) separation between home and waged work, and (iv) women's activity in changing the developing world.

Feminist geographers thus, not only documented gender divisions and the inequalities in women's position in both the advanced and less developed countries but also started explaining the patterns of inequalities. The main source of these works were rooted in the literature from feminism rather than from geography; as a result, the sub-discipline started moving towards a feminist geography rather than a geography of women which require areas of everyday life and of the academic literature that usually are not regarded as geographical (Bowlby, Lewis, Mcdowell and Foord 1989, 158-59). They addressed the issues of the neglect of women, challenged misrepresentations of women and argued that gender has emerged as the axis of social differentiation and inequality in geographical research (Bondi and Davidson 2005, 15).

In the decade of nineties, the feminist geography moved further and opened up questions about ways in which spaces and places are experienced differently by different people and came to be associated with the presence and absence of different groups of people (Nairn 1997; Burgess 1998; Longhurst 2001). This feminist geography thus started rethinking gender and its relationship to space and place. The major proponents of this approach were Doreen Massey and Gillian Rose.

Massey analyzed the dynamic interplay between social relations of class and the spatial organization of production; thus conceptualized the interrelations between space, place and gender. Rose's interpretation of this relationship differs from that of Massey's. She analyses this association in terms of contradictions that can render everyday experiences, especially women's experiences (*ibid.*, 16).

In recent years, the feminist geographers thus got actively engaged in debates that questioned both the unity/singularity of knowledge and the subject of "woman". As a result, a rich literature surfaced on sexuality and gender and the way they perform in different spaces. Moreover, the scope of feminist geography broadened as the feminist geographers started contributing in different fields of geography like economic and labour geography (McDowell and Graham 1996; Friedberg 2001), political geography (Staehcli 1996; Secor 2001), urban geography (Gilbert 1997; Bondi 1998), cultural geography (Katz and Monk 1993; Anderson 1996) and many more (Nelson and Seager 2005, 5).

The above discussion clearly shows four basic themes of feminist geography which emerge and re-emerge to distinguish it as a field of study. First, feminist geography is closely related with political movements and analyzes them. Second, it is an interdisciplinary subfield. Although its roots are grounded in the radical literature, it has engaged itself in major theoretical approaches of geography like in political and cultural geography, in urban and environment research, in the literature on economics and migration as well as in the writings dealing with methodological issues. Third, this subfield studies the relationship between space, place and gender. They stress on how women is the part of multiple oppressions and how these oppressions are impregnated in space and place. Fourth, the feminist geography asserts the importance of women as a subject of study and examines gender as a social and spatial process (*ibid.*, 6-7).

Place of Women in Geography

Geography is an intellectually diverse discipline in which scholars study a large number of subjects. Geographers mainly rely on the synthetic approach for better understanding of the world and the environment-society relations. Feminist geographers among them pay particular attention to the gendered ways in which people interpret places, the access women as individuals and as members of social groups have to places and their ability to act in various ways within those places (Staeheli and Martin 2000, 138-39). Place is one of the most fundamental concepts in the discipline of geography. And, they differ from one another as each is a set of interrelationships between environmental, economic, social, cultural and political processes (Rose 1993, 41).

The feminist geographers claim that the humanist geographers have shown the concern for the way in which ordinary people are subject to various forms of authority. They have not analyzed place in the specific forms of exploitation and oppression that occur in this place. In doing so they neglect women as complex and diverse social subjects and also ignore that place can also be feminized. In this section the author discusses "place" from two points of view; firstly, woman as a subject in the geographical studies, dealt in detail in the next section. Secondly, the position woman occupies in the academia of geography in world with special reference to India.

The question of woman becomes important as she is often related to home; even her work is referred with reference to home. In other words, home becomes a noteworthy place even in the context of man as this place is intensely experienced and is full of significant memories and experiences (Relph 1976; Tuan 1971; Seamon 1980). But, over the period of time, geography has always ignored the domestic, the home, the emotional or anything concerned with women till the very recent advent of feminist geography (Rose 1993, 48). This resulted in the analysis of women's roles and changes in urban and regional structure and activity patterns (Mackenzie 1984, 3). The interesting part of this analysis was that geographers started seeing a relation between women and the place. They started studying and theorizing this relation between process and spatial form on the basis of theory contextualized with women's lives to environmental production and alteration. Many geographers started their work by criticizing the fact that women were either invisible within geographic literature or were analyzed as substitutes for men (Hayford 1970; Burnett 1973; Larimore 1978; Tivers 1978) Thus emerged a "*geography of women*" concerned with socialist-feminist perspective on the relations between women and environment on one hand and a critical engagement with geographic perspectives as a whole and historical materialistic perspectives in particular (Mackenzie 1984, 3).

Another aspect which seeks attention is the insignificant number of women in the profession of academic geography in comparison to other social sciences. The studies done in this regard suggest that there is overwhelming male composition of human geography, especially in the past which is continuing to the present (Warf 2006, 157). Feminists claim that this is one of the basic reasons why the discipline has primarily served the interests of men and worked within masculine paradigms. While some women did engage in exploring society, the knowledge produced by them was not incorporated into the more formal knowledge base of geography within universities (Staeheli and Martin 2000, 138). An example may be cited of Royal Geographic Society (RGS), where till 1915 women were not elected to membership in the RGS and fieldwork as geographic inquiry was limited to a few, elite, white males (Raju 2004, 84).

Even in its demographic composition the discipline has been masculine in nature. A study by Lee (1990) shows that the number of women in permanent academic ranks in United States' geography departments is only 8 percent of geography faculty. Women comprise a quarter of the new doctorates and low ranking professors, but only three percent of full professors, both in large and small departments. Although in 1990s in the United States, the women broke this barrier and entered graduate schools in large numbers. Women became part of the academic staff of many universities but were in junior positions (Poonacha 2003, 2653). The scenario is very similar in Indian geography, where there appears to be resistance not only to the gender studies in geography but also to the entry of women in the academia. This is mainly because the gender discourse is difficult to be placed in the traditional framework of Indian geography. And, over the period of time, women studies, as a distinct discipline outside geography, has developed as an academic statement of a political ideal of education. Table 11-1 provides an insight into the scenario in the geography departments of some of the central universities of India. It clearly shows that the ratio between men and women professors is very large, there are only nine female professors compared to forty-eight male professors. The situation is worse in the slab of associate professors the number is four compared to twenty-eight for men. The scenario seems to be friendlier now for women as the number of women assistant professors is more than compared to the other two designations (sixteen women and twenty-six men). Moreover, nearly one third of these universities (four out of eleven) employ only one woman. The unfortunate consequence of this deficiency is that female students at the bachelor's, masters and doctoral level, lack role models of successful women as professional geographers.

Table 11-1: Female geographers in academic departments in select central universities of India

University	Professor		Associate Professor/Reader		Assistant Professor/Lecturer		Total female faculty members in the Department*
	Males	*Females*	*Males*	*Females*	*Males*	*Females*	
Aligarh Muslim University, Aligarh	02	01	08	00	06	00	01 (17)
Banaras Hindu University, Varanasi	14	01	03	00	04	05	05 (27)
University of Allahabad, Allahabad	06	02	01	00	04	01	03 (14)
University of Delhi, New Delhi	01	00	03	01	02	02	03 (09)
Jawaharlal Nehru University, New Delhi	07	02	06	01	02	03	06 (21)
Jamia Millia Islamia, New Delhi	05	02	02	01	03	03	06 (16)
Mizoram University, Imphal	01	00	01	00	02	00	00 (04)
North-Eastern Hill University, Shillong	04	00	03	00	01	01	01 (09)
Manipur University, Imphal	02	00	01	00	02	00	00 (05)

Visva-Bharati, Shantinekatan	06	01	00	01	00	01	03 (09)
Total	48	09	28	04	26	16	28 (131)

Note: *Only select central universities with geography department have been taken on which information is available on the web. The numbers in parenthesis denote the total number of faculty members in respective department of geography.

(*Source*: Data has been analysed from the information provided at the official websites of the concerned central universities)

Recovering Women in Geography

The reason why women are a particular concern in the study of gender relations in geography is because within the existing patriarchal culture, they suffer from discrimination, and subordination, which prevents equality between the sexes. And in this background, one of the earliest concerns as well as areas of research in this sub-discipline has been the domain of work—the gendered construction of women's primary role as mothers, care-takers and home-makers. The interpretations of work will have to take into account the relationship between work and gender on a notion of "great divide" between home and work. This denotes that the history of work is more than work alone. Conventionally, work is regarded as an area that is clearly demarcated from domestic or social lives, as something people are paid to do, usually for set hours each week (Pahl 1992, 122). Work is often experienced as the opposite of home; it constitutes the "public".

The fundamental transformation in the concept of work in day to day life came with the reorganization of production, which separated the household from the workplace. This detachment laid the foundation for the break-up between consumption and production, women's activities from men's and therefore, *the sexual division of labour*. The division is fundamentally historical. It is the division between production—paid work or economic activity—and reproduction—the bearing and caring of children and the management of the household (Berg 2000). A division exists not only between the domestic and productive work but also within each category (Bullock 1994, 210).

Marx and Engels point out that the first division of labour is between man and woman for child breeding. A theory which claims to be materialist must locate the sexual division of labour in the relations between men and women as social rather than biological categories (Bullock 1994, 99). Brown (1994) also talks of the overall sexual division of labour (Davis 1975; Roger 1981; Oakley 1982; Porter 1982; Massey 1982; Megan 1982; Hufton 1983; Walby 1986; Heath 1994; Hakim 1996; Hanson and Prat 1988; Trat 1998; Clark 2000; Jenkins 2004). This approach examines all the tasks, which have to be carried out in a society and looks to see who actually carries out each task. Many scholars believe that all sexual inequalities are based on this division of labour. Some also point out that this framework has made the home-based women's work invisible and undervalued, whereas market work that is dominated by men gets priority in monetized market economies (Raju 2004, 92). Thus, one can say that throughout the history of geography, home and work have been viewed not only as separate but also as gendered spheres, with the work place epitomizing the male-realm and the residence the female's (Gregson *et al.* 1997, 81).

The research on women and work during the emergence of *Geography of women* was mainly addressing the issues of social justice and individual equality and the answers to questions about women's lower than average incomes, unequal travel patterns and segregation in lower status jobs were not completely addressed (Oberhauser 2000, 60). It was in 1980s that new theoretical and methodological analyses were introduced to answer the unanswered questions. This new approach focused on the economic restructuring that was occurring all over the world, the dramatic shift in employment from the manufacturing to the service sector. This brought a larger number of women in the labour market both in absolute numbers and as a proportion of the female population. A very significant outcome of these studies was that in spite of so many reforms, the labour market still showed the evidences of sexual division of labour. Research demonstrated that in the established professions like administrative, financial and managerial services women occupy the lower status, within all sectors of the labour market. And, some occupations have become more identifiable with women. Women are over represented either in the "caring" and support professions such as teaching, nursing, clerical, counselling and childcare (O'Brien and Williams 2004, 210) or in the multinational factory work where they were low paid, unskilled and non-unionized.

Another aspect which comes under the scope of *geography of women* is how women come to take on identities and think of themselves. How they create places for themselves and how they learn about and practice their identities. This becomes relevant when one observes that women have been used as symbols. The best example is drawn from the Indian society where since the Indus civilization women were seen as being a source of life. They represented fertility and their identity was analogous to Mother Earth, a conception central to the concept of "Mother Goddess". This concept continued to evolve and crystallized into the identity of *Shakt—a* means of acquiring power over the forces of nature. The woman was thus, located in the concept of *Tripurasundari*—ultimate object of sensuous fulfilment (Parikh and Garg 1989, 79). From mythological point of view, the image was that of a self-sacrificing "Savitri". Even today, they are part of this imagination as symbolic representation of the nation. The nation could represent itself as woman. Thus, women became passive symbols of different totalities within the discourse of tradition or religion (Rao 1999).

In the last few decades, feminist geographers brought new dimensions to the study of women. They not only challenged the gender-blindness of geography but also stressed that the significance of women's experiences and activities should be recognized. They drew attention to the links between the domestic (household) and the market both in relation to job market and as part of consumer cycle. In this regard several studies were done to explore the role of women and her involvement in the larger public sphere. The feminist geographers even started mapping women's lives and variations in patterns of their role in the developing countries (Bowlby, Lewis, Mcdowell and Foord 1989, 162).

Thus, the main question, addressed in most of the work done in the last few decades, has been concerned with women's role in the domestic world and her growing involvement in the public sphere. But the questions which have been unanswered are mainly related to women's inequality which is more significant in the discipline of geography. For example issues like discrimination at the work place, wage differentiation, journey-to-work patterns, urban development, role of women as a consumer in globalized world, violence against women, women and crime in city etc, which require analysis of gender roles and gender relations, have hardly been touched. In brief it can be said that conventional geography has neglected women's experiences and gender roles in the economy. The debate on theses issues have just started in the discipline which will bring forward many new theoretical and empirical directions.

The Study of Women as a Method in Geography

Social research turns the chaotic and confusing experiences of everyday life into categories of people in society, categories that reflect prevailing political arrangements (Harding and Norberg 2005, 2009). The social sciences then assign causal relations to people and social-political and economic relations in these categories. But in most of the work done in the social sciences women are ignored as the subject of study. There has been inappropriate contextualization of women over the period of time. The new discipline of *geography of women* has made an immense contribution to the understanding of women and her role in the economy.

This *geography of women* has done this by addressing issues as diverse as the constitution of gendered spaces in employment and restructuring to constructing feminist perspectives on the environment, urban spatial processes and images of urbanity (Kobayashi and Mackenzie 1989; Rose 1989; Pred 1990; Andrew and Milroy 1988; Moss 1993). Thus all theory developed around women has not only redress omissions and misinterpretations in the traditional disciplines but has also created a paradigm shift through which alternatives have been provided to what are viewed as patriarchal models of explanation (Dyrc 1993, 53). Here women are placed at centre-stage both as subjects of research and in creating knowledge. Another interesting aspect of this method is that no new research tools have been introduced rather it has been shown that techniques and research strategies are not dissociable from method in a large sense and therefore influenced by the social context (Rose 1993, 59).

Consequently, the *geography of women* focuses on the need to break through the obsession of spatial science and start to observe how a disintegrated population, broken down by gender, creates environment and is also affected by environments not of their own creation (Breitbart 1984, 73). In other words, it implies an interactive and creative relation between gender and environment, i.e. to involve a focus on the interrelated activities of women and men (Mackenzie 1989, 57).

Conclusions

The above account clearly indicates that the *geography of women* contributes to broader discussions within critical geography and feminist studies through its epistemological and methodological approaches. Some of the works focus exclusively on women and their work, while others deal with gender relations and identities as these are transformed in dynamic relation with the restructuring of work (England and Lawson 2005, 87). Secondly, it has brought the dichotomy of opposites – man and woman, nature and culture, public and private, production and reproduction—into the general theory and practice. As a result, a new discourse has emerged which focuses on the analyses of cultural representation; work on sexuality, subjectivity and social relations; and new developments in the studies of interrelationships of race, class and gender (Bowlby, Lewis, Mcdowell and Foord 1989, 169). This discourse therefore presents challenges to female geographers and they have to make accommodations within the existing frameworks and avoid academic pitfalls. This can be done if the profession continues to attract women into geography especially at the advanced levels and remove barriers which hinder their advancements. Furthermore, gender based biases and prejudices towards females must be uncovered and eradicated. As Palm (1985) states that "by including more women in our field we will enjoy a significant infusion of new talent as well as the prospect of new perspectives on classic geographic problems".

References

Berg, M. 2000. Women's work, mechanization and the early phases of industrialisation in England. In *The Historical Meanings of Work*, ed. P. Joyce, 64-96. New York: Cambridge University Press.

Bondi, L. and J. Davidson 2005. Situating gender. In *A Companion to Feminist Geography*, ed. L. Nelson and J. Seager, 15-31. London: Blackwell Publishing Ltd.

Bowlby, S., J. Lewis, L. Mcdowell, and J. Foord 1989. The geography of gender. In *New Models in Geography: The Political Economy Perspective*, ed. R. Peet and N. Thrift, 157-175. Boston and London: Unwin Hyman.

Breitbart, M. M. 1984. Feminist perspectives in geographic theory and methodology. *Antipode* 6 (3): 72-74.

Bullock, S. 1994. *Women and Work.* London: Zed Books.

Dias, K. and J. Blecha 2007. Feminism and social theory in geography: An introduction. *The Professional Geographer* 59 (1): 1-9.

Dyrc, I. 1993. Ethnography: a feminist method. *The Canadian Geographer* 37 (1): 52-57.

England, K. and V. Lawson 2005. Feminist analyses of work: Rethinking the boundaries, gendering and spatiality of work. In *A Companion to Feminist Geography*, ed. L. Nelson and J. Seager, 77-92. London: Blackwell Publishing Ltd.

Gregson, N. *et al.* 1997. Gender in feminist geography. In *Feminist Geographies: Explorations in Diversity and Difference*, (Women and Geography Study Group), London: Addison Wesley Longman.

Harding, S. and K. Norberg 2005. New feminist approaches to social science methodologies: An introduction. *Signs: Journal of Women in Culture and Society* 30 (4): 2009-2015.

Mackenzie, S. 1984. Women and the environment: Editorial introduction. *Antipode* 6 (3): 3-10.

Mackenzie, S. 1989. Restructuring the relations of work and life: Wwomen as environmental actors, feminism as geographic analysis. In *Remaking Human Geography*, ed. A. Kobayashi and S. Mackenzie, 40-59. London: Unwin Hyman.

Moss, P. 1993. Focus: Feminism as method. *The Canadian Geographer* 37 (1): 48-49.

Nelson, L. and J. Seager 2005. *A Companion to Feminist Geography.* London: Blackwell Publishing Ltd.

O'Brien, R. and M. Williams 2004. *Global Political Economy: Evolution and Dynamics.* Basingstoke: Palgrave Macmillan

Oberhauser, A. M. 2000. Feminism and economic geography: Gendering work and working gender. In *A Companion to Economic Geography*, ed. E. Sheppard and T. J. Barnes, 60-76. Oxford: Blackwell Publishers.

Pahl, R. E. 1992. Defining work. In *Defining Women: Social Institutions and Gender Divisions*, ed. L. Mcdowell and R. Pringle. Cambridge: Polity Press.

Parikh, J. and P. K. Garg 1989. *Indian Women: An Inner Dialogue*. New Delhi: Sage Publications.

Peet, R. 1998. *Modern Geographical Thought*. Oxford: Blackwell Publishers.

Poonacha, V. 2003. Women studies in Indian universities: Current concerns. *Economic and Political Weekly* 38 (26): 2653-2658.

Raju, S. 2004. Gender and geography in India: Why the two shall meet. In *Space, Society and Geography*, ed. S. Banerjee Guha, 83-99. Jaipur/New Delhi: Rawat Publications.

Rao, S. 1999. Women-as-symbol: The intersections of identity politics, gender and Indian nationalism. *Women's International Forum* 22 (3): 317-328.

Rose, D. 1993. On feminism, method and methods in human geography: An idiosyncratic overview. *The Canadian Geographer 37*(1): 57-61.

Rose, G. 1993. *Feminism and Geography: The Limits of Geographical Knowledge*. Cambridge: Polity Press.

Shah, K. K. 2005. *History and Gender: Some Explorations*. Jaipur/New Delhi: Rawat Publications.

Staeheli, L. A. and P. M. Martin 2000. Places for feminism in geography. *Annals of the American Academy of Political and Social Science* 571(September): 135-150.

Townsend, J. G. 1991. Towards a regional geography of gender. *The Geographical Journal* 157 (1): 25-35.

Warf, B. 2006. *Encyclopaedia of Human Geography*. London: Sage Publications.

CHAPTER TWELVE

REDESIGNING SOCIAL GEOGRAPHY OF INDIA: DIMENSIONS, CONSEQUENCES AND METHODS

SOHEL FIRDOS,
CENTRE FOR STUDIES IN SOCIAL SCIENCES

> "…the field of social geography is inadequately developed in India. In fact, it is taking its roots and is likely to flourish as a major thrust area in the coming decades" (Ahmad 1999, 34).

"Geography now comprises a wide range of systematic studies which have one or more of environment, place and space as their foundational concepts (or organizing themes) but whose external links to other disciplines are sometimes at least as strong as their ties with other fields within their home discipline" (Johnston 2005, 11). The umbrella definition of Geography has embraced a flourishing fragmentation, however, comprising a great deal of variation in the extent to which sub-disciplinary communities promote their work as contributing to the appreciation of "areal differentiation". Within human geography, an increasing range of studies focuses on individuals" self- and group identities—what Dear (1997, 232) calls the "personal politics of space-time"—as people and groups continually reconstitute themselves. The role of place in those structuration processes, and hence the creation and re-creation of areal differentiation as the geography of "difference", remains fundamental to the entire project, however.

Fragmented Geography?

The study of Geography is currently fragmented in four main ways:

Substantive differentiation involves sub-disciplinary divisions identified according to their subject matter. At the macro scale, this is exemplified by the widely recognized human-physical split, involving two separate communities, each with some internal coherence but having substantial contact with the other (and the argument for a third major community—environmental geography in some terminologies—is for an activity which rarely combines the intellectual activity of the two, as against their superficial subject matter). At the meso scale, there are the subdivisions of human and physical geography, some compositional (such as social, political and economic geography) and some contextual (urban geography, rural geography and regional geography of x, for example)—on contextual and compositional. And, at the micro scale, these have their different topical areas of investigation, such as urban social geography, each with its own exemplars.

Epistemological differentiation involves divisions according to differing beliefs in the nature of knowledge–what can be known, and how. Buttimer (1993) identified four major metaphors, for example, involving the world as one of:

- a mosaic of forms;
- a mechanical system;
- an organic whole; and
- an arena for events.

The first two are associated with modes of science and social science which emphasize explanation as the product of repeatedly-verified empirical generalizations, whereas the other two are associated with achieving understanding.

The division between the two main types of inquiry is not necessarily a simple split between physical geography, on the one hand, and aspects of human geography, on the other (with spatial analysis in human geography perhaps occupying an intermediate position, disavowing positivism but promoting positive analysis). Realism, for the social sciences for example, has much to offer physical geographers, and others are actively exploring a range of scientific perspectives.

Differentiation in rationale involves divisions among geographers in the *raison d'etre* for their discipline (or segment thereof). Three or four such divisions are sometimes identified), but the basic split is between "pure" and "applied", which are linked to "explanation" and "understanding", respectively.

Community differentiation involves divisions which are they geographical. These vary in scale, and include:

1. the macro scale divisions, which are largely language-based (with that based on English almost certainly the largest);
2. The meso scale divisions, most of which are based on separate countries ("national schools"); and
3. The micro scale divisions, which are most often based on one or more academic institutions ("intellectual schools", such as the Berkeley School based on Carl Sauer between the 1920s and the 1960s).

Becoming an academic geographer involves being socialized and or accultured into one of the many disciplinary fragments that coexist within this coarse four dimensional matrix. Some individuals work in two or more separate fragments, others shift among them (although the "distance" between some are much greater than between others, and the volumes of migration are consequently much less): many remain within one fragment for very long periods.

Some Peculiar Characteristics of Geography

The characteristic of geography as a fragmented academic discipline identified here applies equally well to most other disciplines in the social and natural sciences. But there are additional features that are perhaps peculiar to geography, which has long been widely recognized as different from most other academic disciplines in its degree of overlap with others. Whereas geographers have for long explored along and across their frontiers with other disciplines, historians have tended to remain much more self-contained within their own institutional structures. That overlap is largely asymmetrical, with geographers being much more outward-looking than the academic neighbours with whom they interact, although the extent of imbalance in the "intellectual terms of trade" varies across the fragments: some of the specialized journals generally associated with geography are largely populated by non-geographers (defined by their institutional locations), for example, whereas others are almost entirely populated by geographers (for example, compare the *Journal of Biogeography* with the *Journal of Transport Geography*).

What this means is that the fragmentation of geography has probably produced more centrifugal tendencies than has a similar situation in other disciplines. Many of geography's communities substantially overlap those in other disciplines, as illustrated by attendance at specialized seminars and other meetings, membership of inter-disciplinary groups (few of which have the structures usually associated with learned societies), publications in inter-disciplinary journals established to service the overlap, and citation patterns. Geographers are more likely to join networks established in other disciplines than vice-versa.

Social Geographical Perspectives on "Difference"

There are several ways to approach the concept of difference. While geographers and other social scientists have used divergent definitions of "difference", we understand difference through socially produced markers (such as race, class, gender, and sexuality) and their attendant geographies (colonial geographies, post-colonial geographies, patriarchal geographies, feminist geographies, white geographies, and so on). The relationship between these markers and their geographies is a dialectical one: one constitutes the other with neither being understood outside the context of the other. Difference, then, signals diverse geographies and geographic experiences; it implies that "the social relations of spaces are experienced differently, and variously interpreted by those holding different positions" (Massey 1994, 3). Hence, difference always implies difference-in-place. Moreover, the concept of difference also signals the ways in which non-dominant identities are *socially constructed as different* from dominant identities. Thus rationality has been centred on the figure of the white, heterosexual and patriarchal Western man with all other deemed to be "outside" this orienting figure. To put it in another way, social markers such as race and gender are visible social constructs which mark difference (between whites and non-whites and men and women, for example); these differences are spatially organized and therefore not only visible through the scale of body, but also through material geographies—different people hold different geographic positions (in the home, the workplace, the city, or the suburbs for example). Difference-in-place, then allows us to examine "the hierarchical and unequal relationships among different groups" (Scott 1988, 179). It also gives insights into the ways in which the geographic positions of non-dominant groups challenge geographic hierarchies. For example, how we come to know difference through geography is structured by geographic projects such as imperialism and capitalism. These projects signal how broad geographic patterns (such as spatial and economic expansion) shape our world according to profit, displacement and power. Yet difference also signals critical emancipatory knowledges (such as feminism, post-colonialism, and anti-racism) and experiential knowledges, such as the geographies which we all live in the everyday world, which unsettle broad geographic projects such as imperialism and capitalism. In other words, "different" bodies are not only assigned "different" geographies, they are also actively experiencing and producing space.

One further point is worthy of note in relation to difference-in-place. Not only do various people occupy place in various ways—across time and space—but control over the production of space gives powerful groups the ability to produce difference as well as the right to be in space. Slums, sprawling suburbs, over polluted regions, gated communities, under-developed and over-developed nations, homeless hostels, and native land claims, are just some examples of how geography and geographic knowledge are, locally and globally, tied to practices of spatial unevenness. What these spatial formations reveal is the ways in which geography is mapped according to race, class, and gender specific interests. They also reveal that geographic knowledge—how we "know" and "understand" the external world—is inevitably tied to spatial formations and hierarchies.

We are not asserting, however, that formations of race, class, gender, and so on are spatially rendered in the *same ways* over time and space but, rather, that geographic expression is extremely variable. David Delaney contends that axes of power can combine in multitudinous ways "to produce the richly textured, highly variegated, and power laden spatialities of everyday life" (Delaney 2002, 7). He claims that what is important about these spatialities is how the division of continuous territory into "insides" and "outsides" facilitates the categorization of groups into "us vs. them".

Coming to Know Difference through Social Geography

Understanding of difference has been one of the major concerns of social geographers. According Donna Haraway (cited by Noel Castree, *et al.*, 2005: 43) "knowledge is local, specific and embodied, encapsulates an important way in which difference can be understood. That is, space and place are intimately connected to race, caste, gender, class, sexuality and other axes of power; all geographic knowledges are situated and location matters". For Haraway, situated knowledges are those of embodied, located subject (Holloway 2004, 169-170). Further, she is of the view that "situated knowledges create a conceptual and material space through which non-dominant geographies can be articulated and theorized. Difference, it could be said, can be situated within and around situated knowledges. "So, what is considered different in geographic terrains and why? How does landscape perpetuate or democratize difference? Why does difference matter so much to spatial organization? Do difference and situated knowledges unsettle conventional geographic patterns? That geographic organization is also an organization of difference is probably most obvious in colonial projects, which segregate and hierarchically 'manage' non-white populations for economic and geographic profit". (McKittrick and Peake 2005, 44). In geography, difference is formed in other ways too. Different people "occupy different areas and perform different tasks" at a place as per their status which could be defined in many ways (and this differs of course, from home to home, region to region, religion to religion and caste to caste in case of India). To quote McKittrick and Peake (*ibid.*, 44) again "Who cleans homes? Is a cleaned home a gendered space, a racial space? How are local, regional, national and global spaces differentiated according to social markers? What we are trying to stress is that the particularity of 'identity' and 'self' (*who* is different) implies some sort of *spatial* difference – be it race, sexuality, gender, body, dress, community, nation, citizenship or status: socio-cultural markers make a geographic difference and vice-versa" (*ibid.*).

Differences are neither socially neutral and value-free nor devoid of "layers". In one hand they reflect "both oppressed/oppressor relations, and the complex situated knowledges that challenge these relations", as argued by McKittrick and Peake ((*ibid.*). It is a spatial expression of geographic problems (such as the 'other' who is segregated, incarcerated, profiled) and it is a spatial opportunity to express political, social and economic possibilities. Racial segregation all over the world or religious and caste based segregation in India, for example, is a spatial expression of difference- it separates communities, perpetuates uneven economies, and geographically marks the landscape. However, "….geographies of social segregation also invoke varying responses to cycles of domination….all reconfigure the meanings of places".

"The tensions between difference and geography point to the ways in which, despite progressive resistances to social and geographic domination, the spatial organization of the world is still hierarchical". This hierarchy is, moreover, naturalized; it emphasizes and hides particular processes which profit from what Ruth Wilson Gilmore (2002) calls "the displacement of difference" (cited by McKittrick and Peake 2005, 44). That is, geographic hierarchies—different scales of power and knowledge which can be seen in cities, on streets, in homes, on bodies, across nations—continually privilege some, i.e. those with "traditional right", for example the whites in racial terms or the so-called "upper castes" in the Indian society. Nevertheless, "shaping of bodies by the world around them" could be is simultaneously interpreted as either "favourable or adverse" or both. One can easily find various scales, of body to the globe, interlinked with "broad geographic projects of domination". These, according to McKittrick and Peake (2005, 45), include imperialism, globalisation, transatlantic slavery and the 'carving up' and colonization of Africa, processes of 'white flight' and of 'no go' zones in cities. "These dominant geographies contain within them experiential geographies and critiques, that is, those spaces and places where the 'dailyness' of difference is lived….and resisted.

Geographic Responses to Difference: Nature-Culture and the Body

In recent years geographers have placed lots of emphasis on the geographies of differences. Some of the key concepts geographers have raised in order to think through difference include nature-culture, human species, uneven development, the body, the psyche, and race, gender and sexuality as social constructs. Social geographers, in particular, have drawn attention especially to the "visibility" of difference. For them, "difference is *placed*, in part, because difference is visible". When one sees difference, through common social markers like race, gender, and class, it speaks to historical and contemporary practices of domination/subjugation which could be racial, sexual and economic, respectively. There are also geographies of social differences. "Visible differences, although having different meanings to different social groups, have been used by dominant groups to structure geographic organization, i.e., uneven social relationships are spatialised according to social markers (cf. McKittrick and Peake 2005, 47). As a matter of fact, social differences are manifested in the landscape through spatial organization of settlement, spatial patterns of movement and interaction.

Social Geography of India: Challenges before It

"[The Indian society] has been undergoing wide ranging changes throughout history, although the pace of change has not been uniform in all regions or among all sections of our society. Processes such as education, urbanization and industrialisation, … have all induced the society towards change. The impact of these processes is neither uniform nor abiding, it is mostly fragmented" (Ahmad 1999, 11). The processes of socio-economic and political transformation in India have thrown up new challenges before the social geographers. Social geographers can contribute significantly to understand the contemporary issues by conducting pure academic research and/or may help in policy formulation by investigating the following themes. Firstly, the emerging spatial dimensions of social configurations claiming access to political institutions, public goods and services have drawn little or no attention from social geographers. I propose that these dimensions should form one of the prime concerns of research and teaching for social geographers. Secondly, the neo-liberal economic policies have been creating newer socio-spatial exclusion in the urban centres or accentuating older ones which have not drawn any attention from the practitioners of the discipline. I propose that to make social geography more relevant in the present times and to connect with other social sciences disciplines, we need to study concepts such as "social polarisation", "segregation" and "exclusion" and evaluate these concepts in terms of their relevance in understanding deprivation in India. In order to develop a better understanding of these concepts and their applicability, the social geographers may revisit some of the earlier works, though not limited to, studies such as the caste as a determining factor of spatial exclusion in urban areas. The earlier studies suggested that the residential housing in urban centres was a continuation of the rural model—separate quarters for different castes. In a traditional set up, there were separate residential quarters for Brahmans, Vaishyas, blacksmiths, weavers, washermen, tailors and other artisan castes e.g., Brahmans live in *mohalla* Chauban in Firozabad). The low castes such as the sweepers and the Kanjars generally lived on the periphery of the town. The major research agenda may be: Has the spatial organization of cities based on caste changed or they have remained embedded in space? If the changes have taken place, what are the driving forces behind such changes? What is the role of newer forms of economies in spatial reorganization of city space and how are they unfolding.

We also need to find newer methods or/and modify, re-sharpen the existing ones to understand the phenomenon of socio-spatial exclusion. I am hopeful that inclusion of these themes within the ambit of social geography will make the discipline more vibrant and responding to the social needs of our times.

Social Geographical Perspectives on Public Policy

The first feature of social geography that informs policy-making is an insistence on the relevance of context. Geographers are generally aware that universal explanations or policy prescriptions don't work well—one size does not fit all. Theories, explanations, models and plans that work in one part of the world do not necessarily travel to another. Peck (1999) distinguishes between "shallow" policy researchers who aim to smooth the transfer of policy solutions from one place to another, and "deep" researchers who are more aware of locally embedded solutions. Economists are typical of the former, geographers of the latter. Trudgill and Richards (1997) contrast the policy of world's preference for generalizations with environmental scientists' sensitivity to the context. Furthermore, they caution that policies devised for the large scale need not be successful at smaller scales. Urban or inner city policies that fail to appreciate that the particular mix of disadvantage and decline varies from one city to the next are likely to fail. Human particularly social and physical geographers have shown a consistent appreciation of context-dependence, and this provides an essential corrective to the over-generalization of policy.

Secondly, social geography combines insights and methods from other sub-disciplines of geography, and provides an integrated approach to social problems, be it in urban areas or rural areas. The routine interaction of social geographers with colleagues of other sub-disciplines, diffuse a greater awareness of human geography as a whole. Social geographers and other geographers share a sense of complexity, feedback, context-dependence and non-reductionist understanding, elements that are vital to policy science in social and environmental issues.

Thirdly, social geography has a long traditions of fieldworks, a practical and personal involvement with the objects of study—whether inner cities or spatial organization of settlements in Indian villages. Collection of both census and sample data through both participatory and non-participatory is prerequisite for any research. Ronald Cooke (1992 cited by Rogers 2005, 282) terms this "the landscape imperative", the deployment of field based skills to observe, explore, and monitor real places. The landscape provides hints to social geographers not only about inter-relatedness of processes but also about why spatial contexts matter. Since long, social geographers have been involved in developing depersonalized and abstract models of human behaviour by using computer programmes. However, the increasing popularity of various forms of participatory research methods have helped us to develop not only a better understanding of people's lives from the inside, but has also enabled the researchers to get involved the subjects of inquiry in the policy process itself. "Although there is now an overdue critical analysis of the politics and practices of field work, revealing its power relations and ethical considerations, it remains a valued skill within the discipline" (cf. Rogers 2005, 282).

Fieldwork is closely related to a range of other skills, techniques and methodologies that often comprise the main tools of social geographers. Surveying and cartography like traditional (geographical) skills could be now found adapted to computer based technologies and expanded by a variety of other sophisticated and modernized technologies. For example, commonly known digital information, remote sensing, Geographical Information Systems, Geographical Positioning System, computer aided cartography, and spatial modeling (cf. *Ibid.*). It is well accepted now that what is often broadly termed as "GIScience" has revolutionised analysis, representation and communication of the geographical data and it has immense potential to "become amazingly powerful" in future. Therefore, expecting "exciting possibilities that scientific visualization holds for the traditional geographical concern with cartography" (see, Rogers 2005). Again, communication is found to be "one of the problems of connecting science and policy". As far as geography is concerned, with the newer forms of interactive mapping, geographers enabled better to express their "findings and insights" more successfully and with far greater clarity and appeal than ever.

One of the characteristic features of geography that makes it more useful often goes unnoticed. The foundations of geography were laid down in the second half of the 19th century European universities with the basic aim "to promote geographical education and awareness among the population". This was not restricted to the university lecture halls or schoolrooms; rather it often involved the lecture theatres of the geographical societies as well as meeting places of other societies, groups and clubs. With the changing times and in the new circumstances, the role of circulating geographical knowledge has been assumed and generally performed by print (journals, magazines and newspapers) media and popularly by electronic media (radio and television). It is true that professional geographers are critical of such media for being trivial and "popularizing". But we can neither forget the significance of dissemination of scientific (including geographical) ideas nor ignore the fact that media play an important role in making the scientific ideas easily accessible and communicating about host of geographical issues to masses. It is argued that the ethical communication of science is "very difficult" (Rogers 2005, 283). It is demonstrated, through the example of the science of climate change, that "poor reporting in the media creates uncertainty and ignorance". And, geographers can engage themselves in a "valuable role" of more effective dissemination of scientific findings and their related implications Henderson-sellers (1998). We will have to remember that "an informed citizenry is indispensable to adequate policy formation and implementation". (Rogers 2005, 283)

Social geographers are already doing works that to some extent contribute towards formulating public policies. They have expertise in spatio-temporal perspectives, an understanding of the integration between human and physical processes, solid foundations in field works, interdisciplinary training, cartography, remote sensing and spatial modeling. Along with these, their "long-standing concerns for effective communication" of their findings, place them in a good position towards contributing to the literature of social sciences.

References

Ahmad, Aijazuddin 1999. *Social Geography*. Jaipur: Rawat Publications.

Buttimer, A. 1993. *Geography and the Human Spirit*. Baltimore, MD: Johns Hopkins University Press.

Cooke, R.U. 1992. Common ground, shared inheritance: Research imperatives for environmental geography. *Transactions of the Institute of British Geographers* NS 17: 131-151.

Dear, M. 1988. The post modern challenge: reconstructing human geography. *Transactions of the Institute of British Geographers* NS 13: 262-274.

Delaney, D. 2002. The space that race makes. *The Professional Geographer* 54: 6-14.

Gilmore, R.W. 2002. Fatal couplings of power and difference: Notes on racism and geography. *The Professional Geographer* 54: 15-24.

Henderson-Sellers, A. 1998. Communicating science ethically: Is the "balance" achievable? *Annals of the Association of American Geographers* 88: 301-307.

Holloway, Lewis 2004. Donna Haraway. In *Key Thinkers in Space and Place,* ed. Phil Hubbard, Rob Kitchin and Gill Valentine, 167-173. New Delhi: Sage Publications.

Johnston, R.J. 2005 Geography: Coming apart at the seams? In *Questioning Geography: Fundamental Debates*, ed. Noel Castree, Alisdair Rogers and Douglas D. Sherman, 9-25. Oxford: Blackwell.

Massey, D. 1994. *Space, Place, and Gender*. Minneapolis: University of Minnesota Press.

McKittrick, Katherine and Linda Peake 2005. What difference does difference make to Geography? In *Questioning Geography: Fundamental Debates*, ed. Noel Castree, Alisdair Rogers and Douglas

D. Sherman, 39-54. Malden, MA/Oxford/Carlton: Blackwell Publishing Limited.

Orford, S., D. Dorling, and R. Harris 2003. Cartography and visualization. In *The Student's Companion to Geography*, ed. A. Rogers and H. Viles, 151-156. Oxford: Blackwell.

Peck, J. 1999. Grey geography? *Transactions of the Institute of British Geographers* NS 24: 131-135.

Rogers, Alisdair 2005. A policy-relevant geography for society? In *Questioning Geography: Fundamental Debates*, ed. Noel Castree, Alisdair Rogers and Douglas D. Sherman, 277-293. Malden, MA/Oxford/Carlton: Blackwell Publishing Limited.

Scott, J.W. 1998. *Gender and the Politics of History*. New York: Columbia University Press.

Trudgill, S. and K. Richards 1997. Environmental science and policy: Generalization and context. *Transactions of the Institute of British Geographers* NS 22: 5-12.

CHAPTER THIRTEEN

CULTURAL GEOGRAPHY IN INDIA: NATURE OF ITS MARGINALITY AND FUTURE AGENDA

RAVI S. SINGH,
BANARAS HINDU UNIVERSITY

Introduction

Geography is a vast field of study in which all sub-fields are not treated at par in practice. There has been a continuing debate on what constitutes the "core" of geography and accordingly what all are "hardcore" geographies? Naturally, the debate, on what is geographical and what is not still continues. Long back, Carter (1968, 2) remarked, "Geography, like other fields, is plagued at times by its public image". Even now it is an unresolved issue as a matter of fact. This problem is not theoretical or imaginary; there are real-time instances of discriminations for and against a particular sub-field. Discrimination against the physical geography sub-fields are little known compared to many human geographies which have been rejected as non-geography; of course, at many places complaint against the neglect of the former could be found (as indicated in the last chapter of this book). Such treatments are met mostly by the upcoming fields (of human) geography often rubbished and blamed to be "non-geography" with a common askance "is it geography?" Or, "how is it geography?" Often a young researcher working/worked on a non-conventional theme faces such rude and baseless questions—no matter whether s/he works on crime, health, or sacred places. Themes in cultural geography are often victim of such misconceptions in India and such works are not only rejected but also publicly ridiculed! Here, it needs to be confessed that it is simply impossible to substantiate these facts as they are part of (verbal) interactions which take place either in the course of a viva-voce or interview(s) for selection, seminar and conferences, and also in departmental corridors and on tea/coffee tables. It is very difficult to reason out why such tendencies are found. Perhaps it is because of our jaundiced or myopic academic vision which may develop due to our lack of knowledge of the history of geography (perhaps that is why Sauer (1956) insists on compulsory course in history of geographic thought) or sheer ignorance or personal bias — liking/disliking. In addition to that, the nature of this problem varies from institution to institution, society to society and from country to country.

The conception of cultural geography seems to be quite hazy even in the West as late as the later half of 20th century: "The development of cultural geography as a special phase of geographic study was stimulated in the last half of the 19th century by Charles Darwin's *Origin of Species*. Moreover, support for the new field could be seen in the fact that Buckle's *History of Civilization of England* (1881) devoted more than 100 pages to the discussion of the dependence of man upon his environment" (Dicken and Pitts 1970, 26). There are such similar instances of scholars being not able to distinguish between human and cultural geography as they keep on using "(human) cultural geography". Hence, such ideas could best reflect engagement with ecological relations between the physical earth surface and human cultures in the German tradition and the well-acclaimed study of *genre de vie* of the French school. It will not be incorrect to term such engagement with the ecological relations as indicator of the infancy of cultural geographic tradition. By the end of the 20th century, the mature sub-discipline is marked by distinct theoretical positions and methodologies. Contemporarily there are evidences to the fact that we look at vital human issues—conflicts, global relations, terrorism and also development through culture. Clearly it expresses the pertinence of cultural questions across the world. But, has the Indian Geography contributed to this stream of development? What have been the areas of discourse and dialogue? What are the general lacking and their reason(s)?

In this backdrop, the present effort is devoted to read the 21st century Indian cultural geographic issues and questions critically. The contents to follow are organised into five major sections: the nature of historical roots, recent development (since late 20th century), and the Indian scene, what we continue missing, and tasks ahead with a concluding section containing summary of all discussions presented.

The Nature of Historical Roots

In the history of geography, the first due consideration to study culture is found in the celebrated two volume work of Freidrich Ratzel (1844-1904) titled *Anthropogeographie*. Though his enthusiast followers highlighted the issues Ratzel took up in the first of the two volumes, it was surely incomplete or partial understanding of the Ratzelian perspective of geography in which the main focus appears to be on human culture. Unfortunately, that led to formation of a misconceived popular image of this geographer as a proponent of "environmental determinism" in geography. Almost contemporarily, similar ideas were being nurtured by the leader of French geography and the pioneer of French geographic tradition Paul Vidal de la Blache (1845-1918). The ideas expressed in the course of his inaugural address, at the Sorbonne on the February 2, 1989 made his considered view clear about the role of *genre de vie* in developing *pays* (local homogenous areas). His ideas were based on the second volume of Ratzel's *Anthropogeographie,* as believed by scholars (James, 1972, 246-247). How far the Vidalienne conception owes to the German root may be an issue of specific enquiry; however, it could be noted without much of objection that the French school did maintain distinctiveness as reflected in the works of Vidal himself along with the battery of his pupils who occupied chairs at leading French university centres of geography. Special mention here may be made of Jean Brunches (1869-1930) who was a worthy successor of his master and also one among the leading early contributors to the field of cultural geography.

In the Anglophonic part of the world, noteworthy lead in terms of contributions to cultural geography till recently is taken by the American geographers. The dominant traditions within the American geography since the early days of the 20th century had been influenced by the Ratzalian ecological conception and anthropologic interest of Franz Boas (1858-1942) in ethnographic study of (local) communities (cf. Cosgrove 2004, 134). The academic tradition started by Carl Ortwin Sauer (1889-1975) and the chain of his students and students" students led to the foundation of what we call today "Berkeley School" of cultural geographic enterprise which dealt with primarily "human interventions" and the natural world's "transformations" at various levels. Undoubtedly, Sauer and subsequent generations of his students there at Berkeley established a dominant tradition in cultural geography at the world in general. *Readings in Cultural Geography* (eds. Wagner and Mikesell 1962) contains articles reflecting the conventional problems pursued by the Berkeley school. The major themes, under this dominant tradition, which have been pursued conventionally are domestication (of plants and animals), spatial diffusion of domesticates, technology and material practices, ecologies, modes of life and their impact, etc. These all have been pursued as mediums of landscape changes.

Scholars find cultural geography as the one which bridges the social and earth sciences seeking an integrative view of humankind in its physical environment. Geographers, especially cultural geographers can not study an element without the temporal frame. In the words of Sauer (1963, 360), "The geographer can not study houses and towns, fields and factories, as to their where and why without asking himself about their origins. He can not treat the localisation of the culture, the process of living together of the group; and he cannot do this except by historical reconstruction". That is why it appears convincing that cultural geographic enquiries subsist on collaboration with historians and archaeologists as much as ethnographers and anthropologists (cf. Wagstaff 1987). Recurring themes in cultural geography identified were culture region (with boundaries), diffusion, cultural ecology (and environmental perception), cultural integration, and cultural landscape.

Recent Developments

Various perceptible changes have taken place in the world; some took shape quicker and some others rather slowly. In response to them, disciplinary focus and practices too have modified. As far as cultural geography is concerned both its conception and contents have underwent modification. To illustrate the same, here one may cite *Re-reading Cultural Geography* (eds. Foote *et al.* 1994). Writing the "Foreword" to this volume, Phil Wagner (1994, 3-8) candidly admits shortcomings in the ideas floated in *Readings in Cultural Geography*; simultaneously, one can check the contents and intents (of the editors) of this volume which is indicative of fast changing perceptions and new problems pursued by new generation of geographers practising cultural geography. In this process, during the 1980s the practitioners and promoters of "new cultural geography" heavily came down on the Sauerian project. Despite such criticisms, the ecological-ethnographic tradition has been able to survive rather "re-invigourateed". Thanks to the recasting of new conceptions about the dualism in nature and culture and the post-modern debates. Many issues got raised in the process of debates taken place in the last quarter of 20th century and many new ideas seem to have taken the centre-stage of the intellectual churning which is happening across the globe. For our purpose, in the present context, mention of one of the intellectual strands termed as "eco-feminism" is worthwhile. The eco-feminists have strongly criticised the dualism espoused earlier between nature-culture and instead emphasise the co-productiveness of nature-culture.

Of the many sub-fields of geography, globally speaking cultural geography seems to be assuming greater significance in the 21st century then ever! Going by the strong contemporary trends, it is not an exaggerated remark. The last quarter of the last century witnessed taking over of the debate-space by many shades of critical thought and issues which are essentially speaking culture-centred or at least culture-oriented; in fact more often directly cultural as such. Such an all-pervasive influence seems to have transformed the whole gamut of human geography as cultural. It is no wonder to find cultural geography growing quickly within a shorter time-span.

The end of the world wars heralded a new phase of "modernism". Here I use the phrase "a new phase of modernism" purposely. "Modernism" has been used variedly and has wider connotations as well. The modernism therefore taking place in the post-world war period was not the only one. It brought about shifts in the form of a critique and a response too. The changes initiated in the arts soon gripped other disciplines. At the beginning of the 1980s, a movement which posed questions and doubts towards the grand claims and theories, etc. took shape and later on was termed as "post-modernism". Under this wider umbrella, many positions voicing questioning and openness gave rise to a variety of perspectives—post-structuralism, post-colonialism, deconstruction, feminism, etc. Almost simultaneously, at least in geography emerged criticality. The critical theory owing much to classical Marxism, focused on the historicity of social action made inroads into human geography. The critical human geographers engage themselves to emancipatory politics; progressive social change and development of critical theories with geographical research application and political practice (cf. Painter 2004, 126). Critical human geography precisely speaking involves: opposition to unequal and oppressive power relations, development and application of critical theories, commitment to social justice and contemporary politics. The issues which have seemingly been highly debated within the domain of critical human geography are the praxis in geography, positionality/reflexivity, and internationalism (Painter 2004, 126-128). The changes seen in the domain of cultural geography precisely during the last quarter of 20th century could be broadly identified as

1. New Cultural Geography,
2. Cultural Turn, and
3. New Age Science.

New cultural geography

As indicated earlier, the Sauerian concepts of culture and landscape though being dominant ones, attracted many critics and criticisms during the 1980s especially in British geography. A new genre of thinking influenced by sociological critique of (material) culture and landscapes put forth views in favour of a new geographical concept of culture, concerned more with space and spatiality explained through social relations of production and re-production. On the basis of various texts which had started appearing especially since 1970s, one may distinguish "New cultural geography" as aligning with social sciences and humanities rather than the conventional association with biological and earth sciences. Such a shift meant a lot as far as cultural geographic discourse is concerned. Under its influence, problems of culture and communication, symbolism and meaning, identity and territory, and the likes emerged searching new grounds (cf. Smith and Foote 1994, 27-33). As consequence of this shift, admittingly the range of topics studied and the strength of interpretations widened and multiplied, respectively with direct social connectedness.

New cultural geography could be convincingly viewed as a response to changes taking place in the social composition, cultural politics, emerging voices of "lesser" people, etc. in the Western world especially in the city life. That is how the traditional focus on "homes" increasingly became less attractive/when "homelessness" emerged more prominently. Similarly ethnicity, class, gender, sexuality, consumption, age, body, music and entertainment forced into cultural geographic domain. It is difficult to deny the existence of these issues" prominence and relevance ultimately! Actually, the forceful intellectual debates taking place in the disciplines of social sciences and humanities have created a compelling effect and to me the effects are simply non-escapable for geographers interested in study of culture and cultural patterns.

Cultural turn

An intellectual shift which took place in the 1990s initially in the fields of humanities and social sciences and later in the domains of human geography, brought cultural issues in to the forefront has been termed as "cultural turn". To understand this "turn" in geography, one can see it in three ways, as entered in the *Dictionary of Human Geography* (Crang 2004):

1. it has signalled and promoted shifting disciplinary landscapes,

2. the growing preoccupation with culture-related concepts across all sub-disciplines, and
3. as the intellectual preoccupation with questions of culture is related to wider transformations that make culture a pressing concern in contemporary world.

In this way, cultural turn has been seen as replacement of the political-economic approach emerged in the 1970s and dominant in the 1980s. This approach allegedly neglected the significance of the cultural processes which are believed, under the sway of this new shift, to play a significant role. Cultural turn apparently benefited with and gained strength from the simultaneously growing shades of post-modernist thoughts espousing the cause of culture-centred issues; of course, in a different tone and tune! Concerned human geographers could be seen taking two opposite stands: The advocates praise it as "corrective shift" and the critics see it as "deeply problematic". This stalemate may continue. As far as intellectual practice is concerned, what we need is to take this opportunity for enriching and strengthening geographical skills and not turning away from "non-cultural" problems and perspectives (Crang 2004, 141-143).

Contemporarily, culture may be defined as learned collective human behaviour and learned traits. So, it does not include anything inborn or instinctive with special reference to behaviour. In the same vein, cultural geography could be defined as the branch of human geography that deals with spatial variations found among the culture groups and also the spatial functioning of society. As per Denis Cosgrove (1948-2008), who contributed crucially in reshaping cultural geography through his work on landscape, especially on the relations between landscape ideology, image and practice, the focus of cultural geography is on cultural phenomena which may or may not vary by place and also to explain how human beings function in space (Cosgrove 2004). The concerns and core issues well reflect the growing attraction of cultural geographers towards more on conceptual and theoretical debates than the "worldly shapes" (cf. Lorimer 2005; 2007). Attempts to assess the non-representational theory with reference to cultural geography find it being enthusiastically received by geographers" community which is "thinking hard about … its promise of a politics and ethics reborn, and about the necessity of certain philosophical postures" (Lorimer 2008, 551).

New age science

Human beings tend to return to their roots from time to time. In the recent times there have been attempts to cast light on the scientific backgrounds of the ancient wisdom. That is the archaic and ancient past is (re)interpreted in the scientific manner. Sites of ancient wisdom and age-old practices are unique locales found across cultures and civilisations with which people still feel deeply associated and are involved as well. But, their complete interpretation and explanation is beyond a singular disciplinary boundary and calls for multi-/trans-disciplinary attempts. As a response, we are returning to ancient texts to dig out the scientific knowledge they contain which could be gainfully used to respond to the present day human crises; trying to understand underlying messages in ages' old continuing and live cultural practises. Consequently, areas like cultural astronomy and sacred geometry, etc. have come in to being (e.g. Malville and Singh 1995; Singh 1995a; Singh 2000). Of course, in terms of number and frequency, such studies are limited; however, quite rich and strong in terms of interpretations and the meanings. Growing involvement of modern scientists in this academic enterprise is ever growing and cultural geographers definitely can play considerably larger role provided they develop intimate connectedness with (own cultural) roots!

Indian Scene: Innovations and Recent Trends

As indicated initially in this essay, cultural geography is one of the sub-fields of human geography which are lesser known and least practised in India compared to others. Not only that, even the conception of cultural geography here is sort of archaic in relative sense. Editorial statement in an edited recent anthology on cultural geography, "Cultural geography essentially focuses on the rural context and rural content" (Grover and Singh, 2004, xxiv), is one evidence to prove how narrow, limited and outdated notion prevails with regard to a vibrant sub-field. Talking precisely about cultural geography, predominance of (rural) settlement studies has made it quite a typed field.

The leading cultural geographers, a small band actually, like A.B. Mukerji, Aijazuddin Ahmad, Kashi Nath Singh, Ram Bali Singh, Rana P.B. Singh, etc. were initially trained in settlement geography. However, they later on tried to develop respective specialised fields. Mukerji, the sole representative of the Berkeley School, who had worked under Fred B. Kniffen, cultivated diverse fields for cultural geographic works like rural settlements, house types, caste, cultural ecology and so on. Unfortunately, he could not develop a strong academic genealogy to carry forward his variety of academic tradition even after permanently shifting to Punjab University's geography department at Chandigarh. Aijazuddin Ahmad, a renowned human geographer based at Jawaharlal Nehru University, initially worked on settlements in the arid regions. He collaborated for a fairly long time with Prof. Moonis Raza and worked on issues of cultural, social and historical geography. Naturally, his students pursued a variety of research problems related to tribes, languages, literacy and education, etc. falling in the overlapping area between social and cultural geography. Kashi Nath Singh served as a link between the first and the second generation geographers of the so called Banaras School of Indian geography known for its strong tradition of settlement studies (cf. Rana Singh 1995b, 276 and 2009b, 59; Sopher 1973); initially in urban settlements, under the dynamic leadership of Prof. R.L. Singh, who later on encouraged rural settlement researches too. His famous work on "The Territorial Basis of Towns and Village Settlement in Eastern Uttar Pradesh, India" (Singh 1968) published in the *Annals of Association of American Geographers* for long time remained inspiring milestone for aspiring Indian professional geographers especially from North India. And many anthropologists from USA like Richard G. Fox who so frequently quoted from this work. In the same line, Ram Bali Singh (1971) produced quite a rich doctoral work on Rajput clan settlements. And, Rana P. B. Singh (1974) worked on the diffusion of clan settlements in Saran Plain analysing space-time process. He tried to test mainly Erik Bylund's simulative model and also cultural diffusion model of Torsten Hägerstrand (1916-2004) partially.

With passage of time however Rana Singh gradually focused on new areas of cultural geography with which at least Indian geographers were by and large had no sufficient introduction. His works in areas like geography of religious belief system, pilgrimage, symbolism, sacred geography, scientific cultural astronomy, heritage studies and (modern) literary geography with special reference to Varanasi—the cultural capital of India not only added new issues to the list of researchable areas in Indian cultural geography but also helped in getting a new recognition to Indian Geography and geographers in the West through his collaborative works. On the other hand, themes of cultural geography—urban landscape, religion and pilgrimage, literary images, food preferences, cultural practices and so on—have been visited occasionally by some "dweller" Indian and expatriate Indian geographers. Apart from that later of these, Bhardwaj's (1973) work proved to be seminal one often referred by scholars working on religious practices and pilgrimage in India from across disciplines.

At this point, before jumping over to a brief review of the major trends, it is worthwhile to cast a synoptic look at a few major works. To me, it appears that for the present purpose, one may refer to four major works: *Man, Culture, and Settlement* edited by Robert C. Eidt, Kashi N. Singh and Rana P.B. Singh (1977), *An Exploration of India* edited by David Sopher (1980), *Mythical Space, Cosmology and Landscape* by Singh and Khan (2002) and *Cultural Geography: Form and Process* edited by Grover and Singh (2004) which are in some way representative of the practised cultural geography of own time. The "cultural turn" phase serves as the dividing line between the first two and the last two. True to the fashion of the time, an international compendium of thirty-five essays, is organised into four sections: "man's role in changing geographic space", "patterns of cultural ecology", "system analysis of culture areas", and "settlement system and regional development". These sections'respective titles make the general focus clear. And, the rationale of including the last section which is applied in nature was perhaps that Prof. R.L. Singh, in whose honour this book was done, has the credit of developing the tradition of applied geography studies at Banaras (Varanasi). In a way, it reflects continuity and change in this field. Yet, India related essays especially could not respond to Sopher's (1973) call. Sopher's (1980) volume reflected the continuing research interest of his and his pupils. All ten included chapters in the book were undoubtedly strong and inspiring but perhaps failed in addressing the weak-spots he had indicated himself earlier (Sopher 1973). One may be critical of the approach(es) adopted in their works; still in some sense Singh and Khan (2002) representatively showcase the range of general problems undertaken by Indian geographers in the subfield of cultural geography. No wonder that it has rightly earned acclamation (cf. Singh 2009b, 64). *Cultural Geography: Form and Process* (2004), covering diverse cultural geographic issues, through twenty-nine essays spread over five sections, barely crosses over "simplicity" whereas the 21st century cultural geography is into "critical" phase (see, Singh 2005). Barring a few exceptions, majority of the essays being descriptive do not show signs of vibrancy and hence barely inspire the future studies.

The recurring themes in cultural geography of India as appears from the published works fall under: settlement and house (type) studies (especially rural settlement ecology/habitat, diffusion, distribution and morphology), caste and tribe related studies, and literary geography. However, the overall growth could be termed as sluggish in terms of temporal growth and stunted, if seen in size. On the contrary, India being the home to rich culture, alive through practices and represented in diverse heritage, provides vast opportunities. The enthusiasm of cultural geographers and the momentum of the works carried up are far from encouraging. We can take brief stock of a few innovations and main headways made in the Indian cultural geography as the following.

Geography of religions

Reference to religion in geography could be at least traced back to Immanuel Kant's (1724-1804) *Geographia Religionum.* However, the credit of the 20th century revival, recognised in the geographers' community, goes to the American geographers. In a recent work, Kong (2001) reviewing the progress taken during the 1990s in the geographic research on religion, finds that descriptive approach is now replaced for sure by the analytical one wherein issues like politics and poetics of religious place, identity and community are undertaken for analysis. In India, geographers, though not very prolifically contributing, have shown their general interest in this sub-theme in several ways. Bhattacharya (1961) and Kurian (1961) perhaps were a few first who tried to look at the relationship between geography and religion. However, works in the field of geography of religion date back to at least the 1930s when perhaps the first Indian work (Rao 1937 quoted in Sopher 1968, 392) appeared. Later, publication of Sopher's *Geography of Religions* (1967) gave valid recognition and acceptability to it. A recent work (Stump 2008) has objectively addressed major issues of the geography of religions. Though not stressing the "geography-ness", the adopted approach itself sufficiently takes cares of disciplinary, i.e. geographical, concern and intension (Stump 2008, xv-xvi). Initially the concern had been to understand the diffusion, distribution and patterning of religious groups. An early example is of the doctoral work of Bobb Stoddart (1966). Rana Singh (2009d) brought out the progress taken place in geographical studies on religion and pilgrimage, respectively, in India. It is quite clear that Indian contributions are few and hence cover only a limited number of issues.

Tirthayatra (pilgrimage) has a strong tradition in all (ancient) religions including Hinduism in which it has apparently contributed in integrating one part of Bharat (i.e. India) with others and maintaining unity in diversity internally. Perhaps the first academic work on Hindu pilgrimage was by Agehananda Bharati (1963). But, a geographic work appeared in the form of one of the pioneer works by Sopher (1968) on circulation of pilgrims wherein he not only tried to understand spatiality of the Gujarati pilgrim movement but also sought to explain the role of this process in folk education and integration of a region (here Gujarat) with the rest of India. Later on, a similar study by Bhardwaj (1973) was made on pilgrimage (including pilgrim circulation) with reference to a few select *tirthas* of north India which attempted a quantitative typology. Another important contribution came in the form of an anthropological work (Morinis 1984) dealing with Hindu pilgrimage tradition in West Bengal. In a review article Singh (1992b) has categorised the works on pilgrimage as those dealing with travel genre and place description, works in historical perspective, circulation and pilgrimage, pilgrimage-typology, distribution and organisation of sacred places, the cosmic frame (pilgrimage mandala), anthropological dimensions, and pilgrimage as resource in cultural heritage.

Traditionally the works of geographers on religion focused on very limited themes. Singh (1992a) summarises geography of religions' works into regional religious geographies; demographic and distributional aspects; sacred geography as ecology; time, ritual, and space organisation; religious ecology and animal. Of these, majority of the geographical studies have dealt with the distribution of the religious communities.

Literary geography

The significance of literature in cultural studies in geography is well established (Dhussa 1992; Singh 2004). India, having a vast pool of literature in different languages—Sanskrit, Pali and a variety of regional languages no doubt contains a huge resource to which both dweller and expatriate English litterateurs qualifyingly add. However, literary geographic tradition almost a hundred years' old have not benefited, especially in India. Barring a few exceptions in which the classical texts were explored for their geographic content and it has mainly excelled in the post-1970s with the evolution of humanistic approach in geography. The entire corpus of literary geographic works falls in five major categories (cf. Dhussa 1992):

1. Works dealing with methodological and theoretical issues of the field;
2. Researches on the subjective use of imaginative literature;
3. Works exploring objective (geographic) information in literature;
4. Works searching application of literary works in class rooms; and
5. Studies creating the (imaginative) cultural landscape as portrayed in literature.

A quick survey of the works in this sub-field well reveals predominance of the third variety, mentioned above, in which scholars are found exploring out geographic content in the literature. Vedic, Puranic as well as epic literature have been widely utilised in this vein. Most of such works in the form of post-graduate dissertation and doctoral work nevertheless lie unknown gathering dust in a corner shelf of departmental/university library as they never see the light of publication! with the passage of time, new horizons have opened up wherein human existence, action, experiences and expressions which have direct bearing upon making, unmaking, and re-making of places and cultural landscapes have been given due consideration (cf. Ravi Singh 2003). Now it appears that literary geography is coming out of the Puranic garb and has begun to address post-modern critical issues into which contemporary cultural geographic discourse is engrossed.

Geography of fairs and festivals

Congregation at a place at some (regular) intervals is one of the earliest human practices described as the root of several other cultural developments—enshrinement of holy places, pilgrimages, and urban formations (the non-economic explanation) too. In ancient cultures, where festivals, have important place in cultural life, and strong connection with "power of place" (*genus loci*), people tend to assemble at sacred time at sacred places to perform rituals to get "sacred merit". Otherwise also, fairs and festivals are organised without any precise sacred motives as revealed by the participating visitors (cf. Sopher 1968). People assemblage and performance is still a continuing tradition in rural India; no doubt on decline due to various reasons—socio-cultural to economic—requiring a separate exercise. Allix's (1922) work is an earliest example of understanding fairs geographically. Such attempts on India focused on fairs are very limited, as revealed through published literature. Lodrick's (1984; 1987) works have successfully interpreted a fair of Rajasthan and two festivals related to Lord Krishna celebrated in Vrindavan (Uttar Pradesh). It is a pointer to the fact, which cultural geographers (in India) need to remember, that we can easily understand the basic nature of fairs which may not be always sacred. Singh, Singh and Singh (1987) is the lone example wherein a fair, called Dadri Mela (Ballia District, Uttar Pradesh), is analysed holistically. Another worthy and exclusive work (Singh 1989), emerging out of his doctoral thesis, on the fairs and festivals of Ballia district is a good follow up. It provides a framework of study in which the context of general "Greater tradition" and religious symbolism is searched with a clear focus on the spatiality of rituals and functions performed; and, interlinking of the local tradition with the "Greater tradition". In a way, it is an effort to look at fairs mainly as part of people's belief system. Cultural geographers may use the same as resource for developing an analytical framework for undertaking studies on geographic formation of the regional cultural identity at various levels and scales. It may be added here that the fairs and festivals are also the site and occasion, respectively, of vivid interactions important not for their direct economic value but also for the formation of people's identity and their attitude and perception about the "other(s)".

Festivals, celebrated at mass level throughout India, in one hand are expression of centuries old cultural roots and clear geographic content with bold regional element on the other. One instant instance coming to my mind is of *Bihu*[1] celebrated in Assam, one of the north-eastern states of India. Where direct religious centrality may not be found but link with nature-culture and associated ancient most practices are profuse. Similar examples are ubiquitously found across India still waiting for cultural geographic treatment.

Food: habit patterns and regional preferences

Food habit is a cultural trait with ecological underpinnings. Further, it could well directly relate with diseases, especially those which have to do with dietary quality. Sorre (1962, 445) considered diet important for geographic investigation. He quotes Blache (1922), "Among the connections that tie a man to a certain environment, one of the most tenacious is food supply….which experience has shown to be best suited to human needs in a given climate" (Sorre 1962, 445). Cutler (1962) discussed food sources in the New World in a historical perspective. These two works illustrate that food is directly related to geographic environments and diffusion plays important role in making a particular item of food available to people of a region other than the original source region.

In a country like India inhabited by host of culture groups, immense potential of cultural geographic studies of food exists. Debysingh's (1980) work, though focused on poultry keeping in India, points to the fact that food consumption behaviour and pattern is culturally determined; and, that also changes with time. Similarly, one can find spatial variations at all levels and within the same ecological setting (cf. Singh 1999). Besides the daily consumed food, direct impact of season(s) in food prepared (and consumed) can be noticed. I remember a programme done by the NDTV

India[2] on the progress of (south western) monsoon right since it stroke the Kerala coast and moved on into the mainland. That team was perhaps headed by Ms Naghma Sehar, a geographer by educational training. They also explored what special is cooked and how, during the monsoons, in different regions falling en route of the advancing monsoon. Though their objective was to make popular reporting, one gets a clue that similar exercises are worth undertaking by geographers (as well as other scholars) with an academic objective at various scales. Such studies can ably generate empirical information on food issues and also that how culture is crucial in not only deciding what is edible or not but also how shortage/surplus is a function of culture both temporally and spatially.

We know of the continuing tradition of food habit studies in anthropology; however, there is quite a feeble tradition in geography. Here, it is worth mentioning that a geographical study of food habits and preferences could be well utilised in applied geographic approach too. In geography, we are aware of case studies dealing with carrying capacity of land and nutritional calendar etc. but as a matter of fact they have not gone well beyond the set pattern of quantifying the food requirements and supply (shortage/surplus). A known contribution in this field has been Chakravarti's (1974) work. And, many more are awaited to be done by the future geographers and the younger lot among the present generation of Indian geographers as well. Debysingh's (1980) analysis is almost three decades' old and in the meantime there is apparent change in Indian social norm including those governing the food habit. The 21st century thus is an opportune time to map the changing food behaviour and patterns, and the emerging new economics of food supply-demand and surplus-shortage too.

Dance, music and art

Art and art forms bear immense geographic influences. Karan (1984) explored the interplay of landscape, religion and folk art in a cultural region of Bihar called Mithila associated with the epic *Ramayana*. Of all the art forms, music is found omnipresent across the cultures in the world. And music contains again a rich geographic resource. Ford and Henderson (1974), for example, discussed the image of place in American popular music. Their hypothesis was that "songs both reflect and influence" people's image of place. That image changes in time frame and also varies from people to people. Nash (1975) in a detailed analysis demonstrated the place of music in cultural geography. His "exploratory" investigation that music could be gainfully studied by geographers in several ways—

methods through which sounds are produced, raw materials used for manufacturing instruments, patterns of musical behaviour and consumption, migration and diffusion of music, music production and consumption, as well as the geographical correlates of music like climate, terrain, latitude, altitude, etc. (cf. Nash 1975, 1-14). Gold (1992) has exhaustively captured the bard's singing tradition of Rajasthan which is an oral epic tradition important in Hinduism in this part of North India.

In *Purvi* style of singing in Indian light classical music, heroine, longing for her beloved husband, is often found describing places and peoples of the distant land where her beloved has gone in search of employment, leaving her behind in (their) ancestral village. In this context, one is reminded of legendary figure in Bhojpuri folk music and master of melodrama artist Bhikhari Thakur[3]. In classical music, where greater emphasis is on standardised elements and practices, there is little scope of geographic explanation. The same however is not true in case of folk art forms, here dance and music, as the impact of the region and associated characteristics is apparently found on them. Similarly, their study could be undertaken at various scales. Perhaps, there is no region of India without a typical folk song, music and dance form of its own. These are inseparable part of any traditional festival and function; and, hence, have considerable role in the formation of the regional identity. Being born in the *Bhojpuri* speaking sub-region of the Hindi speaking north India, I know through personal experience that there are different songs for different celebrative occasions: birth related function(s), wedding, worship etc (e.g. see, Tewari 1988). Similarly, different songs for different stages of wedding ceremony are still sung in the countryside. These songs are not sung for fun and entertainment alone. In one hand they have to do with the piousness of rituals being performed and on the other hand they are suggestive of the prevalent aspiration(s) of the society and influences coming from other practices and contemporary media, especially cinema, apart from the embedded luxuriant presence of motifs and metaphors. In the same vein, mention worthy is *Biraha*[4], literally "a song of separation" (Marcus 1989). In the *Birahas* a strong narrative element of local/regional socio-political issues ranging from cruel dowry deaths to political victimisation and state-oppression is often found.

At the regional level too apparent variations are found in terms of the ways of celebration and major instrument used. Famous folk dance forms like Bihu (Assam), Ghumar (Rajasthan), Baul [(West) Bengal], Bhangra (Punjab), and Giddha (Haryana) are a few well known examples beside Odissi (Orissa), Manipuri (Manipur), Bharatnatyam (Tamilnadu), Kathakali and Mohiniyattam (Kerala) and Kuchipudi/Koochipoodi (Andhra

Pradesh) which, are in one hand classical and have simultaneous strong element of regionality, and have somewhat common origin either in temple or royal court tradition. Of course, all of these have drastically changed and evolved during the modern times. In this context, there is yet another example of music, dance, and art (painting) form, which grew out of direct personal initiative and interest of Rabindranath Tagore, known as Rabindra Sangeet (music) and Nritya (dance) which are by every standard regional styles characterising and representing (West) Bengal. Both Rabindra Sangeet and Nritya are inseparable element of (West) Bengal's cultural identity. This is one area which has not been able to attract enough interest of (Indian) geographers reflected through the scanty works on a very few themes only. Here one may cite Singh's (1983) work on the dances of Manipur, one of the north-eastern states of India, perhaps the only one geographical study of dances by any Indian geographer. Though not rigorous in terms of analysis; he has tried to show hoe the Manipuri dances are influenced by the local environment.

What Do We Continue Missing?

An observation made down four decades by Sopher (1973) directly indicated the missing dweller's perception in Indian geographic tradition unlike the West. He made it clear that,

> "It should be a geography in which the popular landscape symbols, mental maps, and "sense of place" of the dwellers, that is, the folk ecosystemic models, are made to interact with and become incorporated in the intellectual constructs of strangers and officers. This is what we sometimes find in the best regional geographies of an earlier European generation. The local perceptions that these draw on are, certainly, often expressed in a cultivated, though not necessarily cosmopolitan, voice – as an interest in local history, antiquities, folk tales (which may richly inform local geography), place names, and topography" (Sopher 1973, 112).

He further insists, "While classic landscape geography may have failed to adjust its scale of investigation to technological changes in circulation, its potential usefulness for Indian studies is not thereby invalidated; as Pinchemel (1968) has recently pointed out once again, landscape is "the expression of a civilization". In the virtual absence of such a tradition in Indian geography, it is almost impossible for us to construe the symbolic information communicated in the Indian landscape, as we should wish to do if Philip Wagner's (1972) recent exhortation is to be attended (113-114). The resources for (cultural) geographers available all around—in the material landscape, mass of visiting devotees at shrines and participants of fairs and festivals, marriages and marriage parties" (cf. Sopher 1973, 115).

Crippled with inherent weaknesses, illustrated through the above quotes, cultural geography in India progresses; albeit, as noted earlier, this growth is sluggish. Singh and Singh (2004; 2008) surveying the progress taken place during the first eight years of the 21st century observe the existence of following typology of the works available:

- Cultural ethnology: text, context and narrative
- Cultural practice: consciousness and performances
- Cultural manifestations: spatiality and astronomy
- Cultural journeys: pilgrimage to sacred places

Going by the present century indications, we need to accept that culture itself is in flux and is exists at various levels in varied forms. Cybernatics and hybridity define urban cultures, not only in developed countries but also developing nations of the world. The studies conducted on India show continuous absence of urban culture studies by geographers. Now, its time to bring Indian urban culture, that is taking shape and form with own peculiarities, under cultural geography scanner. On the other extreme, tribal cultures and culture areas tend to evolve and transform somewhat faster. They too need to be studied longitudinally. The vast pool of texts and traditions offer unexpected potential to be studied in the spatial context.

We, as a matter of fact, continue to miss the chances of utilising the accessible geographic content available in the landscape, texts and traditions. Otherwise, the quotations above would have proved to be redundant for the present piece like writing being attempted after the lapse of complete four decades when David Sopher made his observations. The "sore" of those days perhaps has acquired the form of canceric development. Will it end the life story or we are going to acquire a new lease of life by treating the ailment holistically!

Tasks Ahead

Applying the Habermasian three-tier typology (of science and explanation) to (Indian) pilgrimage studies ("perigrinology"), three approaches could be identified to accomplish the future task in Indian cultural geography (cf. Singh, 1992):

1. Empirical-analytical approach: based on direct experience and observation in a positivist frame.
2. Hermeneutic approach: based on the denial of a separate empirical view is based on perception through a system of meanings which are human constructs developed by each individual process.
3. Critical approach: conceives people as part of complex social-cultural organisation created by them as ways of ensuing individual and generational survival.

It has been observed, with reference to the geography of religion, that the geographers' perception clearly reflects the "closed-ness": "There is no general acceptance in the field of the geography of religion that is to what, why and how a particular aspect to be taken on priority level. Formerly scholars thought to avoid considering personal experience, but now this theme is quite important in understanding environmental aesthetics, ethics and psychic attachment with the place" (Singh 2009d). Things have not changed much as even now a study on geography of religion encounters the same reaction—"Is it geography"? or "how is it geographical"?

It is opportune time to realise and develop the immense potential India possesses: "India, home to one of the world's most ancient civilizations, has a rich intellectual tradition. In this geographically diverse subcontinent of Eurasia, discourses on nature and humanity from the *Upanishads* and the *Vedas* led to the development of various indigenous knowledge systems. Geography thus became part and parcel of human personality and everyday life as well as retaining a central role in Indian intellectual traditions. While these historical texts were regarded as merely scriptures by common people, their richness and variety has recently evoked analytical and critical attention by scholars eager to uncover the roots of Indian geographical knowledges" (Buttimer 2009, ix).

Optimism by the Indianist American cultural geographer Joe Schwartzberg needs serious consideration, "There is, however, much scope from Indian cultural traditions. What is now needed is to enrich geographic research in India thorough a skilful blending, in appropriate cultural contexts, of the rigor of the western scholarly paradigms and greater application of indigenous culturally relevant concepts (Schwartzberg 1998, 253 quoted in Singh 2009b, 7).

Postmodernism and postcolonialism give immense opportunity to cultural geography, like other geographies, in India—returning to the roots. Rana Singh's idea is quite timely, "It is more important to discover something of the meaning of India and to see it through the eyes of the people. We have copied much more rubrics of positivism, which is at the edge of closure in the West. More geographies of people's experiences, memory, and exposition of inherent meaning are required while stressing subjective experience over objective scientific observation. This is not a paradigm shift rather a re-affirmation and re-orientation of the classical ideas. Remember that "there are also places and worlds given meaning by the simple act of human habitation. Some of these are simple worlds: others have deep histories and complex meanings" (Hugill and Foote 1994, 22). India has always consisted and possessed of such geographies, where simplicity has invisible complexity, and visuality has deeper messages. Unfortunately the recent generation of geographers has little sense of feeling this; thanks to our counter partners in the West who slowly realising this and learning to understand it by deeper quest and dedicational skills" (Rana Singh 2009b, 40).

The (new) urban call

Now it is time to realise and recognise that urbanism has deep and wide roots in India and also growing. Naturally, a typical Indian urban culture has developed in all categories and types of Indian urban centres. Nevertheless, cultural geographers in India seem to have undermined this resource's richness for academic explorations. A few attempts have been done focusing built-up landscape, especially in the cities with rich heritage of monuments and royal buildings which are examples of fine masonry pieces, ethnic neighbourhoods, and language. Indian cities are changing very fast to the tune of cosmopolitanisation and also globalisation. These changes, visible in the urban built-up area and architecture, are the outcome(s) of modifying preferences and choice of citizens affecting the economy. No wonder why a city like Varanasi known as the holiest centre of pilgrimage is finding it difficult to save its centuries' old heritage under pressure from booming hotel and restaurant business! The swing to western mode of behaviour of the Banarasis consequently gives rise to malls and multiplexes together with outlets selling garments and apparels of international brands and chains. Rana Singh, through his works, many of which have not been mentioned here due to space constraint, have tried to document the urban cultural heritage (of Varanasi) which can serve as example to inspire similar works for other cities. One of his forthcoming works (Singh 2009e) analyses heritage issue of Varanasi comprehensively. Still many more such studies are needed to analyse the forces, processes and consequences of change, and to develop generalisations. In the years to come, many more Indians will start living in the cities and rural areas too would experience massive urbanisation. These two situations again are sites for cultural geographic explorations. Works by our western counterparts like E.W. Soja and Michael Dear again show directions in which we can bend our studies.

Heritage headways: applied cultural geography

The "cultural turn" has apparently changed the perceived importance and status of cultural geography within human geographic sub-fields. Nevertheless, the "New Cultural Geography" is no more strictly glued to spatial context rather it is more thematic, as indicated earlier and is also the ground of ideological debates on issues like non-representational cultural geography, hybridity, etc. There has been continuous concern about detachment of (Indian) geographic enterprise from a common person's day to day life, i.e. the question of social relevance. Any discipline's works fall within two clear domains: pure disciplinary (theoretical) and applied. Most of the works, undertaken under the broader canopy of cultural geography, fall under the first. We have by far neglected or overlooked the applied aspect.

On the other hand, question of identity has emerged rather strongly in recent times and that naturally underlines newly defined importance of heritage. Expansion of heritage studies, beyond archaeological findings and monumental masonry work(s) and inclusion of both tangible and non-tangible together with sites, unfolds a fertile area for cultural geographers where they can use their expertise to make comprehensive (re-)interpretation and also planning of heritage for contemporary and future use. Our western counterparts have made their mark in this field and a specialised interdisciplinary field of heritage geography is already growing and already around 50 such established centres all over the world are offering Master, Bachelor, PhD degrees. In India too the Delhi Institute of Heritage Studies is an example; yet, it is in the early infantile stage of growth here. It has a direct policy implication, especially in the case of urban heritage which is increasingly threatened despite their role in not urban development alone but also the region in which they are located (cf. Singh 2007). And, many connected issues like pilgrimage-tourism, heritage-tourism, spiritual-tourism, etc. are being dealt together with politics and contestation of heritage, and also environment-heritage-tourism interconnection and interaction. Thus, it is an opportunity, particularly for Indian geographers, to have a tryst with policy making where they have been lacking (cf. Kapur 1998).

Some Preliminary Conclusions

The last quarter of the 20th century saw many shifts and turn. That reflects fertility of the intellectual debates and discourses. It is interesting to note that no major -ism/shift/turn could survive as the centre-stage issue beyond a decade. This vibrancy however could not in general touch dweller Indian geographers much. But, some of those concepts have been incorporated in update syllabi at certain centres. That is quite short of complete pabulum as the ailment is old and infectious. The larger question is do the works done bring out the reality of the problem under consideration? Candid remarks by Sopher (1973) echo in the ears of any person gone through that piece of writing which urges the "dweller" Indian geographer to go back to the roots and see from "own" eyes instead of reading through an adopted magnifying glass which is bound to project any thing but the Indian. Citing the example of Arthur Geddes who had tried to understand the transformation of cultural ecology in the southern West Bengal in the aftermath of Rabindranath Tagore's experimental institution Visva-Bharti at Shantiniketan[5], Sopher's emphasises the virility of the indigenous concepts and ideas to explain the Indian reality. Instead doing that, we went on adopting the western nomenclatures quite short in representing our realities. Observing that "identity" is characterised by many elements which are highly diversified in India, Rana Singh's rightly exhorts:

> "Place matters in different ways, at different scales and in different territories....the attachment to place through its genus loci is an important element in Indian geography, however not seriously drawn attention by dweller geographers. The essence of understanding cultural identities and landscapes are evident at all spatial scales.... In the context of India, relevance of space and place to all human endeavours has been appreciated in different ways with different spectacles. Above all this is the reflection of geographical imagination that rooted in India"s past and may have dreams in future." (Rana Singh 2009b, 38).

Analysing over two decades ago though Bradnock (1986) found cultural geography an "active field of research" in India, cultural geography has ever been a marginal sub-field. And, that is why continues to be neglected in the mainstream Indian geography. The marginality of this sub-field in relation to others is not surprising in view of the proliferation of literature taken place during the recent past. Simultaneously, it suffered and continues to suffer with the definitional non-clarity and vague identity. Often, it is misunderstood with social, historical, and settlement geography. The unavailability of a sincerely attempted general work addressing definitional and methodological issues, scope, approach, etc. related positions in cultural geography in India is a hindrance in the way gaining popularity in my view. Probably because of these problems, brighter lot among those who pursue research particularly among new generation barely get interested and attracted to culture geographic problems. The crisis actually begins at the level of taught courses. Very few centres of post-graduate teaching offer any course on culture geography. Hence, it is imprudent to expect from a student to jump over to conduct research in this field. The mainstream sub-fields' dominance sway attention of professional and students alike at the cost of other sub-fields, which is not academically genuine in any sense.

In conclusion, one may note that the above discussion can not claim to be either complete or exhaustive. On the other hand, many potential areas of study remain barely explored. One example here is of cultural animal geography in which perhaps no work is attempted after Lodrick's (1981) which remains a standard in itself. There still exists tremendous scope for exploratory studies in a variety of conventional and emerging areas as well. That will logically make path for theoretical works as a natural follow up. We can benefit ourselves by availing academic space in sister-disciplines too where our presence continues to be feebly visible (e.g. Singh and Singh 2006; Rana Singh 2007, Rana Singh 2009c; Ravi Singh 2007, Ravi Singh 2009). At the higher plane, cultural geography in India can potently contribute to emerging issues brought to the focus prominently by postcolonial approach (Nash 2002; 2003), e.g. "anti-racism" in place of "race". Rana Singh's two recent publications (2009b; 2009d) show the path for a vibrant Indian cultural geography and its direct use in an awakening of Indian geographers for meeting the challenges of the 21st century. Evidences to the fact of re-vigouration of cultural geography is clear with increasing involvement of the formal international body of professional geographers, International Geographical Union, IGU which constituted a separate study group on cultural geography in 1996 and promoted the Culture and Civilization in Human Development (CCHD) project[6] through the permanent Home of Geography (the official headquarters of the IGU) at Rome, Italy. Cultural geography in India should come up and seek to reincarnate it in the 21st century form even if we do not totally agree by "all is cultural now" under the postmodern influence (cf. Cosgrove 2004, 137)!

Notes

1. Here it stands for the festival celebrated in Assam. Otherwise, "Bihu" is a commonly used term in Assam and its neighbouring states to denote any festival and its celebration.
2. It is a private broadcasting channel based at New Delhi.
3. He is popularly known as the "Shakespeare of Bhojpuri".
4. Name of a two-line metre used in *Bhojpuri* folk songs.
5. See a recent article by A. Bonnett's (2005) "Occidentalism and plural modernities: or how Fukuzawa and Tagore invented the West" *Environment and Planning D: Society and Space* 23(4) 505 – 525, for Tagore ideas.
6. Since 2008 it is attached to the IGU Commission on Cultural Geography (Chair: Berno Werlen).

References

Allix, Andre 1922. The geography of fairs. *Geographical Review* 12 (4): 532-69.

Bhardwaj, Surinder Mohan. 1973. *Hindu Places of Pilgrimage in India: A Study in Cultural Geography*. Berkeley: University of California Press.

Bhattacharya, A.N. 1961. Geography and Indian religion. *National Geographer* 4 (1): 12-17.

Blunt, Alison 2005. Cultural geography: Cultural geographies of home. *Progress in Human Geography* 29: 505-515.

Bradnock, Robert W. 1986/2009. Research in social geography: South Asia. In *Social Geography in International Perspective*, ed. John Eyles, 275-294. London & Sydney: Croom Helm. Reprinted in *Indian Geography: Perspectives, Concerns and Issues*, ed. Ravi S. Singh, 331-346. Jaipur/New Delhi: Rawat Publications.

Carter, George F. 1968. *Man and the Land: A Cultural Geography* (2nd ed.). New York: Holt, Rinehart and Winston Inc.

Chakravarti, A.K. 1974. Regional preferences for food: Some aspects of food-habit patterns in India. *Canadian Geographer* 18 (4): 395-416.

Cosgrove, Denis 2004. Cultural geography. In *The Dictionary of Human Geography* (4th ed.), ed. R.J. Johnston, Derek Gregory, Geraldine Pratt, and Michael Watts, 134-138. Malden, MA/Oxford/Carlton: Blackwell Publishing.

Crang, Phil 2004. Cultural turn. In *The Dictionary of Human Geography* (4th ed.), *op. cit.* 141-143.

Cutler, Hugh 1962. Food sources in the New World. In *Readings in Cultural Geography*, ed. Philip L. Wagner and Marvin W. Mikesell, 282-289. Chicago and London: University of Chicago Press.

Debysingh, Molly 1980. The cultural geography of poultry keeping in India. In *An Exploration of India: Geographical Perspectives on Society and Culture*, ed. David E. Sopher, 105-129. Ithaca, New York: Cornell University Press.

Dicken, Samuel N. and Forrest R. Pitts 1970. *Introduction to Cultural Geography: A Study of Man and His Environment*. Waltham, MS and Toronto: Xerox College Publishing.

Eidt, Robert C., Kashi N. Singh, and Rana P.B. Singh (eds.) 1977. *Man, Culture, and Settlement: Festschrift to Prof. R.L. Singh*. New Delhi and Ludhiana: Kalyani Publishers; National Geographical Society of India, Research Publication No. 17.

Foote, Kenneth, Peter Hugill, Kent Mathewson, and Jonathan Smith (eds.)1994. *Re-reading Cultural Geography*. Austin: University of Texas Press.

Ford, Larry R. and Flyod M. Henderson 1974. The image of place in American popular music. *Places* 1 (1): 31-37.

Gold, Ann Grodzins 1992. *A Carnival of Parting: The Tales of King Bharthari and King Gopi Chand as Sung and Told by Madhu Natisar Nath of Ghatiyali, Rajasthan*. Berkeley, Los Angeles, California, London: University of California Press.

Grover, Neelam and Kashi N. Singh. (eds.) 2004. *Cultural Geography: Form and Process*. New Delhi: Concept Publishing Company.

James, Preston 1972. *All Possible Worlds: A History of Geographical Ideas*. Indianapolis, NY: The Odyssey Press.

Jassal, Smita Tewari 2007. Taking liberties in festive song: Gender, new technologies and a "joking relationship". *Contributions to Indian Sociology* 41 (1): 5 - 40.

Johnston, R.J. Derek Gregory, Geraldine Pratt, and Michael Watts (eds.) 2004. *The Dictionary of Human Geography* (4th ed.). Malden, MA/Oxford/Carlton: Blackwell Publishing.

Kapur, Anu 1998. *Indian Geography: Future with a Difference*. New Delhi: Allied Publishers.

Karan, Pradyuman P. 1984. Landscape, religion and folk art in Mithila: An Indian cultural region. *Journal of Cultural Geography* 5: 85-101.

Kong, Lily 2001. Mapping "new" geographies of religion: Politics and poetics in modernity. *Progress in Human Geography* 25 (2): 211-233.

Kuriyan, George 1961. Geography and religion. *Indian Geographical Journal* 39(1): 46-51.

Lodrick, Deryck O. 1981. *Sacred Cows, Sacred Places: Origins and Survivals of Animal Homes in India*. Berkeley: University of California Press.

—. 1984. A cattle fair in Rajasthan: The Kharwa Mela. *Current Anthropology* 25 (2): 218-225.
—. 1987. Gopashtami and Govardhan Puja: Two Krishna festivals of India. *Journal of Cultural Geography* 7 (2): 101-16.
Lorimer, Hayden 2005. Cultural geography: The busyness of being "more-than-representational". *Progress in Human Geography* 29 (1): 83-94.
—. 2007. Cultural geography: Worldly shapes, differently arranged. *Progress in Human Geography* 31 (2): 89-100.
—. 2008. Cultural geography: Non-representational conditions and concerns. *Progress in Human Geography* 32 (4): 551-559.
Malville, John McKim and Rana P.B. Singh. 1995. Visual Astronomy in the mythology and ritual of India: The Sun temples of Varanasi. *Vistas in Astronomy* 39 (4): 431-449.
Marcus, Scott L. 1989. The rise of a folk music genre: Biraha. In, *Culture And Power In Banaras: Community, Performance, And Environment, 1800-1980*, ed. Sandria B. Freitag, 95-115. Berkeley: University of California Press.
Morinis, Alan E. 1984. *Pilgrimage in the Hindu Tradition. A Case Study of West Bengal.* Delhi: Oxford University Press.
Nash Sr., Peter H. 1975. Music and cultural geography. *The Geographer* 22 (1): 1-14.
Nash, Catherine 2002. Cultural geography: Postcolonial cultural geographies. *Progress in Human Geography* 26: 219-230.
—. 2003. Cultural geography: Anti-racist geographies. *Progress in Human Geography* 27: 637-648.
Painter, Joe 2004. Critical human geography. In *The Dictionary of Human Geography* (4th ed.), *op. cit.* 126-128.
Sauer, Carl O. 1956. The education of a geographer. *Annals, Association of American Geographers* 46: 287 - 299.
—. 1963. *Land and Life. A Selection from the Writings of Carl O. Sauer.* Ed. John Leighly, Berkeley: University of California Press.
Singh, J.P. and Mumtaz Khan 2002. *Mythical Space, Cosmology and Landscape: Towards Cultural Geography of India.* New Delhi: Manak Publications Pvt. Ltd.
Singh, Kashi N. 1968. The territorial basis of towns and village settlement in eastern Uttar Pradesh, India. *Annals of the Association of American Geographers* 58 (3): 203-220.
Singh, R.P.1983. Impact of environment on dances of Manipur. *Geographical Review of India* 45 (3): 16-22.
Singh, Rana P.B. 1992. Geography of pilgrimages in India: Perspectives and prospects. *National Geographical Journal of India* 34 (1-4): 39-

54; also published in *The Roots of Indian Geography: Search and Research*, ed. R.L. Singh and Rana P.B. Singh. Varanasi: National Geographical Society of India.

—. 1995a. Sun shrines, images and alignment at Varanasi. *The Ley Hunter* (Cornwall, UK), No. TLH 123, Summer: 20-26.

—. 1995b. Identity of geography in India beyond 2000: A search of sensitivity for future. *National Geographical Journal of India* 41 (3): 263-281.

—. 2004. *Cultural Landscape and the Lifeworld: Literary Images of Banaras*. Varanasi: Indica Books.

—. 2007. Gaia and ecological awakening: Message of Hinduism for deeper understanding. *Oriental Anthropologist* 7 (2): 213-233.

—. 2009a. Identity of "India" and "Indianness", and geography in India: Sensitive appraisal from the past and move to the 21st century. In *Indian Geography: Perspectives, Concerns and Issues*, ed. Ravi S. Singh, 25-64. Jaipur/New Delhi: Rawat Publications.

—. 2009b. *Uprooting Geographic Thoughts in India: Toward Ecology and Culture in 21st Century*. Newcastle upon Tyne: Cambridge Scholars Publishing.

—. 2009c. Ethno-geography of Sun goddess festival, Chhatha, in Bhojpur region, India: From locality to universality. *Man in India* (Special Issue/March-June) 89 (1-2): 103 - 120.

—. 2009d. *Geographical Thoughts in India: Snapshots and Vision for the 21st Century*. Newcastle upon Tyne: Cambridge Scholars Publishing.

—. 2009e. *Banaras, Making of India's Heritage City*. Newcastle upon Tyne: Cambridge Scholars Publishing. *Forthcoming*.

Singh, Rana P.B. and J.M. Malville 2000. Sacredscape, faithscape and cosmic geometry: A study of holy places of north India. In, *Astronomy and Cultural Diversity* (Proceedings, Oxford VI & SEAC 99 International Conf.erence, June 21-29, 1999, La Laguna, Spain) ed. C. Esteban and J.A Belmonte, 99-106. Santa Cruz de Tenerife: OAMC.

Singh, Rana P.B. and Ravi S. Singh. 2004. Cultural geography. In *Progress in Indian Geography 2000-2004: A Country Report*, ed. H.N. Sharma, 139-146. New Delhi: Indian National Science Academy.

Singh, Rana P.B. and Ravi S. Singh. 2008. Cultural geography. In *Progress in Indian Geography 20040-2008: A Country Report*, ed. D.K. Nayak, 81-88. New Delhi: Indian National Science Academy.

Singh, Rana P.B., S.B. Singh and A.K.Singh. 1987. Religious fair, spatial affinity and economics in India: Case of Dadri Mela (Ballia District). *National Geographical Journal of India* 33(3): 277-304.

Singh, Ravi S. 1999. Ethnic minority groups of Arunachal Pradesh, India: A study in cultural geography. *National Geographical Journal of India* 45(1-4): 112-121.

—. 2003. Place, people and partition: A literary geography study of Rahi Masoom Raza's *Adha Gaon* ("Half Village"). *Annals, National Association of Geographers India* 23 (2): 45-53.

—. 2005. Book Review of *Cultural Geography: Form and Process* (ed. Neelam Grover and Kashi N. Singh). *National Geographical Journal of India* 51(3-4): 97-101.

—. 2007. Mari Mata *Darbar*, a local goddess shrine: A study in enthno-geography of a sacred place. *Oriental Anthropologist* 7 (2): 323-336.

—. 2009. The Goddess Kamakhya temple complex: Symbolism, sacredscape, and festivals. *Man in India* (Special Issue/March-June) 89 (1-2): 121-140.

Singh, Ravi S. and Rana P.B. Singh 2006. Goddesses in Kashi (Varanasi): Spatial patterns and symbolic orders. In *Visualised Space in Banaras: Images, Maps, and the Practice of Representations* (Ethno-Indology Series, vol. 4. Heidelberg Studies in South Asian Rituals), ed. Martin Gaenszle and Jörg Gengnagel, 41-68. Wiesbaden: Harrassowitz Verlag.

Singh, S.B. 1989. *Fairs and Festivals in Rural India: A Geographical Study of Belief Systems*. Varanasi: Tara Book Agency.

Smith, Jonathan M. and Keneth E. Foote 1994. Introduction. In *Re-reading Cultural Geography*, 27-33, *op. cit.*

Sopher, David E. (ed.) 1980. *An Exploration of India: Geographical Perspectives on Society and Culture*. Ithaca, New York: Cornell University Press.

—. 1967. *Geography of Religions*. New York: Prentice-Hall Inc.

—. 1968. *Pilgrim circulation in Gujarat. Geographical Review* 58(3): 392-425.

Sorre, Max. 1962. The geography of diet. In *Readings in Cultural Geography*, ed. Philip L. Wagner and Marvin W. Mikesell, 445-456. Chicago and London: University of Chicago Press.

Stump, Roger W. 2008. *The Geography of Religion: Faith, Place, and Space*. Lanham, Boulder, New York, Toronto, Plymouth (UK): Rowman & Littlefield Publishers, Inc.

Tewari, Laxmi G. 1988. Sohar: Childbirth songs of joy. *Asian Folklore Studies* 47: 257-276.

Wagner, Philip L. 1994. Foreword: culture and geography: Thirty years of advance. In *Re-reading Cultural Geography*, 3-8, *op. cit.*

Wagner, Philip L. and Marvin W. Mikesell (eds.) 1962. *Readings in Cultural Geography*. Chicago and London: University of Chicago Press.

Wagstaff, J.M. (ed.) 1987. *Landscape and Culture: Geographical and Archaeological Perspectives*. Oxford and New York: Basil Blackwell.

CHAPTER FOURTEEN

POPULATION IN GEOGRAPHY: CONTEMPORARY RELEVANCE IN THE INDIAN CONTEXT

APARAJITA CHATTOPADHYAY,
INTERNATIONAL INSTITUTE OF POPULATION STUDIES

Introduction

Population geography in India is in its infancy. To excel in this field one needs to know the variety of aspects dealing with population and geography in totality. Geographers in India, in general, are mostly concerned with spatial description. Majority of them fail to think that geography is not a simple account of variability, it is a more rich and complex subject that demands explanation and innovative methodologies for understanding the human behaviour over space. Perhaps geographers are failing to appreciate the multidimensional aspects of this unique science and hence they are struggling. Understanding population dynamics is the base in population geography. Unfortunately, population geographers are failing to excel for their paucity in comprehensive training in methodology and research techniques, dearth of in-depth knowledge in geography as a whole and population science/demography in particular.

Teaching Content

In the world context, population geography is linked with many parallel sub-disciplines like environment, management, public health, social studies, family studies, urban studies, women studies etc. A cursory look of the syllabus of this subject in any foreign university structurally looks alike as ours in India, yet the depth in teaching, technical aspects of measuring change, model building approaches, and project works are more rigorous overseas. The overall structure of population geography course is given in Table 14-1.

The substantial overlapping between population science—demography and population geography is perhaps the main reason for which population geographers in India are failing to keep pace with the subject's dynamism. Even in the West, Princeton School of population scientists who were mostly trained in sociology and economics with the basics of demography, stolen the geographer's thunder (Weeks 2004). Even Trewartha seemed to have backed away from the broader scope of this subject that he outlined in his presidential address to the Association of American Geographers (AAG) at Ohio in 1953. In 1969 he wrote that "the geography of population as concerned chiefly with one aspect of population study—its spatial distribution and arrangements". From 1950s, many geographers as well as non-geographers with profound training in population science are contributing well in the spatial aspects of population. It is misleading to perpetuate the view that demographers are merely population statisticians. In recent years, population scientists' contributions in explaining regional variation of population components are much more appalling than population geographers, when the latter give mere description and the former is more analytical. Demography and population geography go hand in hand if the "population geographers adopt the demographer's analytical sophistication and the demographers become more spatial". With the passage of time, demographers in India have become more spatial, yet geographers are lagging behind in adaptation of methodological sophistication—be it in statistical modelling or writing narratives or understanding the inner meaning of collected data partly for lack of training and partly for rigidity in appreciating approaches of non-geographers .

A critical look into Table 14-1 gives us enough inputs to talk about the drawbacks of the ways we teach population geography in graduate and postgraduate classes. We always give more emphasis on the distributional aspects of population characteristics, components etc. Yet, we forget to understand that without having sound knowledge on measures and indices and linking theories (which is not limited in demographic transition or optimum population) with practice, especially the current theories of fertility, migration-urbanization and development, a geographer is handicapped in understanding the spatial aspect of any population related issue. Hence theory building by population geographers of the present genera is a day dream. Neither we touch population estimates and projections nor we comprehend population policies critically and whenever we see any numerical juggleries—be it simple arithmetic—we put the ball on the court of population scientists and demographers. Such escapist nature is detrimental for academic proliferation.

Table 14-1: Structure of population geography syllabus and teaching content

Topic	Taught in India	Taught elsewhere
The Fields of Geography and Population	Past development, linkages	Past, present and future, subject philosophy, debates and questions
Data Sources, Measures and Indices	Census	Census, sample surveys, dual registration, data validity, data error and adjustment
The Spatial Distribution of Population	World population distribution, factors	Genesis of present world population distribution, future scenario
Population Growth: Trends and Theories	Demographic transition, optimum population, Malthus, Marx	History, causes of change, short and long term population growth, Future prospects, pre Malthus, Malthus and Post Malthus
Population Estimates and Projections	?	Estimates and projections: mathematical and compound; sub-regional, regional, global
Geographic Aspects of Fertility	Fertility basic concepts, pattern, distribution	Fertility basic concepts, direct and indirect measures, advantages and drawbacks of measures, factors affecting fertility, variation, biological-socio-economic, economic theories of fertility, trend and future
Geographic Aspects of Mortality	mortality basic concepts, pattern, distribution	Mortality basic concepts, measures, standardization, life tables and its use, infant mortality, maternal mortality, causes of death, factors of mortality differentials, pattern
Population Redistribution(Migration) and Urbanization	Migration type, Indian trend, urban distribution, variation	Urban definitional ambiguity, urbanization and urban growth, indices and measurement, world urbanization prospects, problems, country specific pattern, causes; migration concepts, census

		measures, direct and indirect measures, theories of migration, international migration, trend-prospects.
Demographics and Population Composition	Literacy, sex ration, labour-force? Superficial touch	Age-sex structure, population aging, demographic dividend, sex ration and its recent complications, literacy, work force, poverty, social class
Population and the Environment	?	Environment concept, ecology, system, global and local environment, concerns, future scenario, world concerns, policies
World Population: Major Issues for the 21st century	?	Fertility: pro-natalist vs. anti-natalist Mortality: child, mother, world health, HIV, Poverty and malnutrition; Climate change, International migration; slums; gender and women's deprivation, development vs. coercion; population and food security etc.
Population Policies	?	World population conferences, policy of developing and developed world, country specific policy—genesis, progress and future goals.

Note: "?" Indicates vague teaching and no clear cut syllabi

Population geographers in India are weak in statistical techniques which are well studied in foreign universities even at graduate levels. We boast ourselves as map makers, geography being our core subject. Shockingly, how many of us would be able to read and analyse symbol maps by measuring its randomness, association, independence and distributions? Though we often talk about the aspects of spatial distribution and trend, not many of us are in a position to understand the application of matrix (essential for analyzing networks), Lorenz curves (for distribution, dissimilarity, concentration), index number, log scale, moving averages and trend line fitting which are all useful for analyzing time series data. Our concept on probability distribution is abysmal. Neither can we design a probability sampling frame before conducting a research work, nor do we have the depth of testing hypotheses, though most of our students construct researchable hypotheses!

Lack of data, often cited as a drawback of population analysis, is not a point well taken. Indian census is the largest census of the world and the volume of data it generates is excellent to write hundreds of research papers. International migration is the only field that needs more data. Sample registration system generates huge statistics on fertility and mortality. Besides the above two data sources, we do have some excellent large scale data like National Family Health Surveys, Reproductive and Child Health Surveys, Causes of Death Statistics (RGI), data published by Centre for Monitoring Indian Economy, National Sample Survey Organization, Indian Agriculture and Climatic Data set etc. that are good enough to analyse and understand the spatial dynamics. What we need to know is the basic knowledge of standardization, interpolation/projection and statistical controlling while using secondary data. Above all we can always collect primary data for in-depth understanding of any phenomenon. It is sad to say that once again we lack enough knowledge of preparing meaningful questionnaire that connects our study needs or the training of collecting qualitative data. Few universities take into account the importance of teaching research methodology at the master's degree level and very few give the training of data collection in field. Our researchers thus learn how to jot down some socio-economic data without understanding its context, to interpret such data awfully and then to unscientifically establish the connectivity of objectives and information collected. Social geography and population geography go hand in hand as the explanatory part of any population phenomenon needs sound ability of theorization and writing talent. Strong emphasis should be given in field training to every budding geographer. Increasing the frequency of field visit and practical work based on collected data is the need of the hour to make the future of geography bright.

The problem stands in the root. Geography, from the very basic school level, confuses students. It fails to establish an assimilation of the two entities of the earth. "History is about chaps; geography is about maps" goes the old proverb. "Ask anybody about what a geographer does; the answer will probably be much the same: geographers make maps, a most marginal technical input in the world of higher learning. The dominance of GIS is reinforcing that belief, ignoring the fact that we do much more than merely that", remarks Lahiri-Dutt (2005/2009).

A geographer must know the name of all capitals, rivers, mountains….; she is there to remember such things meticulously. A student at school level must be able to throw up innumerable names of places of production, distribution and trade in examinations. Is this geography? Is this the way of connecting man and nature? We have finished geography by the way we teach the subject, the way we structured the books and the way we are examining the students—we have made geography moribund, disable and lacklustre. If you ask any geographer to do soil analysis for improvement of agricultural production, she will say that it is a job of a soil scientist; if you ask her to analyze a social data in the local context, she will promptly put the ball in the court of a sociologist or anthropologist; if help is sorted from a geographer to express some policy views on urban issues, population issues or environmental concerns, she can name a host of experts in urban planning, demography or environmental sciences who can be a better judge in these subjects, alternatively she can make a long list of policies without theorizing those suggestions and without analyzing the efficiencies of existing policies!

At school, college and university, we are repetitively taught to become an expert of "memorizing" names and "data" in human geography and recollecting the same thing in physical geography. Going little beyond, at doctoral level, we learn to copy as usual the "geographic factors" in explaining everything from different books (as journals are infrastructural constraint) and theses done earlier and we also learn to form sentences by simply looking into numbers without going beyond the meaning of raw data. Above all, the comments of the examiners of such doctoral work do not bother the honourable teacher as remarks are "good" as usual from his comrade. Numbers speak, landscape acts and geographers, being the "queen of social scientists" fails to connect man and nature.

Published Research Works

Published works are one of the indicators of advancement of the discipline. There are number of journals that handle issues connecting geography as well as demography/population. To name a few are *International Journal of Population Geography* or *Population Space and Place*, *Urban Studies*, *Asia- Pacific Migration Studies*, *International Journal of Migration, International Migration, Population-Environment, Environment and Planning D. Society and Space*, *Social and Cultural Geography* etc. and all the above international journals have a high impact factor, ranking within first fifteen of the journals published either in the field of population or geography. As mentioned earlier, social geography is also a part and parcel of population studies and hence all the social geography journals are rich platforms for population geographers or population scientists to publish their research works.

The sub-themes of recently published articles by Indians/non-Indians working on India are shown in Table 14-2. The themes cut across population studies and geography (see appendix at the end of this chapter). A variety of papers can be seen in well known journals that are dealing with issues like migration and left behind family; process of migration and network analysis, illegal or cross-border migration, women's migration and adjustment, remittances and migration effects, student's migration, environmental change and management, geopolitical issues, labour market, marginalized population groups, population health, spatial dynamics of demographic components etc. The journals such as *Urban Studies* and *Environment and Urbanization* offer fertile ground to urban planners, architectures, geographers, economists, demographers, sociologists and others to publish their related works. Most of the articles published by Indians in the recent volumes handle with change in urban land use, urban planning strategies, urban infrastructure and amenities, urban projections, private-public partnership in urban development, way of urban living, urban poverty, homeless urban population, slum redevelopment strategies, rurban planning, congestion and pollution, urban violence and security etc. Point to mention here is that most of the research contributions are made by non-geographers or geographers working overseas, though a number of doctoral theses are being awarded by Indian universities without taking into account their meaningful input in the subject (in terms of paper publications and/ or presentations).

Table 14-2 shows only a tip of an iceberg—if I start looking into journals like *Professional Geographer*, *Population and Development Review*, *Population, Demography, Journal of Population Research, Progress in Human Geography, Antipode* etc., a bunch of articles on India can be sorted out, yet the number of Indian geographers publishing their work in peer reviewed journals on geography or social science are abysmal. Critics of geography as a marginal social science identify the drawbacks of the subject as a whole (see Cutter *et al.* 2002; Pandit 2004; Weeks 2004; Silvey 2004; Conway, 2004; Kapur 2004; Lahiri-Dutt 2005/2009; Prasad 2005; Sunny 2006/2009). Population geography, being a part and parcel of the subject, also falls under the same purview. "Answer must be sought within, not outside, of the discipline; from the institutions, the bodies that fund research, from the academia as a whole. Inevitably, questions of quality arise—in teaching, of teachers, and students, the quality of the curriculum and, of course, the quality of university administration and facilities offered by institutions" as rightly opined by Lahiri-Dutt (2005/2009) in searching for the causes for marginalization of geography as a social science.

Table 14-2: Themes of some recent articles by Indian population geographers in related journals

Journal topics	Sub-themes
Urban studies and Migration	Remittances, gender, cross-border immigration, left behind population, refugee, student migration, forced migration, geographies of diasporas, changing urban land use.
Fertility, Mortality and Family	Floating population, cross-border/cultural marriages, child mortality, spatial variation, gender and family.
Development	Poverty and place, housing demand, marginalized groups, policy analysis, gender related issues, health and nutrition.
Estimates and Projection	Small area estimates, housing, urbanization, urban simulation, simulation in population distribution.
Methodology	Spatial data compatibility, geo-demographic classification.
Environment	Environmental refugees, urban sprawl, health and environment, water resource- population dynamism, ecosystem change, environmental migration, agricultural frontiers, climate change, deforestation.

(Source: Based on published articles in *Population Geography/ Population, Space and Place; Population and Environment, Gender, Place and Culture; Environment and Planning D. Society and Space, Social; and Cultural Geography; Annals of the Association of American Geographers*)

The Future

The dualism of geography (physical and human) as felt by Trewartha, is due to the ignorance of man's role as the creator and originator of the cultural landscape. The plethora of issues that are emerging with complications of man's day to day living gives us a fertile ground to sense and act afresh. Population geographers of India need to start thinking beyond the traditional *Do's* and *Don'ts*.

In 1991, R.C. Chandna raised few questions to the population geographers of India. To quote, "there have been only few attempts to theorise.... Model building, hypothesis testing and theory formulation are the real challenging fields for population geographers of the coming years... Macro level studies are now picking but only at the snail's pace and so are the theoretical formulations by Indian population geographers. Challenges are serious warranting equally serious effort." What advancement and efforts have been made for the last seventeen years to improve the quality of studies? As a whole, geographers are extremely smart of escaping accountability mainly for their rigidity to learn new things. We need collaborative research, integration of spatial analysis by applying new techniques, understanding culture by reading, seeing and systematically expressing ourselves more and more—no matter whether a "geographer" or so called "geographic journal" is scrutinizing our work! Some emerging issues with policy relevance where population geographers can thrive well are the population-environment relationship, migration-development and environment, development and displacement, poverty, public health, gender inequalities and census controversies. We need to indulge ourselves in public debate and research that leads to policy formulation. We have to look to the future positively and to do so "why not err on the adventurous side" as opined by Conway (2004).

The relationship between man and nature is becoming complex over the years. The recognition of population-environment complexity by international communities like IUSSP, PERN, and IIASA are encouraging. The future of population geographers is brilliant from carrier perspectives too. They just need little training to understand demographic analysis. The new sources of data, time series information, modes of analysis and research, improvement of research tools, will lead to an ever expanding role for good social scientists in population research, in the business world and in the government realms.

Even if a single geographer of India can do one authentic and meaningful work, s/he can re-establish the importance of this unique subject in our country. It is not that good work is stopped by them. There are still academicians in this very field who are often being quoted by the fraternity of social scientists all over the world. Forget about "-isms" in geography, forget about "geographical" work; think and work which is post-disciplinary, interactive, meaningful and contributory to the welfare of the human society. Approach may differ, techniques may diverge and views may vary. Then only a discipline thrives.

References

Buttimer, A. 1983. *The Practices in Geography*. London/New York: Longman.

Chandna, R.C. 1991. Population geography in India. *Population Geography* 13 (1-2): 1-6.

Conway, D. 2004. On being part of population geography's future: Population-environment relationship and inter-science initiative. *Population, Space and Place*10 (4): 295-302.

Cutter, S.L., R. Golledge and W.L. Graf 2002. The big question in geography. *Professional Geographer* 54: 305-317.

Prasad, Sheela. 2005. Geography in India: Marginal 'space'. *Economic and Political Weekly* 40 (1): 79-80.

Kapur, Anu. 2004. Geography in India: A languishing social science. *Economic and Political Weekly* 39 (37): 4187-4195.

Lahiri-Dutt, Kuntala. 2005/2009. Geography as a marginal science. *Economic and Political Weekly* 40 (7): 689-691. Reprinted in, *Indian Geography*: *Perspectives, Concerns and Issues*, ed. Ravi S. Singh, 242-250. Jaipur/New Delhi: Rawat Publications.

Pandit, Kavita 2004. Postscript: LOOKING BACK Looking Forward. *Population, Space and Place* 10(4): 309

Sunny, Yemuna. 2006/2009. Analyzing current practices in geography education. *Economic and Political Weekly* 41 (3): 270-278. Reprinted in *Indian Geography*: *Perspectives, Concerns and Issues*, ed. Ravi S. Singh, 257-278. Jaipur/New Delhi: Rawat Publications.

Trewartha, G.T. 1953. A case for population geography. *Annals of the Association of American Geographers* 43: 71-97.

—. 1969. *A Geography of Population*: *World Patterns*. New York: John Willy & Sons.

Weeks, J.R. 2004. What did he know and when did he know it? Putting Gleen Trewartha's call for population geography into historical perspective. *Population, Space and Place* 10 (4): 279-284.

Appendix

Bhagat, R.B., Mohua Guha and Aparajita Chattopadhyay. 2006. Mumbai after 26/7 deluge: Issues and concerns in urban planning. *Population and Environment* 27 (4): 337-349.

Bhagat, R.B. 2005. Rural-urban classification and municipal governance in India. *Singapore Journal of Tropical Geography* 26 (1): 61–73.

Chatterjee, Ipsita. 2007. Packaging of identity and identifiable packages: A study of women-commodity negotiation through product packaging. *Gender, Place and Culture* 14 (3): 293-316.

Roy Chowdhury, Rinku and B.L. Turner II 2006. Reconciling agency and structure in empirical analysis: Smallholder land use in the southern Yucatán, Mexico. *Annals of the Association of American Geographers* 96 (2): 302-322.

Guilmoto, Christophe Z. and S. Irudaya Rajan. 2001. Spatial patterns of fertility transition in Indian districts. *Population and Development Review* 27 (4): 713-138.

Dahiya, Bharat. 2003. Hard struggle and soft gains: Environmental management, civil society and governance in Pammal, south India. *Environment and Urbanization* 15 (1): 91-100.

Datta, Kavita. 2007. "In the eyes of a child, a father is everything": Changing constructions of fatherhood in urban Botswana. *Women's Studies International Forum* 30 (2): 97-113.

Dutta, Shyam S. 2000. Partnerships in urban development: A review of Ahmedabad's experience. *Environment and Urbanization* 11(2): 13-26.

Balabdaoui, F., J.P. Bocquet Appel, C. Lajaunie and S. Irudaya Rajan. 2001. Space-time evolution of the fertility transition in India, 1961-1991. *International Journal of Population Geography* 7 (2): 129-147.

Gidwani, Vinay and K. Sivaramakrishnan. 2003. Circular migration and rural cosmopolitanism in India. *Contributions to Indian Sociology* 37 (1-2): 339-367.

Gidwani, Vinay and K. Sivaramakrishnan. 2003. Circular migration and the spaces of cultural assertion. *Annals of the Association of American Geographers* 93(1):186-213.

Zachariah, K.C. and S. Irudaya Rajan. 2001. Gender dimensions of migration in Kerala: Macro and micro evidences. *Asia Pacific Population Journal* 16 (3): 47-70.

Zachariah, K.C., E.T. Mathew and S. Irudaya Rajan 2001. Impact of migration on Kerala's economy and society. *International Migration* 39 (1): 63-88.

Kathleen, O'Reilly. 2007. "Where the knots of narrative are tied and untied": The dialogic production of gendered development spaces in north India. *Annals of the Association of American Geographers* 97(3): 613-634.

Ladusingh, L. and Chungkham Holendro Singh. 2006. Place, community education, gender and child mortality in northeast India. *Population, Space and Place* 12 (1): 65-76.

Lahiri-Dutt, Kuntala. 2006. 'May God give us chaos, so that we can plunder': A critique of resource curse and conflict theories. *Development* 49 (3): 14-21.

Mehra, B.S. and P.K. Mathur. 2001. Livestock grazing in the great Himalayan National Park conservation area—A landscape level assessment. *Himalayan Research Bulletin* 21(2): 89-96.

Noxolo, P., P. Raghuram and C. Madge. 2008. 'Geography is pregnant' and 'Geography's milk is flowing': Metaphors for a postcolonial discipline? *Environment and Planning D: Society and Space* 26 (1): 146-168.

Raghuram, P. 2008. Migrant women in male-dominated sectors of the labour market: A research agenda. *Population, Space and Place* 14: 43-57.

Raghuram, P. and C. Madge. 2006. Towards a method for postcolonial development geographies: Possibilities and challenges. *Singapore Journal of Tropical Geography* 27 (3): 270-288.

Raghuram, P. and E. Kofman. 2002. The State, labour markets and immigration: The case of skilled emigrants in UK's medical labour market. *Environment and Planning A* 34 (11): 2071-2089.

Rajan, S. Irudaya. 2003. Does geography matter in fertility research? *Population Geography* 27: 615-626.

Ramachandran, S. 2004. There are many Bangladeshis in New Delhi, but…: Methodological routines and fieldwork anxieties. *Population, Space and Place* 10 (3): 255-270.

Sheikh, Khatibullah and Shrinivas Rao. 2007. Participatory city planning in Chhattisgarh: A civil society initiative. *Environment and Urbanization* 19 (2): 563-581.

Simpson, Edward and Stuart Corbridge. 2006. The geography of things that may become memories: The 2001 earthquake in Kachchh-Gujarat and the politics of rehabilitation in the pre-memorial era. *Annals of the Association of American Geographers* 96 (3): 566-585.

Sivam, Alpana. 2003. Housing supply in Delhi. *Cities* 20 (2):135-141.

CHAPTER FIFTEEN

MIGRATION STUDIES: MISSING LINKS AND FUTURE AGENDA

ARCHANA K. ROY,
BANARAS HINDU UNIVERSITY

Introduction

Migration is as old as human civilization. Throughout human history, migration has been a courageous expression of the individual(s) to overcome adversity and to quest for a better life. Today, globalisation with rapid advancement in communications and transportation has greatly enhanced the capacity and desire of the people to move around the globe in search of better pasture (Annan 2006). The importance of migration in economic development and social change has been realised over the period of time. Goldscheider (1984) stated that the movement of population has been and remains an essential component of economic development, social change and political organisation. In 1994, International Conference on Population and Development not only acknowledged migration as an integral part of population, but also clearly explained the relationship between migration and development (UN 1995). Also migration is one of the key issues of political and public debate all over the world due to security concerns and human rights. Recent decades have witnessed a number of dramatic changes in the migration landscape due to various reasons. This new era has created opportunities and challenges for societies throughout the world.

According to an estimate by International Organization for Migration (IOM 2005), in 2005 there were some 192 million persons of the global population living outside their country of birth. The figure is doubled what it was in 1980 and equivalent to the population of the world's fifth most populous country, Brazil. However, international migrants represent only 2.9 per cent of the world's population as compared to 2.5 per cent of the total world population of 1960. While the great majority of those who move is still of the internal migrants (individuals or families who migrate within their own country), migration can be voluntary or forced, although the actual experience may contain elements of both. Most people migrate for employment, family reunification or marriage. The demand for labour migrants (i.e., those searching for better economic opportunities) has been a major factor in rising levels of migration to developed countries and urban areas. It is with respect to this group that experts invoke the potential role of migration in development and poverty reduction, especially given the significant impact that financial remittances and other benefits can have on countries of origin. Forced migration and trafficking, on the other hand, encompass the more poignant vulnerabilities associated with international movements—particularly where it involves women and children (UNFPA 2006).

As far as India is concerned, throughout its history, it has received migrants from various parts of the world and has absorbed them instinctively with their culture, language, economic and social status. According to census 2001 there were about 5.1 million migrants who have come across the international border. About 97 per cent of these immigrants were from the eight neighbouring countries (including Afghanistan). Majority were from Bangladesh (3.0 million), followed by Pakistan (0.9 million), Nepal (0.5 million) and Sri Lanka (0.1 million). About two-third of these immigrants from the neighbouring countries had migrated to India at least 20 years back perhaps at the time of partition and later during the Bangladesh war in 1971. In comparison to 1991, there is 31.6 per cent decline in international migration to India (excluding Jammu and Kashmir) in 2001. This is due to substantial decline in the number of recent migration and death of earlier migrants due to old age (Census of India 2001).

While immigration has slowed down, emigration from India has increased considerably. It is the second largest migrant sending country after China, with Indian Diaspora estimated at some 20 million persons, spread across a hundred and ten countries. They are serving their host nations with distinction as entrepreneurs, workers, teachers, researchers, innovators, doctors, lawyers, engineers, managers and even political leaders. In eleven countries there are more than half a million persons of Indian descent and they represent a significant proportion of the population of these countries (Sharma 2002). They not only contribute to the host country but to the country of origin. As per the World Bank's report, India received the largest amount of remittances (US$30 billion) from migrants in the world in 2008 followed by China, Mexico, the Philippines and Poland (Ratha *et al.* 2008).

However, the number of internal migrants exceeds far that of international migrants—they are an established and essential part of the labour force. In India, migration as a result of regional disparity started during the British period, and it further intensified in post-Independence period. Kundu and Gupta (1996) showed that the backward states like Bihar, Himachal Pradesh, Kerala, Rajasthan, and Uttar Pradesh have high out-migration rates in year 1961, 1971 and 1981. While the developed states like Gujarat, Haryana, Karnataka, Maharashtra, Punjab and West Bengal were absorbing these migrants. They registered high life time inter-state migration. Their contribution to the economic, social and cultural life of both their regions of origin and destination appears to be beyond question. As per 2001 census, out of 1,029 million people in the country, 307 million (30 per cent) were reported as migrants by place of birth which was only 144 million in 1961. The states like Maharashtra (3.6 million), Delhi (1.9 million) and West Bengal (0.4 million) have registered sharp rise in migration between 1991-2001. On the basis of net migration (gain and loss of migrants) during last decade Maharashtra stands at the top of the list with 2.3 million net in-migrants, followed by Delhi (1.7 million), Gujarat (0.68 million) and Haryana (0.67 million). Whereas, economically poor states like Uttar Pradesh (-2.6 million) and Bihar (-1.7 million) were with largest number of net migrants migrating out of the state. Thus the direction of migration stream in India is from economically poor states of the east to economically prosperous states of the west.

Most of the studies conducted in India in the last four decades have highlighted various issues related to characteristics, causes and pattern of migration, "push" and "pull" mechanism and use of remittances. This has definitely given a very important perspective to the study of migration and also covered the issue of consequences and outcome of migration in economic terms. However, majority of scholars in the field of migration viewed migrants in isolation from the family and community context to which they belong. Empirical research studies conducted on impact of migration have largely focused on utilization of remittances. If migration is to contribute to economic progress, social cohesion, and justice and peace in the 21st century there is a need to re-examine the way we think about migration.

In this backdrop, the present paper examines the status of migration studies in two contexts, one in changing pattern of migration and the dealing of these issues as reflected by the published literature and another is status of migration studies in the syllabi and curriculum of population studies in general and geography in particular. Further it tries to suggest a framework for the migration studies in the Indian universities.

Status of Migration Studies in India: A Critical Review

Pattern of migration

Migration not only involves the movement of specific number of people from one place to another but also the selective movement of people with particular characteristics. Rural to urban migration has been particularly male selective and furthermore the pattern seems to include rich students and poor workers (Dasgupta 1978). Rural out-migration was found more from lower castes and class and among illiterates or marginally literate (Connell *et al.* 1976; Saxena, 1977; Gill, 1984; Banerjee, 1986; Khandelwal, 1994). Oberai and Singh (1983) found that in rural Punjab, incidence of migration is low among the peasants, particularly those owning 1-5 acres of land. While a study conducted in Varanasi district of Uttar Pradesh noted that the propensity to move is greater from both bottom and the top strata of rural society than from those in between (Yadava 1989).

Another predominant feature in developing countries including India is male out migration leaving their family behind in the villages. It is found that in addition to economic reasons, social factors were also associated for male-alone migration, like desire by older members of the family/community to keep the male migrants tied to the village home which is possible if the wife and children continue to live in the village (Jetly 1987) and the perceived inability of women to go to work in the city (Sen 1960; Jetly 1987); hectic travels and difficult and uncertain working conditions at the destination place make it difficult for women workers to join their male counterparts (Ghosh and Sharma 1995). Sajjad (1998) explained that since marriage brings many responsibilities males are compelled to migrate. Thus, main reasons why women are left behind are economic and socio-cultural conditions. Migration has major implications on the left-behind families, especially on the wives of migrants, but it has not received the attention which it deserves except few studies (Gulati 1993; Sekar 1997; Zachariah *et al.* 2001; Bose 2000; Roy and Nangia 2003).

Complexity of causes of migration

The migration patterns are seen to be a collective expression of millions of individual and family decisions made for economic, social and political reasons, including some forced to seek asylum (UN 1993). In developing countries the labour migration of distressed people is occurring as a "spatial" manifestation of poverty, underdevelopment and spatial disorganisation (Mukherji 1991). However, both internal and international literatures tend to reach at a consensus that migration is primarily motivated by economic reasons. According to Census of India (2001) the female migrants outnumbered their male counterparts in both rural and urban areas—more so in rural areas than in urban areas. For male migrants (12.3 million out of 32.8 million male migrants) "work/employment" was reported as the most important reason for migration while for female migrants it was cited as marriage (42.4 million out of total 65.4 million female migrants) during 1991-2001.

The findings of empirical studies show that rural poor are more influenced by "push" factors while "pull" factors are more likely to apply on the rural rich (Saxena 1977). The "push" factors operating at the place of origin are not only numerous but are also very complex, and vary according to spatial and temporal context. However, it may be assumed that the regional disparities in economic activity and population problems are the basic factors, which initiate migration. High population density, surplus of labour force, high unemployment rates, meagre incomes, the demand for higher schooling, rural-urban wage differentials, distance between village and city, pattern of land possession, caste-conflict and the prior migration patterns have been identified as some of the main determinants of rural out-migration in most of the regions in India (Gosal and Krishan 1975; Connell *et al.* 1976; Population Reports 1983; Banerjee 1986; Sharma 1986; Singh 1986; Yadava 1989; Sajjad 1998).

Another side of the coin is if 30 per cent of the people are moving, 70 per cent of Indians still live at their place of birth and do not migrate. Migration theories perhaps need to examine more closely not only why people move, but why (preponderantly) people do not move; the factors inducing immobility as well as mobility in terms of centripetal socio-economic factors as well as centrifugal ones.

Migration and development

Migration has wide-ranging implications on the economy and society both active (in terms of remittances) and passive (diffusion of ideas, technologies, and social change through knowledge, attitude and practice). Studies in Punjab found that large part of remittances has been utilized for non-productive purposes including conspicuous consumption. However, they help the families staying back in village to improve financial conditions. The process of migration also plays a suitable role in promoting better literacy rate by (Oberai and Singh 1983; Mehta 1990). The studies conducted in Punjab (Oberai and Singh 1983) and Bihar (Kumar 2000) have highlighted that migration has a positive role in agricultural development since it brings about improvements in the productive forces such as physical and human capital and makes it possible for the migrants to continue maintaining small plots of land at the place of origin in Bihar (cf. Haan 1997). Studies of rural areas of Varanasi (Yadava 1989) and rural Bihar (Singh 2000) reveal that families of migrants and commuters are living a better quality life than non-migrant or non-commuter families. Venkatarayappa (1991) concluded that migration considerably improved economic status of migrant families in rural areas. This, in turn, encouraged the change in their way of life such as, dress, housing, start spending more, read newspaper and appreciate and encourage girl's education.

A longitudinal study (1970-1990) conducted in Purnia district of Bihar (Rodgers 2000) reported that as a result of remittances and shortage of local labour, the income and consumption levels of the poorest group of agricultural households had risen markedly. So poverty in these villages had declined and on the whole people were benefited. However, these improvements were occurring not due to economic growth in Bihar but due to remittances from western states like Punjab, Haryana and Delhi. These ideas have been further reinforced by the findings of other scholars (Kumar 2000; Singh 2000)).

Contradictory to the above findings, some of the scholars have critically noted that such migration helps individuals in their survival; it does not lift them upward in terms of social mobility and prosperity (Connell *et al.* 1976; Breman 1985 and Mukherji 1991). A large number of poor families receive small sum of money, which is hardly sufficient for the purpose of both consumption and investment (Shah 1998). Rao (1992) pointed out that due to rural out-migration, villages are deprived of young talent on one hand and pressure is asserted on urban infrastructure on the other.

The migration and development equation has become a foremost subject of research and policy debate. Regarding international migration, now we need to go beyond simple theories of a "brain-drain" from dependent countries to understand much more about remittances and their development impact (Munck 2008). On the impact of internal migration on the development of sending areas too, the evidences are mixed. In some cases predictably they increase inequalities but in some they act as an equaliser. However, changing geo-political and geo-economic landscapes, as well as the "knowledge-economy", are fundamentally changing the processes of migration and globalisation leading to the emergence of new development thinking and strategies. Unpacking of the general debates in the era of globalisation, may better reveal the empirical complexity and social dynamics of the overall process of people movement. There is need to carry out more longitudinal studies in different geographical context for theorisation of migration and development relationship.

Migration and globalization

Globalisation, characterised by a move to more competitive markets is likely to increase disparity between and within nations, providing the stimulus for rapidly increased migration. Since the labour does not enjoy similar freedom of movement like capital, technology and finance; there is high chance for illegal or irregular migration, pushing migrants into more and more dangerous methods of migration and by the involvement of trans-national organised crime (UN 2001). So there is a serious need to respond to such migration issues and rethink about the international immigration laws where there exist substantial demand and supply.

According to Migration News (2009), in contemporary period of global economic meltdown many migrants have lost their jobs and returned to their countries of origin as well as there is also slowdown in the deployment of new workers, especially to the higher-wage European and oil-exporting Gulf countries. Job losses during the 2008-09 recessions have been concentrated in four major sectors—construction, financial services, manufacturing and travel-related services. It means job loss is of both migrants from low-skilled jobs in construction and travel-related services to blue-collar and high-skill jobs in manufacturing and financial services and ultimately offsetting the remittances and resulting into short term financial crisis.

Likewise, several issues of human importance are emerging up due to globalisation. As in the twenty-first century, migration is no longer largely unidirectional and permanent in nature; it is increasingly temporary and circular. However, certain aspects, such as spatial distribution of population, movement of labour from agriculture to manufacturing and tertiary activities and rural-urban migration, have not been in the core of development debate on globalisation. Looking into the quantum of flow of migrants and its complexities, there is a need for studying migration from the prism of globalisation.

Management of migration and (urban) governance

The migration management and governance of migrants has become one of the emerging issues for various state administrations. The countries which have initiated systematic plans for management of migration have mainly worked in context of trans-national migrants to control the socio-cultural, economic, behavioural and legal conflicts between locals and foreign residents. But in some of the countries migrants have been also found responsible for human and drug trafficking and illegal migration which pose more challenges to the local governance.

Regarding internal migration, where there is no clear cut difference between migrants and locals, the issue of governance is looked in the context of urbanisation. Throughout Asia, including India, the wave of people moving to cities will only swell as years go by. According to an article published by Nagrath (2008) in *Businessworld*, about half of India's population would live in cities by 2041. Indian cities are hard pressed to accommodate and provide urban facilitates for their current residents—pegged at 12 million in Mumbai, 14 million in Delhi, 5 million in Kolkata and 4.5 million in Bangalore—it is hard to imagine how they will cope up with the new additions (Nagrath 2008). The influx of migrants has resulted into squatter settlement, crime, low quality of life, low educational level and extra pressure on the urban amenities. Overcrowding and environmental degradation make the urban poor particularly vulnerable to the spread of diseases, from cholera and tuberculosis to HIV/AIDS and other epidemics.

The challenge, then, is of managing this scale and pace of migration from rural areas to urban agglomerations. Migration is unstoppable and growing phenomenon. As more and more people preferring cities as their choice of residence, the challenge lies not in checking migration, but in managing and governing our cities better, to improve quality of life and standard of living to make our cities more conducive and sustainable. The issue of migration management should be an integral part of city plan and decision making so that the planners have some alternatives in advance because sudden measures would be very inadequate and may result in problems rather a solution.

Migration and gender dimension

Until the mid-1980s, migration was regarded as a male-dominated phenomenon in country like India; but, today, women account for almost half of the migrant population globally. Migration is often seen as gender-neutral because it deals with the process of the movement of persons; however, it is in fact gender-related because migration impacts differently on men and women, and on different groups of men and women in their process of movement (IOM 2009). There are three arenas of migration and gender study: first, women as a regular migrant for employment; second, trafficking of women; and third, status of left-behind women when male migrates, as a survival strategy of the family.

The incidence of women migration for taking up jobs is increasing within India as well as to other countries. Today's movements include an increasing proportion of women taking advantage of opportunities for largely informal work in more affluent societies. The vast majority suffer from the multiple disadvantages of being migrants and being women in the labour market. The experiences of men and women as migrants differ, and most of the differences are due to the role, behaviour, and relationships that society assigns and expects from a woman in a particular country or place of destination. That needs to be studied intensively.

Other issues which need to be investigated under gender dimension of migration are particularly related to problems of children and women trafficking. Trafficking is the dark underside of migration, damaging communities of origin and destination as well as the individuals concerned. Increasingly restrictive immigration policies in Europe and North America are driving more and more would-be migrants into the hands of agents and traffickers. Trafficked women find themselves forced into prostitution, sex-tourism or commercial marriages, or into unpaid or poorly paid domestic, agricultural or sweatshop labour (UNFPA 2006). This also applies to Indian society, but there is no data on such cases therefore more studies are required.

Regarding left-behind families of migrants there is a need to look at male migration from women's point of view: to what extent their work load increases when their men are away, how many economic benefits they get and do they raise their socio-economic status in the society. It is pertinent to know how do women cope with the increased workload and how does it affect their well-being and general health. In the recent years it is being noticed that the more people move the faster AIDS and other STDs can spread because migrants become carrier of such infections and they expose their wives at the place of origin to such infections whenever they visit home (Carr-Harris 1993; Decosas *et al.* 1995; Roy and Nangia 2005). Therefore it would be vital to know whether rural women know about such infections or not and how many of them are suffering from the symptoms of RTIs/STIs, particularly in the areas from where male migration is very high. There are very few studies which have covered such aspects of left-behind women in India.

Migration and issues of health and human rights

The migrants are often associated with the "dark" side of development in terms of poor health due to abuse of drugs, unsafe sexual habits, inadequate drinking water and sanitation and poor living conditions (Munck 2008). Poverty and poor health keeps the urban poor in vicious circle of culture of poverty. There are number of cases reported that migrant workers have infected by sexually transmitted diseases (Halli *et al.* 2007), contagious (tuberculosis, hepatitis) and non-contagious diseases, unwanted pregnancy, mental and psychological health, injury due to violence, etc.

Studies have found that the socio-economic needs of seasonal migrants, who are highly vulnerable population as a result of poor awareness of their rights and consequently their inability to demand their entitlements for the basic services being provided at the place of destination. Moreover, they are socially isolated at the destination and their employment uncertainty makes them further vulnerable (Barbora *et al.* 2008; Pattnaik 2009). Most of the poor migrants are a neglected lot at their destinations as they have no say and many a time they are bound to be exploited, harassed and misused contractors or employers. Sustainable livelihood and a reasonable stay arrangement and association with co-workers later on may provide them some security. But their lucrative jobs and change of places keep them always at loose footings. In order to deal with the issues of migration there is need of inclusive policies for their rehabilitation and inclusive growth of city.

Migration and security

Terrorist attack of 9/11 has redefined state securities and it has necessitated the importance of global cooperation and information sharing. New security measures and agreements have an impact on migration as well. While the flow of persons across borders has always been monitored with respect to state security, the current focus is on pre-checking the personal profiles and making more rigorous efforts to prevent suspected persons from crossing borders (IOM 2009). The role of information communication technologies (ICTs) has been found vital to deal with such situations, thus there is a need of global policy for information sharing.

In the context of India too international migration is very important particularly when India is facing challenge of terror attack being operated from outside the border. The issue of security is not only related to international migration but also to the internal migration. Migration, of a huge number of people in a city or state, results into change in its original demography. It blends societies and their cultures. But some time it may result into conflicts among various interest groups. The recent violence against north Indian migrants in Mumbai by Maharashtra *Nav-nirman Sena* activists is an extreme case in this series. The violence in Assam and other states have also been reported. The issues of son of soils versus outsiders have been raised by several regional political parties. The migrants are also been viewed as a threat to the cultural identity and security of the locals. Where a nation is defined by ethnicity, then minority ethnic groups must supposedly assimilate into the dominant group or run in the risk of being marginalised or, ultimately, be subjected to "ethnic-cleansing" (Munck 2008).

This issue needs to be examined in depth particularly to change the image of South Asian countries by forming a cooperative union like EEC. This is the right time to rethink on this issue exhaustively in the context of overall growth of the region as union and power negotiation at the global level.

Issues of displacement and refugees

According to Lama (2000) there are four types of displacement in India, namely (a) politically caused including secessionist movements (conflicts in North East and Kashmir); (b) localised violence (caste conflicts, religious fundamentalists); (c) identity based autonomy movements (Bodoland, Khalsaland, *Amchi* Mumbai); and (d) environmental and development induced displacement (dams, universities, industries, roads etc and; flood, drought, tsunami, earthquake etc.)

According to an estimate by internal Displacement Monitoring Centre (iDMC, 2008), over one third of the India's 608 districts were affected by insurgent movements in 2007. The armed movement by Maoist groups known as Naxalites was the most widespread insurgency in the country and had displaced over 100,000 people of Chhattisgarh. Clashes between a tribal groups and immigrant settlers in Assam caused the displacement of over 200,000 people, and communal violence in Orissa displaced tens of thousands. In Nandigram region of West Bengal, displacement followed conflict arising from the government's plans to evict the local community and use the land for a development project popularly known as Nano Project. Some groups displaced for years continued to be trapped in relief camps or relief colonies (see Table 15-1). Over 30,000 ethnic Bru internally displaced persons (IDPs) were living in deplorable conditions in camps in Tripura while state authorities in Mizoram refused to allow them to return home. India's largest IDP group, the over 250,000 Kashmiri Pandits (from Jammu and Kashmir state), were also unable to return even after 18 years of their displacement. More than 20,000 people in Gujarat who had fled their homes during communal violence in 2002 could not return to areas of origin for fear of further violence.

Another aspect of forced migration is case of refugees. India is asylum for the refugees from its neighbouring countries like Tibet, Afganistan, Sri Lanka, Bangladesh etc. mainly caused by civil unrest and political repression. As per the estimate by UNHCR there were 11,286 refugees in 2007 which exclude the Tibetans.

Table 15-1: Internal conflicts and displacement

State/Region	Who are displaced	Time of displacement	Estimated No of displaced
Jammu &Kashmir	Kashmiri Pandits	1990s onwards	J&K 15,000 NCT Delhi 3,600
	Hindu	1998	12,000 from Doda, Poonch and Rajouri districts
	Others	2001 war	1,800 families
North-East	Santhal tribals	Inter-ethnic clashes 1996 and 1998	200,000 people,
Assam	Immigrant Muslim	Inter-ethnic clashes Bodos October 2008	212,000 people
Manipur	Hmar and Paite	Naga-Kuki clashes in the 1990s	Large number of people displaced
Tripura	Bengali immigrants	since 1980 to June 2008	120,000 8,700 families

Mizoram	Bru/ Reang Tribe	Bru-mizo clash in 1997	35,000 in Tripura
Nagaland		Isak-Muivah and Khaplang frictions	No figures
Chhattisgarh	Villagers	Since 2004	49,000 to Dantewada and Bijapur 65,000 to AP, Ma-harashtra and Orissa
Gujarat	Majority Minority Muslims	Communal violence in 2002	100,000-250,000 people in March 2008 estimated over 21,000 people in camp
Orissa	Majority Christian	Communal violence in Kandhamal in 2008	Estimated 50,000 people
West Bengal	villagers	SEZ at Nandigram Jan-March 2007	1500

(*Source*: IDP India Overview; http://www.internal-displacement.org/countries/india)

Estimating IDPs is a big challenge particularly in the context of conflict-induced IDPs in India as there is no government monitoring agency and humanitarian and human rights agencies have limited access to these IDPs. What sorts of crisis IDPs face related to drinking water, job, sanitation, health, education? In India very few researches have been conducted on the spatial variation of distribution, causes, adjustment and problems of refugees and IDPs. The views of locals should also be incorporated. In many cases like climate, migration and conflict the relationship between cause and effect is very complex and need special attention so that a concrete policy can be formulated.

Geographical Perspective of Migration

Migration is a trans-disciplinary issue hence scholars from a range of disciplines look into the matter from the different perspectives. "Migration", referring to semi-permanent or permanent change in place of residence by a person or community for certain time period, is one of the fundamental components of population dynamics which involves two important concepts of geography i.e. time and space. According to Salt (1986, 166) migration study has become the corner stone of the geographical interest in population. Hassan (2005) emphasised that migration has been one of the favourite topics of research and teaching in population geography. Undoubtedly, the aspect of population geography, which developed in the early decades, was migration analysis—an issue which never attracted to the demographers in the same way as it was in the case of fertility, nuptuality and mortality.

The major constrain in studying migration issue is availability of accurate and detailed data which is also related to definition of migration. The most convenient and broadly applied definition is given by the UN multilingual dictionary which defines "migration as a form of spatial mobility between one geographical unit to another, involving permanent change in residence". Here it involves two aspects: (i) permanent change of residence for one or more years, and (ii) shifting from one predefined administrative unit. However, this definition has certain limitations, all the movements are not made with intention to settle permanently like refugee, guest workers or labour migration; the seasonal or circular migrants are not taken care by the definition; migratory status of people is also affected by the reclassification of the geographical area.

The major sources of data on migration are census, population registers and surveys. In most of the countries like India, census remains the comprehensive source of internationally comparable information on internal migration and international migration. However, it has certain limitations as, census is conducted at an interval of 10 years hence, short term, seasonal or circular migration made during inter-census period can not be recorded; it gives limited information about migrants like age, sex, literacy, occupational information collected at the place of destination. In the European countries, Latin America, China and Japan, the population reports are another source of demographic characteristics though, very few countries collect data on migration information (like Sweden, Norway, etc.). Highlighting its importance in collection of migration data, United Nations Statistics Division states "If complete, population registers can produce data on both internal and international migration through the recording of changes of residence as well as the recording of international arrivals and departures" (UNSD). Periodic surveys like NSS in India (55th round, 1999-2000; 49th round 1993; 43rd round, 1987-8; 38th round, 1983) also provide important source of data on internal migration and related aspects. There are more serious problems in getting data on international migration. There are few sources which provide information such as the United Nations Statistics Division that collects statistics on international migration flows and on the stock of migrant population through the Demographic Yearbook data collection system. International Labour Organization collects and compiles international labour migration statistics; the United Nations High Commissioner for Refugee (UNHCR) collects and compiles data on asylum seekers and refugees; the United Nations Population Division prepares estimates of migrant stock for each country and undertakes regular studies on national policies on international migration and assesses the impact of international migration on development.

Most of the geographers prefer to choose census for migration study due to wide coverage of geographical area and availability of data on regular intervals. Most of the studies conducted in the past are focused on analysis of trend, pattern, causes and economic impact of internal migration. However other related aspects like families of migrants at the place of destinations and origin, socio-politico-cultural differential at the place of origin and destinations, cause and effect of migration play significant role which need to be covered only by empirical studies, are not taken care of by most of the researchers in general and geographers in particular.

As far as analysis is concerned, maps have been major tools for illustrations. No doubt maps are very successful in showing flow of migration but very less effort has been made for statistical estimation of migrants, statistical models and multivariate analysis of the data to understand the complexity of migration and its overall impact on development of the migrants, their families and society. Most of the geographers have not undertaken any holistic study, in other words, rather they have become cartographers of demographers.

The problem of migration falls under both urban and rural planning and that's why it is beyond the reach of a specialist like urban or rural planner. The problems of urban slums (due to high influx of migrant labour) are as important as sustainable livelihood of the rural poor in backward regions. Here the role of geographer becomes pertinent as the discipline of geography which is not only concerned with physical environment but also deals with human's interaction with its surroundings.

Geography provides better perspective of understanding to the problem of migration.

Thus, the geographers need to address this problem in present context seriously keeping holistic view on the problems and micro level interventions for sustainable development of communities. With the help of some important modern tools like remote sensing and geographical information system (GIS), geographers can take lead by working out different possible solutions by mapping the problems correctly. This can help in carving their professional niche in a better way in the present competitive world.

Migration Study in Syllabi of India Universities: A Critical Review

Study of spatial mobility has been an important matter of concern for geographers but it has also been equally dealt by sociologists, demographers, economists, anthropologists and historians. This section aims to examine how the issue of migration is dealt especially in the syllabi of higher education. It is clearly evident that issues of fertility and mortality have got separate space in the syllabus of population studies (Table 15-2). But migration has been taught either with urbanization or human ecology moreover there is hardly any serious attempt to link them. Another point is that in some of the universities' courses advanced study of fertility and mortality studies (Annamalai University) are taught, while migration study has been still at the basic or conceptual stage. Reason might be that India has been going through second and third phase of demographic transition during second half of the last century, hence the fertility and mortality study have been major thrust area of research; and migration was not paid much attention. Migration being the factor of population redistribution that's why only causes, pattern and trend of migration have been studied widely. Interlinking migration with gender and changing scenario of economy, environment and policy is missing. Particularly in the era of globalisation when human beings are taken as a capital, their movement plays significant role in economic and political dynamism and needs special attention. With rapidly changing world economy, social system and increasing complexity of the phenomenon, migration needs interdisciplinary approach of study.

The Indian syllabi need to be restructured in order to develop better understanding among the students about the migration phenomenon. It certainly requires some comprehensive knowledge other than conventional boundaries of Geography. The demographic component has tremendous potential for future research for the geographers. There is a need of systematic planning of the course curriculum starting from the 10 + 2 /intermediate level of school education to the research level. In order to incorporate new dimensions in the curriculum some suggestive measures for redesigning syllabi may be proposed as the following:

Senior Secondary/+2/Intermediate Course

Concept and definition of migration; streams of migration, causes of migration under pull & push theory taking India as a case for study. As NCERT has introduced one chapter dealing with migration in India in class XII book, similarly, other states secondary boards/intermediate councils should also introduce at least one chapter).

Under graduate course

In undergraduate course migration is dealt in various papers like in human geography, population geography and in geography of India. A comprehensive section could be introduced appropriately giving consideration to the following:

1. Conceptual: mobility and migration, migrant and non-migrant, in-migration and out-migration, immigration and emigration, place of origin and destination;
2. Causes, pattern and trend of migration in India
3. Types of migration in dichotomy (international vs. domestic; permanent vs. temporary; voluntary vs. forced migration)—these concepts should be explained in the context of India:

    I. *Permanent, voluntary and international migration*: Indian Diaspora in Africa, Europe, Caribbean Islands;
    II. *Temporary voluntary International migration* i.e. guest worker and migrant on the work permit to USA, Europe, Australia;
    III. *Forced migration*: politically motivated population transfer from Pakistan and Bangladesh during partition of India; Chakma refugee from Bangladesh, Tibetan asylum seekers, Tamilian refugee;
    IV. *Domestic voluntary migration*: Economic migrant like permanent, seasonal and circular migration from less developed regions to more developed regions of India; and

V. *Domestic forced migration*: politically motivated transfer freedom fighters from Indian mainland to Andaman. Displacement caused by environment and development programmes.

4. Brief history of international migration

Post-graduation course

Source and limitation of migration data; direct and indirect measures of migration; theories of migration (classical and neo-classical), qualitative and quantitative method of migration study; migration policies.

Ph.D. programmes

The aim of the course should be to engage students in exploring a range of scholarly debates concerning both theoretical elaborations and empirical studies within the broad field of migration. It should combine surveys of general theoretical and empirical developments with more specific analyses of various cases relating to Indian situation. This will enhance the students" skills, both with regard to their comprehensive understanding and their ability to carry out in-depth analysis. Both quantitative and qualitative researches are equally important. It also requires various research methodologies to study the intensity of problems of migration which may be the part of social science research like survey, case studies, participatory research techniques, ethnographic research, applied medical examinations, anthropological research and archival research.

Some of the important areas for the future research in migration can be identified as generating research collaboration among researchers; integrating the migration studies in India with those studies carried out in other local and global settings; connecting the migration debate in India and in "near abroad" such as neighbouring areas in SAARC region; initiating inter-disciplinary research to address sensitive issues like illegal migration, human trafficking, ethnic conflicts, human rights and security; and developing critical perspectives based on empirical knowledge on current migration policy, besides key issues in theoretical and conceptual framework.

Table 15-2: Presence of migration study in syllabus of Indian universities: A comparative view

University	Course Name	Papers	Position of migration
Annamalai University	MA in Population studies	***I Year*** - 510-Population Growth, Structure and Characteristics - 520-Population Data Sources, Evaluation, Adjustment and Graduation - 530 - Fertility - 540 - Mortality - 550 - Population Ecology, Urbanisation and Migration ***II Year*** - *610* - Research Methodology & Statistical Techniques - 620 - Population and Socio-Economic Changes - 630 - Population Theories - 640 - Population Policy and Programmes - 650 - Population Education & Communication - 651 - Health and Nutrition Education	**550–Population Ecology, Urbanisation and Migration**

Bharatiar University	MPhil/PhD in Population Studies	- Advanced Studies in Mortality and Morbidity, - Advanced Studies in Nuptiality, - Advanced Studies in Fertility, - Migration and Urbanization and - Advanced Studies in Reproductive Health	**Paper: Migration and Urbanization** **Unit II: Migration** Basic concepts definition, its importance in the study of population. Types of migration. Internal, international, temporary and refugee. Sources of data and limitations Trends and differentials. International migration. **Unit III: Measures of Migration** Place of birth approach, place of last residence approach, duration of stay approach, combinations of the three, indirect methods of estimating migration: growth rate method, vital statistics method, life table and census survival ratio methods. **Unit V: Theories of Migration** Theories of migration, Peterson's typology, Regenstein's laws of migration, Stouffer's model of intervening opportunities and competing migrants, Zipf and Kant gravity models.
Banaras Hindu University	BA Geography	Human Geography Population Geography	Types and determinants, Types and determinants of migration, Trend of internal migration in India, International migration.

	MA/MSc Geography	Population and Settlement Geography (Major specialisation group)	Population geography: Fertility, Mortality and Migration Migration theories: Ravenstien and Everetts Lee

V. Conclusion

Migration has been an important issue for social scientists but they have addressed the problems in isolation presenting partial pictures of migration. With increasing complexity in the nature of migration due to neo-liberal globalisation, there is a need to recognise and integrate various dimensions of migration studies like, human trafficking, illegal migration, feminisation, health, national security, ethnic violence, forced migration, migration management and policy formulation. At present there are a lot of gaps in migration studies. Migration which affects both the area of origin (out-migration) and the destination in-migration) has been given little importance in framing any national policy in India (Singh, 2000). The role of geographer becomes very pertinent to internalise the issues of migration study in holistic manner and further redesign the course curriculum accordingly starting from intermediate level to PhD level.

Acknowledgements

The author acknowledges Dr. Sarfaraz Alam and Dr. Upendra Nath Roy for their valuable contribution in the paper. I also acknowledge my heartfelt thanks to Dr. Ravi S Singh, the editor of the book, for his encouragement kind cooperation.

References

Annan, Kofi 2006. The Secretary-General (UNO) address to the UN General Assembly on International Migration and Development, , September 14, New York.

Banerjee, B. 1986. *Rural and Urban Migration and Urban Labour Market: A Case Study of Delhi*. Bombay: Himalaya Publishing House.

Barbora, Sanjay., Theime Susan, Siegmann Karin Astrid, Menon Vineetha and Gurung Ganesh. 2008. Migration matters in south Asia: Communalities and critiques. *Economic and Political Weekly* 43 (24): 57-65.

Bose, Ashish. 2000. Demography of Himalayan villages: Missing men and lonely women. *Economic and Political Weekly* 35 (27): 2361-2381.

Breman, J. 1985. *Of Peasants, Migrants and Paupers: Rural Labour Circulation and Capitalist Production in Western India*. Delhi: Oxford University Press.

Carr-Harris, J. 1993. Eco-health in the rural environment. *Health for the Millions* 1 (2): 2-6.

Census of India. 2001. *Migration Data Abstract on Data Highlights*. New Delhi: Census of India (Ministry of Home, Government of India).

Connell, J., Biplab Dasgupta, Laishley Roy and Michael Lipton. 1976. *Migration from Rural Areas: The Evidences from Village Studies.* Delhi: Oxford University Press.

Dasgupta, Biplab. ed. 1978. *Village Studies in the Third World.* Delhi: Hindustan Publishing Corporation.

Decosas, J., F. Kane, J.K.Anafri, K.G. Sodji and H.U. Wagner. 1995. Migration and AIDS. *Lancet* 346 (8978): 826-828.

Ghosh, P.P. and A.N. Sharma 1995. Seasonal migration of rural labour in Bihar. *Labor and Development*, 31(50): 118-136.

Gill, Indermit 1984. Migrant labour: A mirror survey of Jullunder and East Champaran. *Economic and Political Weekly* 19 (24-25): 961-964.

Goldscheider, Calvin. 1984. *Rural Migration in Developing Nations: Comparative Study of Korea, Sri Lanka and Mali.* London: Westview Press.

Gosal, G.S and G. Krishan. 1975. Patterns of internal migration in India. In *People on the Move*, ed. L.A. Koziniski and R.N. Prothero, 193-206. London: Methuen & Co.

Gulati, Leela. 1993. *In the Absence of their Men: the Impact of Male Migration on Women.* New Delhi: Sage.

Halli, Shiva S., James Blanchard, Dayanand G. Satihal and Stephen Moses. 2007. Migration and HIV transmission in rural south India: An ethnographic study. *Culture, Health & Sexuality* 9 (1): 85-94.

Hann, Arjan De. 1997. Migration and poverty: A case of Bihar. Paper presented in the international conference on Bihar in the World and the World in the Bihar, Dec 16-19, ADRI, Patna.

IDMC. 2008. India: New and Protracted Displacement ongoing in absence of Formalised Response http://www.internal-displacement.org/countries/india (Accessed 23 March 2009).

International Organization for Migration. 2005. International migration trends: Facts and figures. *World Migration Report*,http://images.gmanews.tv/html/research/2007/12/world_migration_report_2005.htm <accessed on 24Feb 2009.

—. 2008. *World Migration Report* http://iom.ch/jahia/webdav/site/myjahiasite/shared/shared/mainsite/published_docs/studies_and_reports/WMR2008/Co_WMR08.pdf. (Accessed 21 March 2009).

—. 2009. Migration & gender. http://www.iom.int/jahia/Jahia/about-migration/developing-migration-policy/migration-gender. (Accessed 23 March 2009).

Jetly, S. 1987. Impact of male migration on rural females. *Economic and Political Weekly* 22 (44): WS47-WS53.

Khandelwal, S.V. and Chorghade S.V. 1994. *Migration of Unskilled labourer (A Study of Causes, Permanency and Assimilation in Nagpur)*. New Delhi: Classical Publishing Company.

Kumar, Ajay. 2000. Growing in and out of poverty, people's own Perception: A case study of Bihar, E-seminar on Poverty, website: http//www.bihartimes.com. (Accessed 1 June 2001).

Kundu, A. and Shalini Gupta. 1996. Migration, urbanization and regional inequality. *Economic and Political Weekly* 31 (52): 3391-3398.

Lama. Mahendra, P. 2000. Internal displacement in India: Causes, protection and dilemmas. *Forced Migration Review* 8: 24-26.

Migration News. 2009. *Remittances and Recession* 15 (1). http://migration.ucdavis.edu/MN/more.php?id=3500_0_5_0). (Accessed 18 March 2009).

Mukherji, Shekhar. 1991. The nature of migration and urbanization in India: The search for alternative planning strategies. In *Dynamics of Population of Family Welfare*, ed. K.B. Pathak and K. Srinivasan, 203-249. Bombay: Himalaya Publishing Corporation.

Munck, Ronaldo. 2008. Globalisation, governance and migration: An introduction. *Third World Quarterly* 29 (7): 1227-1246. URL: http://dx.doi.org/10.1080/01436590802386252. (Accessed 30 September 2008)

Nagrath, Sumati. 2008. Urbanisation: Managing the chaos. *Businessworld***,** 11 July 2008. http://www.businessworld.in/index.php/economy-and-banking/managing-the-chaos.html. (Accessed 28 February 2009).

Oberai, A.S. and H.K.M. Singh 1983. *Causes and Consequences of Internal Migration (A Study in the Indian Punjab)*. Delhi: Oxford University Press.

Ratha Dilip, Sanket Mohapatra and Zhimei Xu. 2008. Outlook for Remittance Flows 2008-2010: Growth expected to moderate significantly, but flows to remain resilient. *Migration and Development Brief 8,* Migration and Remittances Team, The World Bank, November: 3. http://siteresources.worldbank.org/INTPROSPECTS/Resources/334934-1103150151 65/MD_Brief8.pdf. (Accessed 15 February 2009).

Rao, Sudha V. 1992. Population and Development. *Margin* 24(2): 166.

Rodgers, Gerry. 2000. Growth in Western India reduces poverty in Bihar. E-seminar on Poverty. http//www.bihartimes.com. (Accessed 15 May 2001).

Roy, Archana K. and P. K. Nangia. 2003. Left behind women in Bihar. In *Migrant Labour and Human rights in India*, ed. Gopal K. Iyer, 73-84. New Delhi: Kanishka.

Roy, Archana K. and P.K. Nangia. 2005. Impact of male out-migration on health status of left behind wives—A study of Bihar, India. Paper presented in XXVth International Population Conference of International Union for the Scientific Study of Population (IUSSP), July 18-23, in Tours, France. Available on web http://iussp2005.princeton.edu/download.aspx?submissionId=51906

Sajjad, Haroon. 1998. *Employment of Landless Labourers*. Delhi: B.R. Publishing Corporation.

Saxena, D.P. 1977. *Rural Migrattion and Cultural Change*. Bombay: Popular Prakshan Private Ltd.

Sekar T.V. 1997. *Migration and Social Change*. Jaipur: Rawat Publication.

Sen, S.N. 1960. *The City of Calcutta*. Calcutta: Bookland Publication.

Shah, Farida. 1998. *Rural-Urban Migration and Economic Development*. Jaipur: Printwell Publishers & Distributors.

Sharma, H.L. 1986. Distribution of number of migrants from a household. PhD thesis, Banaras Hindu University, Varanasi.

Sharma, J.C. 2002. Inaugural Address on Indian Diaspora. First international conference on Indian diasporic experience: history, culture and identity, January 22–24, Hemchandracharya North Gujarat University, Patan. http://www.ngu.ac.in/center/Dias/index.htm. (Accessed 27 February 2009).

Singh, Indu B. (2000). Bihar, yet to reap the bonus of her backwardness. E-Seminar on Poverty in Bihar. http://www.bihartimes.com. (Accessed 15 May 2001).

Singh, J.P. 1986. *Pattern of Rural-Urban Migration in India*. New Delhi: Inter-India Publications.

United Nations 1993. Demographic and health survey results: Madagascar. *Population Network News*, No. 4, July, Part IV.

United Nations 1995. Programme of action adopted at the international conference on population and development, Cairo, 5-13 September 1994. *Population and Development*, Volume 1, 30. New York: United Nations.

United Nation Population Fund. 2006. A passage to hope: Women and international migration. *State of World Population*.

http://www.unfpa.org/swp/2006/english/chapter_1/index.html. (Accessed 24 February 2009).

United Nations. 2001. Globalization, migration and trafficking: Some thoughts from the south-east Asian region. Occasional Paper No. 1 under UN Inter-Agency Project on Trafficking in Women and Children in the Mekong Sub-region. http://www.un.or.th/TraffickingProject/Publications/globalisation_paper.pdf (Accessed 15 December 2008).

United Nation Population Fund. 2006. *State of World Population: A Passage to Hope—Women and International Migration.* New York: UNFPA.

—. 2008. Negotiating culture: Poverty, inequality and population. *State of World Population.* New York: UNFPA.

Venkatarayappa, K.N. 1991. Influence of urbanization of rural areas. *Indian Journal of Social Research* 35(2): 15.

Yadava, K.N.S. 1989. *Rural-Urban Migration in India: Determinants, Pattern and Consequences.* Delhi: Independent Publishing Company.

Zachariah, K.C., E.T. Mathew and Rajan S. Irudaya. 2001. Social, economic and demographic consequences of migration on Kerala. *International Migration* 39(2): 43-68.

Population Registers (http://unstats.un.org/unsd/Demographic/sources/popreg/popregmethods.htm)

CHAPTER SIXTEEN

RECENT TRENDS AND FUTURE DIRECTIONS IN URBAN GEOGRAPHY

VIJAYA P. KHAIRKAR,
UNIVERSITY OF PUNE

Introduction

Within urban geography the last decade has seen an explosion in qualitative work in terms of the types of methods utilized, the topic addressed and potential theoretical consequences (for review see Findlay and Graham 1991; White and Jackson 1995; Lawson 1998; Findlay and Li 1999; Graham 1999; McKendrick 1999; Mc Hugh 2000; Graham and Boyle 2001; Boyle 2002; Gober and Tyner 2004; Silvey 2004). The purpose of this paper is to highlight the particular promises, challenges and practically using different methods in urban geography. I begin by addressing how different techniques and research methods answer ongoing problems within the sub-discipline for non-traditional ways of thinking about the problems related to urban geography in the Indian context and articulate further how all the techniques I have considered here are valuable for urban geographers. Then I outline two other promising outcomes of the methods, including the potential and the problem areas, spaces wise distribution, and the potential for the empowerment of researchers, planners, governmental as well as non-governmental organizations and local bodies. I use the rest of the paper to provide a review of some of the methodological details of different techniques, with the idea of encouraging more urban geographers to engage with these techniques. Throughout, I contend that different techniques have the capability to further our understandings of urban processes, and to connect this research to ongoing discussion within the broader discipline. Typically the techniques are an ideal method for both exploratory and confirmatory purposes. In other words, these techniques are useful for gaining background information, clarifying ideas, developing questions and understanding the seriousness of the particular problems processes and patterns. Geographical studies of urban problems and regions in India have made remarkable progress since 1991.

Need for Studying Urbanization

The globalization of urbanization is only a recent phenomenon. Most of the developed countries have already achieved a very high level of urbanization. But the process of rapid urbanization has started only recently in the developing countries. As this process is still unfolding in the developing countries it is revealing special features. Therefore, the study of different aspects of urbanization is very important in order to have a proper understanding of the urbanization phenomenon as well as policies to deal with it.

Level of urbanization is closely linked with the level of development. For example, the level of urbanization is much higher in the developed countries compared to the developing countries. Generally, both fertility and mortality conditions are found lower in urban areas than in rural areas. Many demographic and social scientists believe that urbanization will be one of the key factors responsible for lowering the relatively high rates of natural increase found in many developing countries. This is likely to occur not just through the lower fertility rates among the increasing urban proportion of the population, but also through the diffusion of urban fertility values to rural areas. Viewed in the historical context, cities have always acted as centres of innovations, scientific, technical and industrial development. They were the leading centres of power, higher education, arts and culture. They have also served as vehicles of dispersion of many of these ideas and have played a crucial role in the overall process of development.

One important World Bank study (1975, 1 and 45-56) had forewarned long time ago that between 1975 and the year 2000, the cities of the developing countries will be expected to absorb 70 per cent of the projected population increase of 1.3 billion, people most of them poor. By any measure this is a gargantuan task. Although cities are absorbing large numbers of people, they are doing so in a manner that is both inefficient and inequitable. Few cities are prepared for the vast increases that are clearly foreseeable in the next two decades. Unless there are changes in national, regional, and urban policies, the growth of urban areas will not make its maximum possible contribution to social well-being. The study thus talks about the gigantic tasks ahead in the face of very rapid urban growth, especially in developing countries, huge explosion of urban poor, proliferation of urban informal sectors, its inefficient and unjust production systems, overflow of urban unemployment and under-employment, consequent urban decay and breakdown of urban services, and about the required policy changes for ensuring social well-being of the poor. Harris (1992, ix-xxiii) also pinpoints some of these problems, and calls for better urban management. Rodgers (1989, 1-17) has unfolded growing urban poverty and distortions in the labour markets in the metros of the third world.

Most of the metropolitan cities of this region are growing at a very faster rate and most of them are expected to double themselves by 15 to 20 years, according to United Nations estimates (UN 1993b, 1-11 and 110-150). Rapid urbanization and problems of urban squalor, poverty, and unemployment has become main problems in many developing countries (Oberoi 1987, 1-20; Singh 1989, 1-4). More alarming than urban growth is the urban slum population in mega cities that is increasing at still faster rate (McNeil 1995, 1).Huge proliferation of filthy slums and degradation of human life among the teeming millions of slum dwellers in the mega cities are one of the greatest human problems of this century and this problem will tremendously aggravate in the 21st century (UN 1991b, 85-95). For instance, in India, it is feared that the total urban slum population will double from its present 80 million to 160 million within the short span of just 10 to 15 years (NIUA 1988, 46-65; Ministry of Urban Development (India), Urban Commission, 1988, Vol. I, 1-32 and Vol. II, 1-46).

Another recent study has noted that "A decline in rate of growth of urban population has occurred from 46 per cent in 1971-81 to 36 per cent in 1981-91 this requires an explanation that is complex" (Chapman and Pathak 1999, 135-155). These cities are imperfectly related to their respective hinterlands and failed to generate development waves to the rural countryside (Berry 1966, 1-25). These cities face burgeoning urban problems regarding extreme housing storages, water scarcity, sewerage, transport bottlenecks, air pollution, financial deficits, and management deficiency (UN 1986a; 1986c; 1986d; 1987a; Singh and De Souza 1980, 3-56). Urban decay, urban atrophy, and declining level of human consumption are acutely associated with present-day urbanization (Desai and Pillai 1980). "What we need now are fresh ideas, expert advice, social and political will and above all a determined government action" (David 1996, 6). Additionally, very efficient fiscal and administrative management of the large cities like Mumbai (Delhi, Chennai, and Kolkata, as well) are the most essential requirements of contemporary urban planning (Ghorpade 1998, 361-380).

Recent urbanisation in India leads to a new phenomenon of developing countries in general and India in particular. India has the second largest urban population in the world and the highest number of mega- cities (Kaushik 2003). The characteristic feature of urbanization in India is its concentration in a few regions and especially in large cities. These metropolitan cities grow with an enormous and non- controllable speed which has generated a marginal area both of towns as well as countryside which is called rural-urban fringe. The fringe area is the outer sub-urban zone mostly beyond the cities' administrative limits. It is a transitional zone that could be recognized recently on social grounds by the presence of rural and urban groups which are nowadays much diffused. Normally the area is still partly rural and many of the residents live in the country but are not socially and economically part of it. So it is a wide rural area into which residential development is intruding and new industrial sites and other urban uses are in the process of development. Therefore, the main characteristics are the presence of both agricultural land as well as industries that have sporadically cropped up (Verma 2006). Although Ford (1999) states that the distinctive population geography of this growth zone and its unique set of problems and issues remain under- researched, there are some researchers that examined the processes and problems of the fringe area in selected cities of India and all over the world. These are, for example, Ford (1999), Firman (2004), Henderson (2003), Kalra (2005), Keys *et al.* (2007), and Vij (2005). But, according to Firman (2004), most of the empirical studies focus on the developed world. Furthermore, he stresses that contemporary urban transitions in the developing countries are different from those that the developed world experienced a century ago. He points out that in contrast to the situation in the developed countries at that time, the transition processes in developing countries nowadays occur at a lower level of per capita income and in a totally different global economic situation. But still, even if the processes leading to the development of a rural- urban fringe may vary, many problems occurring in these regions remain the same. In the context of the national capital territory of Delhi, Kalra (2005) identifies six stages in the process of transformation from a rural area to an urban village:

1. Rural stage
2. Rural- urban linkage
3. Transformation of occupation
4. Change of land- use
5. Urban village
6. Urban village redevelopment

As urbanization is a never ending process and fringe zones undergo great changes, the challenges vary over time. These variations due to the development of the region and the specific problems caused by this development can be shown with the help of Kalra's (2005) stages of transition.

According to the second stage, "Rural-urban linkage", transportation and infrastructure connecting the village to the city develops. In this stage the cities come closer to the villages with the help of better transport infrastructure and new modes of transportation. So, now there are streets for individual transportation, but also bus lines and trains that connect the village and the city. This is important for the village people as now they have the possibility to work in the city. Therefore, the commuter belt is often used to define the outer boundary of the fringe in that stage. Moreover, village people now have better access to social facilities like hospitals, buses, police stations, colleges, markets, etc. (Kalra 2005). But a medal has always two sides: in the shadow of all these advantages hides the problem of increasing pollution of the air and the environment in general. Besides this, there is a new linkage concerning infrastructure like electricity, water and waste. But here lies one of the most severe problems of the fringe: in case of an electricity or water shortage which is pretty likely in emerging mega-cities, the fringe is the area where the supply is shortened first. Hence, people of the fringe often have to deal with insufficient water supply, sharing of electricity with neighbours and power outtakes. Even if the area is still widely rural, industries develop sporadically along the improved main line of transportation. They are attracted by lower property prices and service taxes than in the city. This leads to the third stage of transformation which is the change in occupation.

Traditionally agriculture has a low value in the job market and is low paid (Mazumdar 2005). So those who have an additional employment besides agriculture in industry or the service sector have higher income levels. Mostly they are the more literate and educated families in the village (Misra 1987). Also some of the farmers begin to start their own business such as transportation activity or educated farmers have been employed in the service sector like primary and nursery teacher or police departments. For those, the increasing income level leads to an increased standard of living (Kalra 2005). This has a great impact on the way of life in more than one way: assured income, education, modernization and change in attitudes. But as the area is still widely rural many families totally depend on agriculture. Moreover peri-urban agriculture is considered to be a temporary job opportunity for rural-urban migrants until they find a formal job (Mazumdar 2005). So there is an increasing gap between a high and low income level and different ways of life which always lead to tensions. With the new industrial and commercial sites, residential development and other urban uses, the change of land use is obvious. So there is a conversion of land from agriculture and open space to non-agriculture uses (Gowda *et al.*1986). Whereas in villages industries tend to pop up, there are also large areas of residential development, the so called suburbs that displace open space or agricultural land (Keys *et al.* 2007). Often farmers are more or less forced to sell their land in order to realize large building projects. Henderson (2003) regrets the absence of research exploring the conflict experienced by farmers in this context. He states that as the fringe contains different competing actors, land use conflicts are likely to occur as seen at the fringe of Perth. Through this example he came to know that the city's administrative bodies do not pay attention to the farmers` interests and therefore they often tend to lose the fight. But as the growth in urban land use is hardly controllable the city does not grow outwards in well- defined patterns. This leads to an incoherent landscape (Verma 2006) which is the main characteristic of the fourth stage.

Moreover the unplanned development of the area brings lack of public facilities, health centres and schools. It also leads to an insufficient providence of sanitary and water disposal services which has severe impacts on health as well as the environment (Vij 2005). The urbanisation of former rural areas also leads to great changes in social and cultural fields. The most obvious change is simply the enormous population growth in peri-urban regions. Ford (1999) states that these regions are the fastest growing regions in the USA, Canada, the UK, Europe and Australia in terms of population. This enormous growth is a combined effect generated in the inner city, in the growth zone itself and in the outlying rural area. One reason for the growth is the stagnation of the city core as people prefer to live in the outskirts in search for better living conditions. The two processes related to this phenomenon are suburbanization and counter-urbanization. Especially in developed countries these are the dominant reasons for the growth of peri-urban zones. In developing countries they are accompanied by another process of utmost importance: rural-urban migration which is also called rural exodus. In this process, the fringe area often is the destination before entering the core area. As most of the migrants are young adults in childbearing age groups they also tend to settle permanently which causes a process called population retention (Ford 1999). Whereas the local agricultural population sticks to their former social life, commuters participate within the city's daily urban system and therefore tend to adapt or keep the urban lifestyle. Especially with increasing income the way of life changes fast. This can be seen for example by the number of cars in a village as an indictor for modernization and by the increasing percentage of literate people which shows the level of education. Moreover, there is a change in attitudes such as loss of traditional values in religion, decline of traditional norms of living in a joint family system and an increasing trend towards nuclear family households (Kalra 2005). This is critically seen by those who do not participate in this trend which are normally less educated or old people. Furthermore, since not everyone experiences an improved standard of living there are still severe problems concerning housing conditions and sanitation which often lead to health problems. Quality of housing and health status are basic determinants in the quality of life for people. Especially lack of clean water, sanitation and health care are the major causes of low life expectancy and high rate of infant mortality. Therefore public health is an important indicator of human welfare. But medical infrastructure is not accessible to a wide section of residents as it is concentrated in cities and out of reach for many people of the fringe. Lower income groups are normally hit hardest as this group is not able to afford the way to a hospital or health centres. Also in the village itself it is them who do not have protected water and proper habitat conditions (Misra 1987).

The last stage in the transformation, urban village redevelopment, implicates changes in the morphology of the area. For example older rural houses become replaced by modern buildings like skyscrapers so that any rural appearance of that area vanishes. After this last stage of transition, the former village cannot be recognized as such anymore, but is totally adjusted to the city. The expanding city constantly absorbs its fringe area and creates a new fringe further from the city centre, argues Ford (1999). Hence, there is a continual outward movement of the fringe area, which shows again the procedural component of urbanization and the ever changing challenges of the fringe zone over time. As shown above, the people of the fringe area have to face severe problems in the fields of economy, land use development, and all kinds of difficulties related to social issues. They get very few help in improving the situation as urban authorities ignore the problems since the fringe area falls outside their administrative limits (Vij 2005). In order to respond adequately to the challenges of the peri-urban growth, a clear understanding of contemporary dynamics is essential. The examination of local conditions and population characteristics within the specific location is crucial if the impacts of processes there are to be understood and planned for. Particularly policy-makers need to understand the dynamics and impacts of the development in the fringe. They need to understand that the fringe varies over time and not all fringe areas around one city can be handled the same way. As a result the challenges for planning vary over space and time and require carefully targeted policies that are sensitive to local conditions (cf. Ford 1999).

The purpose of this paper is to highlight the particular promises, challenges, and practicality of doing focus group research in urban geography. The following methods in details are discussed with reference to the study urban phenomena in geography:

1. The role of qualitative techniques.
2. Quantitative techniques used in urban geography.
3. Remote sensing used in urban geography.
4. Geographical information system used in urban geography.

The Role of Qualitative Techniques

In the study of urban geography, the uses of qualitative methods are not that old, but the authenticity is important and it is more time consuming. You can not get clear picture with handling few cases in detail. But to understand the problem in depth, the qualitative techniques (QTs) appear important in studies aiming to see the urban youth problems, adolescent problems and urban problems with which the social stigma is attached. The issues and problems of urban geography are complicated. Many times getting data is difficult and tedious. In that case, researches have to use qualitative technique. In QT, the data need be extracted with patience. Researchers have to visit study area again and again, sit for hours together with people. Through conversation, personal and group discussion data is collected and projected with quotation marks and with the fictitious name of the respondent in the research. In this way, with the help of QT first hand information can be collected through respondents own language and intensity. It gives, the Planners an opportunity to understand the severe-ness of the problem. The critical problems like, infant mortality, HIV/AIDS, domestic violence, reproductive health, etc. can be dealt with the help of qualitative techniques.

Quantitative Techniques Used in Urban Geography

Normally, geographers used the theoretical approaches of the study. To get perfect result and to compose the reasons or true problems in last two decades, the statistical techniques played an important role in the geographical studies. Even today, the statistical methods are needed and used in computing and analysing the large data for different cities with multitude of variables. To see the availability of social amenities in rural as well as urban areas, one can compute the different indices. To measure development, the human development index, gender development index, poverty index, physical quality of life index, composite index can be calculated and with the help of these indices the seriousness of the problem can be gauged and it becomes easier for planners in governments, non-government organizations and local bodies to prepare the appropriate development plans. Quantitative research is the systematic scientific investigation of quantitative properties and phenomena and their relationships. The objective of quantitative research is to develop and employ mathematical models, theories and/or hypotheses pertaining to natural phenomena. The process of measurement is central to quantitative research because it provides the fundamental connection between empirical observation and mathematical expression of quantitative relationships. To prove the results properly, it needs some concrete answer. To get authentic answers, quantitative techniques help. Urban geography is concerned with human beings; the answer of people may be biased and error may occur in the answer. Sometime many factors act simultaneously where we can not reach at a particular conclusion. In such circumstances, quantitative analyses face some limitations. Many a times, social aspects can not be dealt with quantitative techniques; in that case, researchers have to use some suitable qualitative technique(s).

Use of Remote Sensing and Geographical Information System in Urban Geography

Remote sensing (RS) is a terminology coined recently. It is the science of making inferences about objects from measurements, made from a distance without coming into physical contact with the objects index study. That is, RS refers to any method which can be used to gather information about an object without actually coming into contact with it (Joseph 2005, 1).

With the help of RS techniques we can study the aspects which are important but due to lack of accessibility we can not reach; to solve these problems RS is useful. In India, since the last fifty years, the government is trying to solve problems like infant fertility, neo-natal mortality, and maternal mortality. By and large they are solved but there are some pockets which are still lagging behind in all walks of life. Use of the RS techniques, with high resolution maps, can solve the problems which still exist in the areas deprived in terms of overall development.

This need is driven by continued urbanization. In 2000, about 3 billion people, representing about 40 per cent of the world's population, lived in urban areas. According to the United Nations, urban population will continue to rise substantially over the next several decades, and most of this growth will be in developing countries. The number of million plus cities (i.e. cities with populations of more than 10 lakhs) will increase to 100 by 2025 CE. Thus, there is critical need to understand urban areas to help improve and foster environmental and human sustainability of cities around the world. Over the past decades, the majority of remote sensing work has been focused on natural environments. Applying remote sensing technology to urban areas is relatively new. With the advent of high-resolution imagery and more capable techniques, urban remote sensing is rapidly gaining interest within the remote sensing community. Driven by advances in technology and societal needs, biannual international symposia on remote sensing of urban areas (since 1997) and remote sensing and data fusion (since 2001) have been very successful (Weng 2007, 1).

Urban development and migration from rural to urban areas are common global phenomena. The expansion of the urban areas is a great problem, and if not tackled properly can lead to haphazard urban growth. Remote sensing data with its unique characteristics of synoptic view, repetitive coverage and reliability has opened immense possibility for urban area mapping and its growth (Chandra and Ghosh 2006, 124).

Geographical information system (GIS) is another recent technique in geography, "A powerful set of tools for collecting, storing, retrieving at will and displaying spatial data from the real world" (Burrough *et al.* 2000). With the help of GIS we can prepare maps as per the requirement; the database can be prepared and analysis as well as query can be done with the help of GIS techniques. Decision support system is computer based software used for taking decisions for an ill-structured problem. This tool can be used for effective generation of information on the decision problem from available data and ideas. To enhance the research in urban geography GIS technique can be used effectively. Data attachment and data analysis is more accurate with this tool. As well as the spatial analysis of any problem is possible. For instance, the household-wise availability of toilet facility in rural areas can be seen with this technique after the preparation of household map of a village.

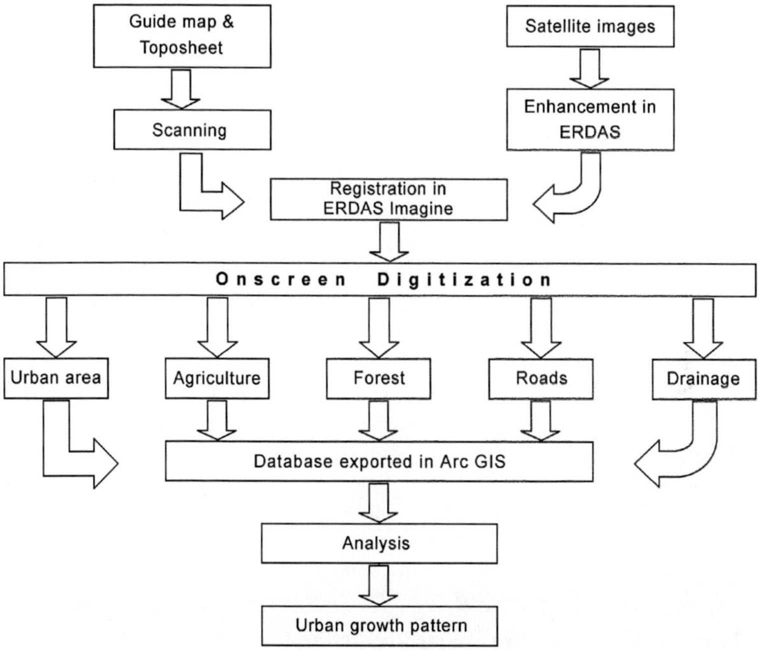

Fig. 16-1: Flowchart for the study of urban growth
(*Source*: Soni 2003)

The Role of Focus Group in Urban Geography

The above techniques are useful for answering certain questions to solve the sector-wise problems of the society; but, the people's participation in solving problems is more important. According to Morgan (1996) focus group is a research technique that collects data through group interactions on a topic predetermined by the researchers. Although once a focus group is set, the conversation between the researchers and the researched often take unexpected turns. The group discussion yields the primary sources of data and group interaction is expected to produce insights that would be difficult to obtain through individual interviews or large scale surveys.

The issues which are connecting urban population, and could not be solved easily, can be dealt with the focus group discussions like, people suffering from HIV/AIDS, people's visit to red light areas, sexual behaviour of adolescent boys, problems related to reproductive health, etc. in urbanism.

Future Directions

Geographers need to pay attention to the importance of the constructed nature of male and female identities and the way that gender is played out through the body, households, communities, labour markets and regional projects of neo-liberal restructuring through which migrants move (Lawson 2000; Price 2001; Domosh and Seager 2001; Silvey 2004b). Still, much work remains to be done in urban geography research to move outside of an essentialist frame of analysis.

Migration, the quality of life and sustainable development of cities need to be studied. The city growth, availability of basic amenities in fringe and changes occurred during last half a century, especially in the post-liberalisation era, should be the focused objective of future studies. All these aspects of urbanization with respect to population should be dealt with the help of RS, GIS and focus group study.

Currently, the processes of world population growth and urbanization are found to be closely interlinked. As a result, it has become increasingly difficult to separate their independent causes and consequences. A continuously larger share of the world's population growth is emanating from and redistributing itself towards the world's urban areas; therefore, many of the problems, frequently referred to as population growth problems, are urbanization-related problems. Therefore, the priority should be given to the research in urban geography.

In this era of globalization, low productivity and depressed social status in rural areas, migration towards cities became important to get better and sophisticated job opportunities. In India, metro and mega cities are flourishing and increasing rapidly but in future there is need to concentrate on the development of small and medium towns in India in terms of infrastructure, social amenities and industries. Researchers should concentrate on the study of these aspects. Migrants' important socio-economic and cultural problems should be given priority as far as urban geography is concerned. Due to globalization, intra, internal and international migration has become important with special reference to youths. Not only small and medium towns but also the receiving regions, which are metro and mega cities, are becoming the hub for industrialization and important centres for education. It therefore becomes imperative to study the availability of amenities in the cities.

Unlike in developed economies, where rural-to-urban migration is connected with an upward shift in the labour force from the agricultural sector to the urbanized industrial sector, labour migration in the developing economies, like India, occurs from rural peasant sector to urban informal sectors, as spatial manifestation of underdevelopment, poverty, and spatial disorganization of the economy (cf. Mukherji 1981, 1-43).

To accommodate growing population and activities in the cities, migrants and core dwellers shift to the area around the city, which is a sprawl of the city. The cheaper rates of land, fresh air, and non-contaminated water are the reasons for people to stay in areas around the city. Many a times government also declares the merger of these villages in the main city, but the problem of infrastructure and amenities remain poor in such areas. If researchers conduct studies on the man-land ratio, availability of amenities and need of the people a truly systematically manner; then, planners can prepare proper city development plans including the surrounding urban fringe areas. Researchers and planners should provide some alternate arrangement for basic amenities like education, health, water, sanitation, transportation etc. following their systematic and authentic research. The cities thus can be developed sustainably for the next fifty to hundred years.

Another serious problem in urban areas in India is rapid growth of slums. During 1984-85, some two million residents of notified slums (i.e. slums that would be granted tenure) had benefited from provisions of new facilities like water supply, sewerage, storm water drains, community baths and latrines, multi-purpose community halls, street lighting, and internal road development. Delhi Development Authority (DDA) is continuing to construct large resettlement colonies for homeless squatters. Attempts are also being made to regularize squatter communities and to gradually upgrade them. In cases where tenure could not be granted, squatters were moved to resettlement colonies on the periphery (UN 1986c, p.25). But even now, after 20 years, these problems persist, and frequent news of violent agitations against demolition of slums in Delhi appears in the newspapers. This remains as a burning issue (Times of India, 5 December, 2000). It needs to greater attention by geographers as well as other researchers.

References

Berry, B.J.L. 1966. Commodity Flows and the Spatial Structure of Indian Economy. Research Monograph No. 111. Chicago: Chicago University Press.

Chandra, A. M. and S. K. Ghosh. 2006. *Remote Sensing and Geographical Information System.* New Delhi: Narosa Publishing House.

Chapman, Graham and Pushpa Pathak. 1999. Indian urbanization and the characteristics of large Indian cities revealed in the 1991 Census. In *Urban Growth and Development in Asia,* Vol. I & II, ed. G.P. Chapman, A.K. Dutt and R.W.Bradnock, 135-153. London: Ashgate.

David, M.D. 1996. Bombay's urbanization. In *Urban Explosion of Mumbai: Restructuring of Growth*, ed. M.D. David, 1-6. Bombay: Himalaya Publishing House.

Desai, A.R. and S.D. Pillai. ed. 1980. *Slums and Urbanization*. Bombay: Bombay University Press.

Domosh, M. and J. Seager. 2001. *Putting Woman in Place: Feminist Geographers Make Sense of the World.* New York.: Guilford Press.

Findlay, A. and E. Graham. 1991.The challenge facing population geography. *Progress in Human Geography* 15: 149-62.

Findlay, A. and F. Li. 1999. Methodological issues in researching migration. *Professional Geographer* 51: 50-59.

Firman, T. 2004. Demographic and spatial patterns of Indonesia's recent urbanisation. *Population, Space and Place* 10: 421- 434.

Ford, T. 1999. Understanding population growth in the peri-urban region. *International Journal of Population Geography* 5: 297-311.

Ghorpade, K.R. 1998. Mumbai Metropolitan Region, India. In *Metropolitan Governance and Planning in Transition: Asia-Pacific Cases* (UNCRD Research Report Series, No. 31), ed. Josefa S. Edralin, 361-380. Nagoya, Japan: United Nations Centre for Regional Development.

Gober, P. and J. Tyner. 2004. Population geography. In *Geography in America at the Dawn of the 21st Century*, ed. G. Gaile and C. Willmott, 185-199. Oxford: Oxford University Press.

Gowda, P. B. and P. D. Mahadev. 1986. Impact of metropolitan centres on rural transformation: A case study of villages around Bangalore. In *Contributions to Indian Geography: Urban Geography*, ed. P. D. Mahadev, 135-146. New Delhi: Heritage Publishers.

Graham, E. and P. Boyle. 2001. Editorial introduction: (Re)theorizing population geography—Mapping the unfamiliar. *International Journal of Population Geography* 7: 389-394.

Harris, Nigel. ed. 1992. *Cities in the 1990s: The Challenge for Developing Countries*. London: University College London Press.

Henderson, S. 2003. Regulating land use conflict on the urban fringe: Two contrasting case studies from the Australian poultry industry. *Australian Geographer* 34: 3-17.

Joseph, G. 2007. *Fundamentals of Remote Sensing* (Second Edition). Hyderabad: University Press.

Kalra, R. 2005. Rural-urban transformation in the National Capital Territory of Delhi. In *Urban and regional development in India*, Vol. I, ed. B. Thakur *et al.*, 197-216. New Delhi: Concept Publishing Company.

Kaushik, S. 2003. Trends of urbanization in India: Some insights from the census 2001. *Geographical Review of India* 65: 181-188.

Keys, E., E.A. Wentz and C.L. Redman. 2007. The spatial structure of land use from 1970- 2000 in the Phoenix, Arizona, Metropolitan Area. *The Professional Geographer* 59: 131-147.

Lawson, V. 2000. Arguments within geographies of movement: The theoretical potential of migrants' stories. *Progress in Human Geography* 24: 173-189.

Mazumdar, M. 2005. Role of urban and peri-urban agriculture in metropolitan city management. In *Urban and regional development in India*, Vol. I, ed. B. Thakur *et al.*, 150-196. New Delhi: Concept Publishing Company.

McKendrick, J. 1999. Multi-method research: An introduction to its application in population geography. *Professional Geographer* 50: 40-50.

McNeil, Mary 1995. Bombay: One City-Two Worlds. *The Urban Age*: *The Human Environment of Cities* 3 (2): 3-26.

Misra, H. N. ed. 1987. *Rural Geography*. New Delhi: Heritage Publishers.

Morgan, D. 1996. Focus groups. *Annual Review of Sociology* 22: 129-152.

Mukherji, Shekhar. 1981. *Mechanisms of Underdevelopment, Labour Migration and Planning Strategies in India*. Calcutta: Prajna.

National Institute of Urban Affairs (NIUA) 1988. *State of India's Urbanization*. New Delhi: NIUA.

Oberoi, A.S. 1987. *Migration, Urbanization and Development*. Geneva: ILO.

Price, M. 2001. The kindness of strangers. *Geographical Review* 91: 143 - 150.

Rodgers, Gerry, ed. 1989. *Urban Poverty and Labour Market*. Geneva: ILO.

Silvey, R. 2004a. On the boundaries of a subfield: Social theory's incorporation into population geography. *Population, Space and Place* 10: 303-308.

Silvey, R. 2004b. Power, difference and mobility: Feminist advances in migration studies. *Progress in Human Geography* 28: 490-506.

Singh, Ajit 1989. Urbanization, Poverty and Employment: The Large Metropolis in the Third World, ILO Working Paper. Geneva: ILO.

Singh, Andrea M. and A. de Souza 1980. *The Urban Poor, Slum, and Pavement Dwellers in Major Cities of India*. New Delhi: Manohar.

Skop, E. 2006. The methodological potential of focus groups in population geography. *Population, Space and Place* 12: 113-124.

Soni V.K. 2003. *GIS for site suitability in urban planning,* M. Tech. Thesis, Department of civil Engineering, IIT Roorkee, Roorkee.

Times of India 2000. "Delhi". December 5, Bombay edition.

United Nations 1986a. *Population Growth and Policies in Mega Cities, Calcutta*. New York: UN.

—. 1986c. *Population Growth and Policies in Mega Cities, Delhi*. New York: UN.

—. 1986d. *Population Growth and Policies in Mega Cities, Bombay*. New York: UN.

—. 1993b. *World Urbanization Prospects*: *The 1992* Revision. New York: UN.

Verma, L.N. 2006. *Urban Geography*. Jaipur/New Delhi: Rawat Publications.

Vij, M. 2005. Changing Rural Land Values at the Urban Fringe of Bhopal. In *Urban and regional development in India*, Vol. I, ed. B. Thakur *et al.*, 231-243. New Delhi: Concept Publishing Company.

Weng, Qihao and Dale A. Quattrochi, ed. 2007. *Urban Remote Sensing*. Boca Raton, Florida: CRC Press.

White, P. and P. Jackson 1995. (Re)Theorizing population geography. *International Journal of Population Geography* 1: 111-123.

World Bank 1975. *The Task Ahead for the Cities of the Developing Countries*. Bank Staff Working Paper No. 209. Washington, DC: World Bank.

CHAPTER SEVENTEEN

URBANISATION, URBAN GROWTH AND URBAN MANAGEMENT IN INDIA: SHIFTING PARADIGMS

ARUN K. SINGH,
BANARAS HINDU UNIVERSITY

Introduction

The most striking feature of India's urbanization is its long history of around 5000 years. Indian urban centres have been playing significant role in shaping the cultural landscape. While talking about the causative factors underlying the urbanization processes, Ramachandran (1989) remarked that during the prehistoric period the urbanization was synonymous with the rise and fall of civilization (cultural process); from ancient to the British period, it was related with the rise and fall of the kingdoms and dynasties (political process); and, in the post independence period it has been associated with industrialization and economic development (economic process). Besides, rural areas have also been converted into urban bodies by the competent authority from time to time (administrative process). So, urbanization is a function of various forces operative over space at different points of time.

Despite a long history of urbanisation, India still remains relatively less urbanized country with 27.78% of the population living in urban areas as compared to the world average and other developed/developing countries. At the world level the degree of urbanization is 42 %. In developed countries like United Kingdom, Germany, Japan, Brazil, United States and France the level of urbanization is above 70% while in the developing countries it is 37%. As compared to developed or developing countries the level of urbanization in India (27.78%) seems to be less but the absolute Indian urban population is gigantic (285.4 million) next only to China and the contribution of urban centres in the national economy is immense. It is estimated that the contribution of urban sector to the GDP has increased from 29 % in 1951 to 65 % in 2001 and 90% of the revenue comes from the urban sector. Cities are considered to be the engine of economic growth and provider of jobs, services and assurance for better quality of life.

Pattern of Urbanisation

Pattern of urbanisation in any spatial unit is measured in a number of ways. The most common and the best-understood method is the percentage of urban population to the total population. Table 17-1 provides trend and pattern of urbanisation for the last 50 years (1951-2001). Till 1991 Maharashtra was the most urbanized state of India. In 1981 Tamil Nadu followed Maharashtra and in 1991 Gujarat replaced Tamil Nadu. The result of 2001 census presents a different picture from the urbanisation point of view. People living in towns and urban agglomerations account for 27.78 %. Tamil Nadu surpassed both Maharashtra and Gujarat and became the most urbanized state. Himachal Pradesh (9.8%) is the least urbanized state followed by Bihar (10.5%). The least urbanized states have the characteristics like hilly and mountainous topography, larger proportion of tribal population and its inaccessibility with respect to metropolitan cities (Ramachandran 1989). The regional pattern of urbanisation has remained quite stable for the last 50 years.

Table 17-1: Level of urbanisation in Indian States and Union Territories (1951-2001)

States	1951	1961	1971	1981	1991	2001
1. Andhra Pradesh	17.4	17.4	19.31	23.25	26.84	27.08
2. Arunachal Pradesh	N.A.	N.A.	3.7	6.32	12.21	20.41
3. Assam	4.2	7.0	8.82	9.88	11.09	12.72
4. Bihar	6.8	8.4	7.97	9.84	10.40	10.47
5. Chattishgarh	N.A.	N.A.	10.38	14.69	17.40	20.08
6. Goa	N.A.	N.A.	26.44	32.46	41.02	49.77
7. Gujarat	27.2	25.8	28.08	31.08	34.40	37.35
8. Haryana	17	17.2	17.66	21.96	24.79	29.00
9. Himachal Pradesh	6.4	6.3	6.99	7.72	8.70	9.79
10. Jammu Kashmir	14	16.7	18.59	21.05	22.76	24.88
11. Jharkhand	N.A.	N.A.	16.01	20.29	21.25	22.25
12. Karnataka	22.9	22.3	24.31	28.91	30.91	33.98

13. Kerala	13.5	15.1	16.24	18.87	26.44	25.97
14. Madhya Pradesh	12	14.3	18.58	22.34	25.27	26.67
15. Maharashtra	28.8	28.2	31.17	35.03	38.73	42.4
16. Manipur	0.5	8.7	13.19	26.44	27.69	23.88
17. Meghalaya	9.7	15.3	14.55	18.03	18.69	19.63
18. Mizoram	N.A.	N.A.	11.36	25.17	46.2	49.5
19. Nagaland	1.9	5.3	9.95	15.54	17.28	17.74
20. Orissa	4.1	6.3	8.41	11.82	13.43	14.97
21. Punjab	21.7	23.1	23.73	27.72	29.72	33.95
22. Rajasthan	18.5	16.3	17.63	20.93	22.88	23.38
23. Sikkim	N.A.	4.2	9.37	16.23	9.12	11.1
24. Tamil Nadu	24.4	36.7	30.26	32.98	34.2	43.86
25. Tripura	6.7	9.0	10.43	10.98	15.26	17.02
26. Uttaranchal	N.A.	N.A.	N.A.	N.A.	N.A.	25.59
27. Uttar Pradesh	13.6	12.9	14.02	18.01	19.89	20.78
28. West Bengal	23.8	24.5	24.75	26.49	27.39	28.03
29. Andaman and Nicobar Islands	N.A.	N.A.	22.77	26.36	26.8	32.67
30. Chandigarh	N.A.	N.A.	90.55	93.6	89.69	89.78
31. Dadar Nagar Haveli	N.A.	N.A.	00	6.67	8.47	22.89
32. Daman Diu	N.A.	N.A.	N.A.	N.A.	46.86	36.26
33. Natioanal Capital Territory	N.A.	N.A.	89.70	92.84	89.93	93.01
34. Lakshdeep	N.A.	N.A.	N.A.	46.31	56.29	44.47
35. Pondicherry	N.A.	N.A.	42.02	52.32	64.05	66.57
INDIA	17.3	18.2	20.22	23.73	25.72	27.78

Note: N.A.: Data not available.

(*Source*: Census of India 1951, 1961, 1971, 1981, 1991, 2001)

The level of urbanisation has been increasing steadily. The western and southern states are more urbanized than the northern and eastern states. All the four southern states Andhra Pradesh, Karnataka, Tamil Nadu and Kerala and two western states Maharashtra and Gujarat have the level of urbanisation above the national average whereas in northern India only Punjab and in eastern India only West Bengal have higher level of urbanisation than the national average. The state-wise variation in the level of urbanisation resembles the pattern of industrial and agricultural growth (NIUA 1988). In spite of the fluctuating trend Delhi National Capital Region and Chandigarh have witnessed the level of urbanisation around 90%.

Growth of Towns and Urban Population

One of the major features of Indian urbanisation is heavy top urban hierarchy and the fast metropolitan growth. This phenomenon can be studied with the help of distributional pattern of urban population and urban centres under different population size categories. Indian census classifies the urban centres into six categories i.e. class I city (population 100000 and above), class II towns (population 50000-99999), class III towns (20000-49999), class IV towns (10000-19999), class V towns (5000-9999), and class VI towns (less than 5000 population). In 2001 there were 4368 towns of which 393 were class one cities. There has been a constant increase in the number of such towns as well as their share in urban population. The pattern of urbanization in India is characterized by continuous concentration of population and activities in large cities (Kundu 1983).

Table 17-2: Category wise number of towns in India (1951-2001)

Category of Towns	1951	1961	1971	1981	1991	2001
Class I	76	102	148	218	300	393
Class II	91	129	173	270	345	401
Class III	327	437	558	743	947	1151
Class IV	608	719	827	1059	1167	1344
Class V	1124	711	623	758	740	888
Class VI	569	172	147	253	197	191
Total	2843	2365	2590	3378	3768	4368

(*Source*: Census of India 1951, 1961, 1971, 1981, 1991, 2001)

The Class I cities possessed more than half of India's urban population in 1961 and presently these towns account for more than two third (68.5%).Number of class II,III and IV towns have also gone on increasing but contrastingly the proportion of urban population has either remained fluctuating or declining. The demographic share of class IV, V and VI towns have significantly come down since 1951. Presently, these 2433 towns (55% of the total urban centres) are accounting for only 10% of urban population. This changing pattern of urban concentration is seen as an outcome of the faster growth rate of larger cities and fluctuating or declining growth of smaller towns, city size jump as well as changing census parameters over census years.

The emergence and dominance of metropolitan cities (population above 1 million) is another dimension of India's urbanisation. In 1901, there was only one metropolis Calcutta; in 1951 the number increased to 5; in 1981 it further increased to 12; and, in 2001 the number of metropolitan cities is 35. Similarly, the proportion of urban population living in these cities has also increased from 12.1% in 1901 to 19% in 1951 and 37.8 % in 2001.

Table 17-3: Percentage of population under different categories of towns (1951-2001)

Category of Towns	**1951**	**1961**	**1971**	**1981**	**1991**	**2001**
Class I	44.6	51.4	57.0	60.4	65.2	68.5
Class II	10	11.2	11.0	11.6	11.0	9.6
Class III	15.7	16.9	17.0	14.4	13.2	12.4
Class IV	13.6	12.8	10.0	9.5	7.8	7
Class V	13	6.9	4.5	3.6	2.1	2.3
Class VI	3.1	0.8	0.5	0.5	0.3	0.2
Total	100	100	100	100	100	100

(*Source*: Census of India 1951, 1961, 1971, 1981, 1991, 2001)

In-depth analysis of over all urban population in various categories of towns and cities, their spatial, economic, social and political dimensions and dynamism depicts distinct features of post-Independence urbanization in India .They are expressed in terms of :

- Rapid increase in population of class I cities;
- Significant increase in the number of class one towns and decrease in class VI towns;
- Accelerated rural-urban migration leading to over-congestion and heavy pressure on urban infrastructure and amenities;
- Building of new capital cities like Chandigarh, Gandhinagar, and Bhubaneshwar;
- Construction of new industrial cities and townships like Bokaro, Bhilai, Rourkela, etc.;
- Physical expansion of large cities to form urban agglomerations;
- Deepening of urban poverty and unprecedented increase in squatter and slum population;

- Introducing city planning to ensure orderly development of towns and cities;
- Increased contribution of urban areas in gross domestic product;
- Urban poverty will overtake rural poverty in percentage terms if the same trend continues;
- Reducing biasness related to exclusiveness and elite centric urban behaviour;

Impact of Urbanization

The impact of urban growth on space, environment and quality of life has been tremendous. Large segments of urban population do not have access to basic amenities like drinking water, sanitation, basic health services and education. The impact of urbanisation may be considered in the context of urban water supply, sanitation, solid waste management, level of environmental pollution, extent of urban poverty and urban crime etc.

Water supply

As per the set standard 40 lpcd water is required in rural areas, 75 lpcd in the towns having no sewerage system, 135 lpcd for urban areas with sewerage system and 150 lpcd in mega cities. According to the 1991 census, 81.4 % of urban household had access to safe drinking water but 40 million persons (18.6%) were reported to be without access to safe water supply. Reports claim that in metropolitan cities like Delhi, Calcutta, Mumbai and Hyderabad, the water supply is more than 200 lpcd but it hides the intra-cities inequities in terms of access, quality and quantity of water supply. India's three largest cities (Mumbai, Delhi and Chennai) are worst in terms of the duration of daily water availability varying between 4 to 5 hours. There are about 80 class I towns in Tamil Nadu, Andhra Pradesh, West Bengal, Rajasthan and Maharashtra which have got per capita supply of less than 75 lpcd. The daily per capita supply of water in Bangalore is about 75-80 litres and in Chennai it is about 70 litres. In Delhi per capita water availability is 200 lpcd but about 30% of the city's population receive only 25 litres or less. Another problem of drinking water in urban sector is that of leakages which accounts for 25-50 % loss in water supply system thereby creating further shortages. Low water pressure and intermittent supplies allow back-syphonage and contamination. Since around 60-70 % of drinking water is drawn from surface streams, their pollution by discharge of domestic and industrial wastes is a direct threat to public health.

Sanitation

Sanitation covers arrangements of drainage of rainwater and effluents, collection and disposal of garbage and removal of human excreta. Proper sanitation is a necessary condition for improvement in general health condition and a better quality of life. But, in the past water supply has received much attention as compared to sanitation aspect. Nearly three-fourths of the population living in cities has no access to any human waste collection and disposal system. The sewerage system exists only in 60 Class I Indian cities. Less than half of the total sewage is collected and only 30-40 % of the collected waste is treated properly. Only 15 % of the urban population has access to private toilets. More than half of urban population particularly in small and medium towns resort to open defecation. A national programme of low cost sanitation aimed at elimination of manual scavenging has been taken up. This seeks to replace about 6 million dry latrines by sanitary latrines in 3600 towns. Low cost sanitation is not considered a total substitute. Due to inadequate sewerage and lack of water treatment facilities, pollutants enter ground water, rivers and other water sources causing water-borne diseases.

Solid waste management

Economic development, urbanisation and improved living conditions in the cities have led to the increase in the quantity and complexity of generated waste. Indian cities and towns are estimated to generate about one lakh metric tones of solid waste every day. The per capita waste generation varies between 0.2 kilograms (small towns) and 0.6 kilograms (large cities) per day, and the current urban MSW generation is estimated to be approximately 0.49 kilograms per capita per day. The collection efficiency is 70-90% in metropolises and less than 50% in smaller towns. Mumbai generates about 3200 tonnes of garbage of which about 97% is collected. The collection varies from about 90% in Delhi and Chennai, about 75% in Bhopal, Calcutta and Hyderabad, 70 % in Kanpur and 68% in Bangalore. Most of the solid wastes that are collected end up in open dumps, sanitary landfill or blocking drainage system threatening both surface water and ground water quality. Solid wastes create one of the most visible environmental problems in low-income areas. These problems are directly linked to inadequate planning, finance and management capacity at the local level.

Urban local bodies (ULBs) spend about Rs. 500 to Rs.1500 per ton on municipal solid waste for collection, transportation, treatment and disposal. About 60 to 70 % of this amount is spent on collection, 20-30 % on transportation and less than 5 % on final waste disposal. Hardly any attention is given to scientific and safe disposal of waste (CPHEEO Manual 2000).

Environmental pollution

Air pollution is the most common form of environmental problem which occurs due to the intense anthropogenic activities. The three major activities which contribute towards the air pollution are domestic, commercial and industrial. Industrial and vehicular emissions are the most important sources of air pollution. Trends indicate that transport in the cities is the key factor responsible for 60-70 % of the local air pollution. Due to the improvement in socio-economic status, desire for more ease and comfort, growing materialistic lifestyle and inadequate as well as inconvenient mass transport system in cities have led to considerable increase in number of motor vehicles.

Table 17-4: Growth of registered motor vehicles (lakh)

Period	Mumbai	Kolkata	Delhi
1974-75	2.21	1.26	2.35
1984-85	4.41	3.11	8.41
1993-94	6.08	5.45	22.39
1998-99	8.60	6.64	30.33

(*Source*: CMIE, 1995 and HUDCO, 2001)

In the last fifty years (1951-2001) the growth in number of vehicles has been 11% per annum. Data available on the number of motor vehicles during1974-99 reveal that there is more than ten-fold increase in Delhi. Similarly, Mumbai and Kolkata have witnessed three and four times increase in the number of vehicles, respectively. These vehicles contribute about 80% of carbon monoxide pollution, 90% of oxides of nitrogen and 65% of hydrocarbons. According to the estimates made by the National Environmental Engineering Research Institute (NEERI, Nagpur), in Delhi and Mumbai the major sources of air pollution are the vehicular emission.

Poverty

Poverty in India is usually measured in terms of calorie consumption level which has been fixed as 2100 calories pcpd in urban area and 2400 in rural areas. With the increase in the level of urbanization and the proportion of urban population concentrating in the metropolises, the problem of urban poverty has become very significant. In the early 1950s, urban poverty in India was viewed as a fallout of rural poverty. This was the first generation of urban poor. Now, these areas have second or third generation of urban poor who have inherited poverty. A study by National Institute of Urban Affairs (1987) found that 68% of their samples of urban poor were women and children mainly belonging to SC/ST and minorities. A study for the Planning Commission by NCHSE, Bhopal concluded that 18 occupational categories account for the bulk of the workers belonging to the poor household and 50 occupational categories account for 90% of the total number of workers. They are engaged in transport (rickshaw, auto, head loaders), constructions (brick layers, painters, carpenters and casual workers), business and manufacturing in micro units.

According to a survey, 32% of Indian urban population lives in slums as against 22% of the world average. Most of them living below the poverty line. Calcutta is the poorest city where 89% of the population lives below poverty line. It is followed by Bangalore (73%), Indore (68%) and Mumbai (55%).

Evidence shows that other manifestation of poverty in terms of access to housing and basic services is equally serious. In 1991, nearly 18.6% of persons in urban areas were without access to safe water supply and 36.1% without any form of basic sanitation. Housing deficit in urban areas is about 7.57 million units, and is particularly much more severe for economically weaker sections of population. Majority of the poor largely get accommodated in slums and informal settlements. Attempts to ameliorate the living conditions by providing subsidies for access to urban infrastructure services have not been effective because of the existing socio-political practices.

Urban violence

Urban India, particularly the large and metropolitan cities, is increasingly becoming a melting pot for different strata of society. A constant increase in crime is reported in such cities attributed to a variety of factors like unchecked migration, illegal settlements, increasing sense of relative deprivation, disparities in terms of various socio-economic opportunities and also uneven income distribution, etc. It is seen that the incidence of total cognisable crimes in 1998 reported in Delhi was highest (59,011) followed by Mumbai (30,017) and Bangalore (29,265). Amongst other cities, Kanpur reported increase of 44.3 per cent over the figures of 1997, followed by Madurai (23.9%). Of the total murder cases reported, Delhi had the highest (523) sharing 18.1 per cent amongst all cities followed by Mumbai (365). Pune city recorded significant increase of 52.3 per cent murder cases over previous year. Again, Delhi stands first in terms of reported 728 robbery incidences. The cities of Jaipur (1,561) and Pune (1,195) reported more riots cases than any other city. Delhi reported the highest rape incidence (365).

The root causes for these problems are accumulation of population in cities particularly in big cities and failures to formulate and implement policies for planned urban and rural development. There is also a lacking of coordination among town planners, policy makers, urban administrators, academician, legal experts, NGOs and CBOs as well as community leaders on the management of cities. Due to these lapses, cities often face social unrest and management crisis. We cannot leave our cities depend on market forces for their growth and development because these forces cannot ensure a balanced and sustainable development. Market forces do not respect social equity and environmental concerns if they are left free to operate themselves. At the same time in the age of liberalisation and globalisation, we cannot plan our cities without considering the market forces and their role in urban management. Since urban management is multi-dimensional in nature, it needs multi-disciplinary approach. Meeting the needs of India's soaring urban population and consequent problems will continue to be a strategic policy matter in the field of urban management. Geographers with a better understanding of the physical and cultural landscape as well as with inter-disciplinary training, spatial vision, social welfare orientation and ecological mind set can contribute significantly in the formation of strategies for sustainable development of cities.

India's Urban Vision and the Agenda of Urban Management

Government of India's urban vision is expressed as "to promote sustainable, efficient, well managed and regionally balanced urban settlement system in the country. A rural-urban continuum should replace the existing rural-urban dichotomy and facilitate emergence of an articulated human settlement with a well defined regional hierarchy that integrates and strengthens the urban centres and surrounding rural areas. The policies, strategies and actions should focus on improving the functioning of urban local bodies, improving urban planning, urban management and governance, reducing poverty, protecting environment, promoting healthy and affordable houses, cost effective and efficient infrastructure and service systems with an overall objective of improving the quality of life in the cities" (*The State of Indian Cities 2001*).

Urban management may be perceived as a set of instruments, instructions, activities and tasks which assures that a city function accordingly and appropriately, basic urban services are provided to the city dwellers and the various private, public and community stakeholders are performing their roles and responsibilities in a harmonic manner. It facilitates co-ordinated efforts and believes in the efficiency, efficacy and equity in the distribution of resources and services. Urban management refers to policies, plans, programmes and practices that try to assure that population growth is matched with available infrastructure, housing and employment. Access to these services depends very much on the initiative of the private sector which are affected by policies of the public sector and by functions that only local governments are eligible to carry out (Claudio 2003).The structure, roles, tasks and functions of local governments and the way responsibilities are assigned to different agencies and stakeholders affect the quality and efficiency of urban management. Equally important is the capacity, quality and quantity of human, material and financial resources without which the urban management practices will turn to fiasco.

Urban Management in India: Paradigm Shift

The history of urban management and governance in modern India dates back to the year 1688 CE with the formation of Madras City Corporation. Its role was to levy taxes for construction of jail and school buildings and for taking necessary measures for protection of the residents and taking responsibility for the payment of salaries to its employees. The Charter Act of 1793 led to the establishment of municipal administration in the Madras, Bombay and Calcutta with the Governor General having the power to levy taxes on lands and houses and to provide for the maintenance of roads, scavenging and police. The second phase of urban governance began with Viceroy Lord Ripon in 1882 who pioneered the promotion of local self-government in India.

The fundamental principle for urban management was to make these bodies representative in nature and empower them to raise their own revenue. The Government of India Act 1919 strengthened the process of democratization of local bodies in India at the grass-root level and provided opportunity for local leadership. The Government of India Act 1935 paved the path for democratic decentralization in the urban local bodies. Nomination of members in local bodies was abolished and attempts were made to give executive functions to urban local bodies.

In the post-Independence era, constitutionally the management of urban affairs being a subject of the concurrent list was put under the domain of the state. Balwant Rai Mehta Committee report (1956) provided the frame for a three-tier structure of governance for strengthening political pluralism at the grass-root level and enabling the local leadership to deal with local problems. Following this, a number of state governments adopted recommendations of the Committee and enacted state Acts for local government in rural areas but the urban local governance was left behind as the Constitution of India had made no specific mention about it. Ashok Mehta Committee (1978), Singhavi Committee (1986) and Sarkaria Commission (1988) were constituted to find ways of strengthening local governance by providing an effective institutional framework at the local level. The Sarkaria Commission had categorically asserted the need for local bodies in rural and urban areas. There has been a deliberate effort on the part of the state governments to weaken the grass-root level democracy and minimize the role of urban local bodies. Practically speaking, prior to the 74th CAA, the existence of urban local bodies as well as their functionaries at various levels were at the goodwill of the state and for all practical purposes they functioned as an extended arm of the district administration.

Since the last decade of twentieth century, it has been widely accepted that cities are the engine of growth contributing around 65% of the total GDP. They can play significant role in macro economic development if planned, managed and governed properly. A series of conferences like United Nations Conference on Environment (Rio Summit 1992), Habitat II Conference (Istanbul 1996) and Campaign for Good Urban Governance (2000) have accepted that managing the urban affairs are not the domain of the state but of the citizens and there is a need for innovative partnership. Traditional functioning of government needs to be tackled in non-traditional ways. This changed mindset helped to promote the principles of decentralization, civil society participation, public-private partnership and the concept of "Inclusive City". The management of urban affairs in India requires pro-active interventions at local level, well trained functionaries, policy instruments and availability of the funds which will determine the direction and destiny of the urban future.

74th CAA: A Watershed in the History of Decentralised Urban Management

74th CAA popularly known as the first generation of urban reforms gives urban local bodies (ULBs) a constitutional status, ensuring their regular election at regular interval and initiating participatory governance by empowering people and ensuring their commitment and involvement in planning and decision-making process.

Within the given scenario the present day urban management and governance needs to become a collaborative efforts of various actors and sectors including government and non-government agencies, public private partnership, NGOs and CBOs and Weaker section of the society. This sets a paradigm shift of urban management from centralized to decentralized form of urban management, from rowing(direct delivery of services) to steering (shaping the public policy) and attaining participative decision making with the specific goals of equity, efficiency, transparency and accountability. The Act recognizes urban local bodies as the principal representative platform for urban population and visions a changed perspectives of development i.e. development through decentralization and people's participation.

The Act assigns certain functions to the urban local bodies that relates to the preparation of micro plans for economic development, environmental protection, social justice as well as for implementation of various development schemes. These functions are appended in the 12th Schedule and list the following eighteen functions:

- Urban planning including town planning;
- Regulation of land use and construction of buildings;
- Planning for economic and social development;
- Roads and bridges;
- Water supply for domestic, industrial and commercial purposes;
- Public health, sanitation, conservancy and solid waste management;
- Fire services;
- Urban forestry, protection of the environment and promotion of ecological aspects;
- Safeguarding the interests of weaker sections of society, including the handicapped and mentally retarded;
- Slum improvement and up-gradation;
- Urban poverty alleviation;

- Provision of urban amenities and facilities such as parks, gardens, playgrounds, etc.;
- Promotion of cultural, educational and aesthetic aspects;
- Burials and burial grounds; cremations, cremation grounds and electric crematoriums;
- Cattle ponds, prevention of cruelty to animals;
- Vital statistics including registration of births and deaths;
- Public amenities including street lighting, parking lots, bus stops and public conveniences; and
- Regulation of slaughterhouses and tanneries.

The inclusion of 12th Schedule in the 74th CAA is to demarcate and specify the functions of local authority to facilitate their autonomy and plan according to the need of the area in a proactive and vibrant manner. This is especially useful for newly urbanized and urbanizing areas, i.e. small towns and cities with poor quality of life and higher level of poverty.

Towards responsive spatial planning

In order to frame a sound urbanisation policy to achieve the urban vision a planner or policy maker has to take into cognizance of the fact that rural villages and other urban settlements in the geographic space form the integral parts of the entire fabric of human settlements. These two geographical entities have a dynamic socio-economic relationship where urban areas create demand for rural produce and in turn provide goods and services to rural hinterland. Hence intra- as well as inter-urban along with intra-regional urban development strategies need to be emphasized by creating proper forward backward linkages. District administration has to take into account the development of rural and urban areas within their jurisdiction. The 74th CAA provides a framework for physical and economic development of the urban area through micro level participative planning. It envisages three tier planning structures in states; (i) local level planning by *panchayats* and municipalities, (ii) regional level planning by district planning committee and metropolitan planning committee respectively and (iii) state level planning by the state government. Article 243-ZD dealing with the district planning committee, makes spatial planning a mandatory function of both the urban and rural local bodies and is required to consolidate the rural and urban plans to prepare draft district development plan.

Geography, Geographer and Urban Management

The status of a subject is closely related with the existing value system of the society, the degree to which it helps the society in resolving the contemporary issues and the number and level of positions held by the students in intellectual circles, government, politics and elsewhere (Misra1982) .Human societies face different problems at different point of time, their value system changes over time so the nature and spirit of the subject needs to be dynamic if it desires to survive and remain relevant. Geography by its very nature is the study of man and his relationship with the environs. Mercer and Powell (1972) remarked that by their very nature, all men are geographers, because geography is concerned with the earth as the home of mankind. It is recognized as the science dealing with the causal relationship among the spatial, biological, physical and human perspectives which searches to explain the world of reality and tangibility (Simmons 1976). This growing thrust of scientific orientation leads everybody to feel himself to be the geographer of his own space (Sautter1975) who is supposed to understand and interpret the physical and cultural components of his surroundings and its influence on his well-being. Hence, a geographer can play a multifarious instrumental role in different capacity in handling the urban problems, managing the urban affairs, making the city inclusive and sustainable by integrating the different disciplines (economics, sociology, engineering and architecture) and their professional approaches into a viable whole.

The policy makers like ministers and the senior bureaucrats with geography background would be helpful in synergizing and finalizing the plans of different regions of the state based on need based priorities and gets the resources make available and utilized in time. The planning exercises require continuous problem identification, need assessment, data collection, analysis, interpretation and updating of data. It is a continuous and ongoing process. Geographer as a planner having knowledge and understanding of physical environment can help in identifying the area specific problems, needs and suggest the viable solution. Spatial planning, an important ingredient of 74th CAA, is emphatically spelt out in 12th schedule in the form of District Development Plan and Metropolitan Plan. Planning is a technical issue that requires specific theme based training. A geographer with professional training of remote sensing, GIS and computer based cartography can be a resource person in developing Master Plans, city development plans and regional plans.

H.H. Barrows (1923) defined Geography as a science of human ecology and focused mainly on human adjustment to physical environment.

Geographer as human ecologist can contribute significantly in integrating and analyzing the interaction between physical and cultural landscape of the area. In many of the development-displacement projects like Indira Sagar Project, Sardar Sarovar Project, Tehri Dam Project, geographers as human ecologist are preferred choice and assigned the task to analyse the association and influence of man and nature over each other and its implications on the natural surroundings. Geographer as subject expert and resource person can help in developing target group specific modules, evolve appropriate pedagogy, develop content oriented reading material and pursue research work in urban management and governance.

NGOs/CBOs have been engaged in organising the community and enabling them to manage their own affairs, understanding of local issues and equipping them eminently to achieve the objectives. As an NGO worker, a geographer can inform and sensitise the community about national planning goals and the relevance of 12th schedule functions of urban local bodies. Take the initiatives in inviting the officers and people's representatives of the area to share the problems, potentials and impediments of the area and seek their commitments for solution. Geographer as social worker and activist too can sensitise and mobilize the community towards the issues related with general well being and better quality of life, service delivery system and sensitise the people towards urban problems and policies. This section of society in assistance with an NGO can be instrumental in developing need based micro plans and proper integration of the urban and rural planning. A geographer is supposed to understand his physical and cultural surrounding. As a researcher he can undertake the physical survey of the area, appraise the available physical and cultural resources and create a resource inventory and urban observatory which could be used by planners, policy makers, academicians and administrator for the overall management of the area.

Most of the urban local bodies have limited resources and their revenue base is so weak that their maximum budget allocation often goes to establishment cost and the local authority is unaware of their existing untapped resources. Geographer as people's representative in the capacity of members of PRI, councillors of ULBs, MLAs and MPs of the respective areas can influence and motivate the people towards resource mobilization, revenue generation and making the urban management an affair of the mass. By innovative ideas and their application he can broaden the resource base of the urban areas and this could be used for developing infrastructure of the city.

An administrator in urban management usually performs the duties of Project Officer in District Urban Development Authority (DUDA) or the

Chief Municipal Officer in urban local bodies. In both the cases they have to develop the plan related with urban poverty, revenue generation, resource mobilization for their respective urban local bodies and improving urban infrastructure for the city dwellers. A geographer with exposure of urban geography, social geography and having some managerial skill can promote and facilitate convergence with the sectoral departments at district/city/town levels. Similarly he can liaise with various agencies at district, state and central level and can influence the policy makers with their reality based proposals and plans.

Way Forward

In the above paragraphs geographers are expected to play a diversified role in urban management and governance issues. To perform these roles properly and professionally they need specialized academic input and training and capacity building at different stages. Initially a geographer intending to work in urban planning and management sector should have an essential background of settlement geography with two broad components: rural settlements and urban settlements. Contents of the curriculum should focus on evolution of settlements, their histogenetic and morphogenetic processes, spatial and demographic characteristics, local resource base, rural-urban linkages and interrelationship. This will help in understanding the rural-urban settlement systems in a complimentary and integrative manner and importance of these settlements in the over all development of the country.

At the later stage while opting for geography as a career the focus should be on issues affecting day to day life of an urban dweller, participatory method of survey, research and planning, sophisticated techniques of map making. A paper entitled "Geography of Urban Development and Governance" could be introduced for this purpose. A brief out line of the course content is suggested below:

I. **Knowledge based themes**: Theoretical bases of urban land use, urban resource base, state of urban infrastructure, urban poverty and urban environment, structure of urban administration, legislative provisions governing urban areas and programmes and policies launched by the govt.

II. **Skill based themes**: Developing Master Plan for a city, Metropolitan Area and District Development Plan with the help of existing base map and modern techniques like computer assisted cartography, remote sensing, GIS and GPS.

III. **Project planning, formulation and management in urban sector**: This part of the component would focus on the concept of project cycle, identification and analysis of the problem and the need, accordingly determination of objectives, strategy finalization and analyzing the role of different stakeholders in achieving the target. Project would focus on urban services (water supply and sanitation, solid waste disposal and management, urban infrastructure) and urban resource base (evolving tools and techniques for revenue generation and resource mobilization).

Concluding Remarks

With the enactment of 74th CAA, the urban local bodies attained the constitutional status and people's representatives of these bodies are entrusted with the responsibility of managing the urban affairs. Their proper training and capacity building is the daunting task for policy makers. HUDCO-HSMI took the initiative by establishing HUDCO Chair in different Administrative Training Institutes during the IXth and Xth plan which were expected to work in the area of training and capacity building for urban sector by analyzing training need, developing training module, preparing reading material and imparting training to the target groups. This paper is an outcome of my previous five years experience of working in the HUDCO Chair, RCVPN Academy of Administration, Madhya Pradesh, Bhopal as a trainer and training manager. This assignment brought me closer to the urban administrator, people's representatives and other stakeholders like NGOs and CBOs and provided opportunity to share ideas on different urban issues and their viable solution and suggestions. I personally see a great employment opportunity and huge job market for geographers in the changed scenario. Before the enactment of the 74th CAA the development of city master plan was the responsibility of town and country planning directorate of the respective states which has limited number of functionaries. Consequently, the preparation of master plan was a very slow process and its implementation was far from reality... Now this responsibility is given to the respective urban local bodies which have neither expertise nor any technical wing in urban planning and management. Knowledge, skill and training of geographers can be useful for developing master plans of the cities and towns. This will not only encourage the youth to opt geography as a career but also keep the subject relevant and useful for the society. In this way the role of geographer can be seen as Enabler, Developer and Facilitator. Similarly, I also perceive the role of geographer as an academician, administrator and public figure. These two sets of role are not exclusive, nor could a line of demarcation be drawn between them. They are coterminous with overlapping characteristics.

References

Alderfer, H.F. 1964. *Local Government in Developing Countries*. New York: McGraw Hill.

Barrows, H. H. 1923. Geography as human ecology. *Annals Association of American Geographers* 13:1-14.

Clarke, G. 1991. Urban management in developing countries: A critical role. *Cities* 8(2): 93-107.

Claudio, A. 2003. The meaning and scope of urban management: an introductory note. In *Urban Management Practices in Secondary Cities in Egypt: The Case of Belbeis,* ed. Azza Sirry, 1-3. Cairo: Elias Modern Publishing House.

Ganeshwar, V. 1990. Basic service approach to urban development: Indian experiences. *Cities* 7(4): 333-340.

Gupta, R.C. 1998. Managing the metropolitan cities of India. In *Million Cities of India,* Vol. II, ed. R.P. Misra and K. Misra, 746-760. New Delhi: Sustainable Foundation.

HUDCO-HSMI. 2001. *The State of Indian Cities 2001*. New Delhi: HUDCO-HSMI.

Jha, R. and Nasreen Siddiqui. 2000. *Towards People Friendly Cities*. Mumbai: UNICEF Maharashtra State Office.

Kumar, B. and R.B. Singh. 2003. *Urban Development and Anthropogenic Climatic Change*. New Delhi: Manak Publications.

Kundu, A. 2005. *Urban Development and Urban Research in India*. New Delhi: Khama Publishers.

Mattingley, M. 1994. Meaning of urban management. *Cities* 11(3): 201-205.

Misra, R.P. 1982. Introduction. In, *Contributions to Indian Geography*. Vol. I, ed. R.P. Misra, 1-14. Delhi: Heritage Publishers.

McGill, Ronald. 1995. Urban management performance: An assessment framework for third world city managers. *Cities* 12 (5): 337-351.

Mercer, D.C. and J.M. Powell 1972. *Phenomenology and Related Non-positivistic Viewpoints in Social Science*. Victoria (Australia): Monash University Publication in Geography.

Mishra, H.M. and Arun K. Singh. 2003. Urban decentralization: An impact assessment with special reference to Madhya Pradesh. *Governance* 3: 1-14.

NIUA. 1998. *India's Urban Sector Profile*. Research Study Series No. 61. New Delhi: National Institute of Urban Affairs.

Pathak, C.R. 2002. Urbanisation pattern, process and policies: A case study of India; In, *Society, Development and Environment*, ed. Ranjan Basu and Shukla Bhaduri, 225-252. Kolkata: Progressive Publishers.

Palnitkar, S. 2004. Capacity building as an element of governance. Paper presented in national seminar on Sustainable Urban Management, May 28-29, at RCEUS, Lucknow.

Ramachandran, R.1989. *Urbanisation and Urban Systems in India*. New Delhi: Oxford University Press.

Sautter, G. 1975. Some thoughts on geography in 1975. *International Social Science Journal* 27 (2): 231-249.

Simmons, I. 1976. To what extent is the geographer's world a real world. *Area* 8 (2): 82-86.

Singh, R.L. 1982. Role of geography as an integrating science. In *Contributions to Indian Geography*, Vol. I, ed. R.P. Misra, 31-55. Delhi: Heritage Publishers.

Sivaramakrishnan, K.C., Amitabh Kundu and B.N. Singh. 2005. *A Handbook of Urbanisation in India*. New Delhi: Oxford University Press.

UNCHS (Habitat) 2001. *Cities in a Globalising World. Global Report on Human Settlement 2001*. London and Sterling,VA: Earthscan.

CHAPTER EIGHTEEN

RETHINKING REGIONAL DEVELOPMENT

BIKRAMADITYA KUMAR CHOUDHARY,
BANARAS HINDU UNIVERSITY

Introduction

Relationship between developmental process and geographical space has been a universal factor since the conceptualisation of development itself at least in the modern sense. Developmental history, in the sense it is defined in mainstream, started since human has learnt about accumulation and the history traces back to the beginning of agriculture when it became possible to store the output and use it later. The sense of space, initially as a physical entity and later as cultural landscape remained attached with the process of development. The initiation of industrialisation and its reliance on fund resources made the case for space more distinctive and scholars started becoming dogmatic about the spatial characteristics. The loosely understood space got more precisely defined in late nineteenth and early twentieth century when theories of development came into existence with the conception of external demand and comparative advantage. Still, the concept of regional development remains a widely debated and loosely implemented concept in the developmental discourse across the world with India not being an exception. Several reasons are cited for this such as the weakness of the implementing agencies, bureaucratic character of the executive organ/s of the state, lack of active participation of the people and so on. However, very few questions have been raised on the way the so-called regional development polices are formulated with a clear idea of benefiting the power group at various levels in different societies.

The present paper's attempt is to deconstruct some of these myths and analyse the conception of regional development using examples from India. In this backdrop, this paper will try to explain the continuity or uniformity in the theories behind "regional development" as Rees (1989) stated "to a large extent, the more theories change, the more they remain the same" and the poor implementation of such theories in India. This paper would also try to reconstruct the theory behind state's strategy and action towards minimising inter-regional and intra-regional disparities that has given rise to higher level of concentration of economic activities in the country.

The role of the academics especially the spatial scientists (geographers) is paramount in this context and this paper will try to present a case. This considers the relevance of space when development is considered as entitlement and capabilities of resources and its sustainability.

What is Regional Development?

Regional development has been understood in different context differently with unanimity that it deals with the process of development at spatial scale. Historically regional development has come into developmental literature in United Kingdom, where the concept was understood as the provision of aid and other assistance to regions which are less economically developed or became depressed due to change in production technology (Mackay 2003). This definition of regional development addresses the practicality of the situation and is based on the experiences from United Kingdom; where poor laws have had been important aspect of policy making to assist the northern regions. In the aftermath of great depression of 1929 and 1930 that affected most of the capitalist economy, the concept of *laissez-faire* was opposed and active state intervention in restructuring of economy was getting accepted. The conception of self-correcting nature of economic development against concentration was thoroughly opposed and role of control and management was agreed upon. The new economic order based on Keynesian demand management, Roosevelt's "new deal", and Stalin's model of centralised planning, especially top-down approach, started restructuring the economic space across the world at national and sub-national order (Lipietz 1992). The debate on the issue began with a general acceptance that the process of development does not start everywhere at one point of time and that means that some places are behind in the process. The reasons for this (non-)uniformity is advanced as natural conditions and historically created differentiation leading to comparative advantage based on resource endowment (Losch 1964; Myrdal 1957; Perroux 1964). The concept of resource as defined by Zimmermann (1951) notes that "resources are not they become". Therefore, the argument about natural conditions become less important and the conditions created by historically differentiated treatment to the region becomes more important. The ways resources are identified at certain point of time become dominant factor that shape the economic space. However, the definition of any concept and of regional development depends on the theoretical conception of the person who defines the process and examines the cause.

The understanding of regional development depends on the way development is defined and Crush (1995) has correctly formulated in the introduction of the book *Power of Development* that development depends on who defines for whom. The term "development" has always been accompanied with complexity right from its origin in the eighteenth century. It grew up with its root sense of unfolding and was provided a fresh lease of life by the emerging new evolutionary ideas of the nineteenth century. In Victorian England, it came up from a mere cultural and theological response to be looked upon as a part of progress. Escobar (1994) defines rather finds development discourse as an invention, which took place in the form of historically produced discourses of development in the post-1945 era. This was governed by the similar principles of colonial discourse but had its own forms of representation. Development is about the forms of knowledge, the power that regulates its practices and the forms of subjectivity fostered by its impulses. As its origin lay in the western world it remained Eurocentric for a considerable period and emphasised on the "capitalist form of growth", or in the "growth of national income". The Third World Countries adopted the term in specific forms like—"championing economic growth", "catching up", and "providing governable subjects" and so on.

Economic developments do provide a paradigm that can be used to help the understanding of the development of better living standard. Of course, the importance of individual level of satisfaction, personal freedom and well-being cannot be denied while defining development. However, such incorporation gives the freedom to scholars to perceive development in different ways in different cultures. Amidst the talk about alternative definition of development (Tulpule 1996; Nash 1963), it is necessary to stick to the mainstream classical conception of development that is said to be Eurocentric and broadly depends on the stage of economic growth. This argument does not mean to suggest that social, political, cultural upward mobility should not count in developmental debate; rather it is to reassert the reality that the social, political and cultural condition of the individual or community depends on the material condition of the individual or the group. The level of satisfaction can be one good indicator (Sen and Nussbaum 2000). However, level of satisfaction and level of unfulfilled desires can not be compared between the groups exposed to limited sets of material conditions. Accepting the economic dimension of development as major determinant, the debate gets concentrated around two major aspects viz. the way how to measure the level of economic growth and at what scale.

The measurement of economic development has remained centred around "typical index approach" of Nash, that access the economic conditions based on a normative value. It can be debated that who is deciding that normative value which is used to identify the gap and the value is derived with what interest. Still, when each unit is accessed in terms of gap between the normative standard and the existing standard, there is scope to say that which units are lagging behind. The problem that is confronted in the whole development debate that needs to be sharpened is the arguments that explain these gaps and suggestive measures. The reason for existing low levels of economic performance and lower level of standard of life is often advanced as complimentary to each other what Myrdal (1957) has summed up as "poor are poor as they are poor"; later on explained as vicious cycle of poverty. The existing gaps are the result of market-oriented policy of the governments across the world. The existing evidence from China reveals that before implementing open-door policy the coefficient of variation across provinces was lower compared to the scenario after adopting the open-door policy favouring the coastal areas compared to rest of the country (Demurger *et al.* 2004).

The second major aspect of measurement of development remains concerned with the question of scale which is less debated by economist and wherever debated it is putting economic development across units of area at various levels leading to the idea that regional development may be regional, national or international in nature. The implications and scope of regional development may vary in accordance with the definition of a region, and how the region and its boundaries are perceived internally and externally. More than this, the concept of regional development needs to be more than simply putting economic development across regions, as economic development starts in a regional setting rather regions getting attached to economic development. Trade economists, however, invariably regarded national economy as spaceless and even international trade as only an exchange system based on comparative advantage. The fact remains that by changing the scale the type of problem changes and so the adopted mechanism and also the expected explanations. Most of the problem of underdevelopment is considered as local and it is argued that the local population generally does not show much concern for the macro affluence or macro level economic crisis. Chakravorty (2005) noted that relative deprivation is a local issue; an unskilled factory worker in Calcutta is rarely concerned about the profit of industrialists in Bombay, just as a landless labourer in any district of Bihar or Orissa is unlikely to care about the income level of landlords in a district of Maharashtra. However, this fact is correct only in a situation when mobility: physical and virtual is restricted. With mobility, the population gets exposed to the larger reality and also shows concerns, as after globalization even the person on the margin of the economy and society gets affected with the decisions that are taken at the global level. Nature of problem in different location is different due to different kind of political and social conditions that shape the economic structure and also the level of material condition of household that decides the bargaining power of the individual and group in larger political and economic set-up. Nature of developmental activity also has different kind of result in different reasons. At national scale, better road transport infrastructure is a boom for economy as commodities can be transported at faster rate to the regions that demand those commodities. Infrastructure also generates the multiplier effect to the regional economy by creating new growth impulses. However, when infrastructure is created only to fulfil the indirect multiplier effect for the cement and stone industry without identifying the commodities to be transported, the region in which it is created faces more problems than the benefits. The question of scale not only needs to be looked at in the simple manner in identifying the problem and the type of understanding of the people. The need is also to study the scale with the conceptual idea of creation of certain type of administrative units and the reason of such creation, as Brenner (1997) referred this fact that the spatial targets like states, districts, or smaller units are the divergent territorial scales at which spatial tactics like planning are deployed.

Regions need to be defined in terms of resource endowment, available technology to harness the resource potentials and the hindrances that cause underutilization of certain kind of resources. Regional development than needs to be understood as the process that aim at development of natural resources that has played a significant role in the expansion of many regional economies (Gunton 2003). The studies that primarily based on the conception of comparative advantage and the nature of external demand has failed to really address the question of different levels of development either at the national level or sub-national or regional levels and a paradigmatic change in needed to address the existing or rather increasing regional inequality across the nations and amongst the nations.

Role of Regional Dimension in Development Process

The great inequality amongst nations and in between regions of nations in terms of economic performance and the distribution of benefits of such economic development is an undisputed reality. There are glaring examples of various kinds of disparities at international, interstate and interregional levels. The concentration of development in terms of accumulation and affluence in a few parts of the world or in different parts of a nation continues to remain the law of nature across the globe. Even in the most developed countries like USA or the industrial countries of Western Europe, the same scenario has been observed a few decades ago and once again regional inequality is on increase in these countries (Brenner 1997; Mackay 2003; Ullman 1964; Wang and Arnold 2008). It is also true that in the poor countries or poor regions such disparities are larger and in the beginning of developmental process these inequalities are supposed to be larger (Myrdal 1957). The reasons advanced for such inequality have been multifarious ranging from racial explanation to environmental determinism to resource endowment and in economic terms comparative advantage, and in more docile form of racism as "zeal to achieve" (Berberoglue 1992; Chakravorty 2005; Hilhorst 1998; Mackay 2003; Perloff and Lowdon 1964; Rees 1989). These explanations are based on the earlier stated two conceptions namely "the way development is defined" and "the question of scale" in regional sense. In economic sense, explanation keeps on changing as the in-built contradictory economic explanation, i.e. the prominence between "open market" and "state intervention" is still far from settling. Alternative explanation for the existing disparities between regions and across nation-state was based on theories; like "Amin's theory of unequal exchange", "Frank's concept of dependency", "Wallerstein's world-system approach" and so on (Larrian 1989). These theories try to explain the reasons that decide the lesser negotiation power with the peripheral regions at international level means the dependency of the developing countries on the affluent centres. On the basis of these theories, conditions of inequality inside the national economy can also be explained through the concept of existing post-colonial social structure, comprador bourgeoisies that influence the monopolistic nature of state in allocation of fund to different sector and in the selected regions.

The three major economic crises during last eighty years or so have been the outcome of the emphasis on "open market" that has affected the global economy primarily led by the capitalist economy. The crises were followed by the regime of active state intervention as the self-correcting mechanism of open economy either in terms of creation of demand or in terms of decentralization did not yield desired results. The experiences so far shows that the "polarization effect" of Hirschman or the "cumulative causation" of Myrdal has been the only reality and the process of economic development has not gone to the stage of either "trickle down" or "spread effect". Experiences in the twentieth century have shown that the nature of economy is such that before it could start decentralising; the stage of the "great boom" come to seize and the economy again starts consolidation leading to centralisation once again.

India is a country with wide natural variations that include lofty Himalayan Mountains, great Gangetic Plain, vast Thar Desert and the Deccan plateau. These natural diversities often provide a shield for the state for not being able to eliminate existing or rather increasing regional disparities. It is most common to relate non-development to lack of resources. However, nature of resources is that they are not resources unless man's technology can use them. A gaze on noted historical developments tarnishes such notion. The example from China strengthens this point clearly. The current "inaccessible regions" of China used to be the centre of economic activities till the 11th century and the current thriving regions of China used to be the areas where offenders were sent to exile. The change in the technology from land route trade (silk route) to trade via sea route changed the landscape of the country. Similar example can be taken form United Kingdom, where the first phase of industrialization started in the North and North-western provinces having the distinction of core of industrial revolution. But, once there was a shift in technology, the region came in the category of depressed region and still remains depressed (Thomas 1981).

Initiatives in India for regional development started with a vision document of National Planning Commission chaired by Jawaharlal Nehru, who later became first prime minister of the country. The need to develop the poor regions and weaker sections of the society was identified. This realization was tried to translate into action during Third Plan period (1962-1967) in Independent India. The efforts in India remain confined to the decentralisation of industries to the industrially backward regions. It was broadly accepted that a country or region can develop only if there is sufficient scale of industrialisation. However, the problem with the conception remained that industries did got established in the region but the population of the region remain broadly untouched with that, as there was not sufficient effort for skill formation to the people that was required for securing a participation in the than set-up industries. Further, no efforts were made to identify the critical aspects of people's life in these backward regions where intervention was needed either in terms of skill formation or in terms of provision of technology so that the people can improve their capabilities to harness the potentials. The latest Five-Year Plan, i.e. Tenth Plan is in operation, also reflects that the state is committed towards to poor region and poor people. The Tenth Plan states:

> Although the issue of regional balance has been integral component of almost on each five year plan, there has been perceptible increase in regional imbalance over the year. We have also been conscious of the fact that national targets do not necessarily translate into balanced regional development. The potentials and constrains that exist at state level vary significantly. Therefore for the first time we have broken down the national targets to state level in consultation with state government.

This statement makes it clear that the past efforts to ameliorate regional disparity have not succeeded due to one or the other reasons; and, the Tenth Plan aims to achieve higher level of equity amongst regions. However, one needs to be careful when reading such documents. The objective of the Tenth Plan has been outlined "to achieve higher growth rate of economy". The objective to attain higher growth rate and distribution of these to poor regions have been historically proved contradictory to each other and almost all theory of regional development has shown that in the initial stage higher growth rate of economy is needed and this can be achieved when taking benefit of agglomeration effect and only in the later phase when growth rate establishes the process of decentralisation began. However, in the plan document the two contradictory goals are outlined in the same sentence without resolving the conflict through suggestions or arguments at least to show how they can be complementary and not contradictory. The setting up of state level targets in the plan has hardly shown any expected impact on reduction of disparity among the states. Another important feature is the intra-state differences in terms of quality of life and economic performance. Maharashtra, an economically developed state, also has the district of Ghadhchiroli; the state of Gujarat also encompasses the district of The Dangs.

Regional Development: A Forgotten Agenda

The process of development is said to begin with the creation of external demand and certain kind of surplus and having a comparative advantage. These all in combination are pre-condition for a classical as well as neo-classical conception of development. In this scenario, relative deprivation of region is bound to increase, and it has been suggested time and again that if state intervenes or at the later stage of development process, when decentralization begins; these disparities are liable to decrease. This has been clearly shown by the famous U shape model of Williamson. However, the problem begins at the conception that whether the state intervene to improve the health of the deprived region and if there are economic and political compulsions for not to intervene what would be the later stage of (capitalist) development. Once again, I cite example from United Kingdom, where up to the mid-1970s, there was a trend towards greater regional equality in the distribution of income and wealth in the country, but after that for next 25 years regional inequalities increased along with other forms of inequalities (Mackay 2003). Similar situation has been reported from USA where, income inequality and concentrated disadvantage have been on the rise since the 1960s as the affluent and poor are becoming more concentrated than ever (Wang and Arnold 2008, 259). Most of the Third World countries started development with foreign aid that had the noble intension of helping developing countries to provide their people with the material opportunities for using their talents, of living a full and happy life steadily improving their lot. However, the result of such activity was emergence of more accentuated form of colonial structure of dual economy. Schumacher (1973, 135) wrote it quite clearly, "One of the unhealthy and disruptive tendencies in virtually all the developing countries is the emergence, in an even more accentuated form, of the "dual economy", in which there are two different patterns as widely separated from each other as two different worlds. It is not a matter of some people being rich and others being poor, both being united by a common way of life: it is a matter of two ways of life existing side by side…".

In India, the conditions after Independence have been similar to what these lines read. The idea of development, in India, was based on the Nehruvian conception of modernity that was based on technological advancement and industrialisation. The country began development with the goal of attaining self-sufficiency in basic structure incorporated regional dimension in planning documents since Third Five-Year Plan in principle. There were allocations for certain kind of programmes to ameliorate poverty, relocate industries, and create certain kind of infrastructure and so on. As the emphasis was on creation of industrial centres in different parts of the country, the process and technology that would have harnessed local resources for the benefit of the people of the region were largely ignored, resulting in two India side by side. There have been several writings on these lines; here two distinct India can be identified: one urban India voicing concern for clean air, less congestion, open space and so on and the other rural India that is still struggling for two squares of meal on an assured basis. The conditions became more alarming since mid-1980s and early 1990s when a paradigm shift was witnessed in Indian economy in tune with the U turn in USA and European countries when boom started (Mackey 2003; Wang and Arnold 2008). Regional dimension of developmental activities were forgotten even in the plans, and definite outcome was almost zero allocation for depressed areas, even there was no modification or efforts to re-identify depressed region or poor regions. Arguments were that once the country achieve a "high rate" of growth, "all problems will be the reality of past". As expected or say suspected by the leftist academicians, this did not happen. Before a steady high rate of growth could be achieved, the boom busted and all efforts are once again got diverted towards "consolidation" of economy. The tragedy of the poor region remains that they wait endlessly for the "spread effect" and "trickle down" to take place but before it the crisis began. These poor regions suffer when the centre reap the fruit of the economic development but when the crises begin these are the regions, which suffer worst. This is not by accident but this is by design, as highlighted by economists across the First World and the Third World (Krugman 2008; Patnaik 2008).

Two dominant often cyclic processes are noted with regards to regional development on the basis of the experiences in the field of knowledge creation and in the field of interventionist implication. In the field of knowledge creation, there is spell of vociferous argument that there is increasing disparity between regions and geography is responsible for it and the balancing argument from the other quarter is that state's intervention does not yield enough result as geographical conditions (e.g. accessibility, weather, or terrain) is not favourable for such developmental activity. Both arguments are not different; only accuse the victim. In the field of active intervention, the efforts have been of two kinds: one the depressed regions are given additional aid or relocation of industrial activity in favour of depressed /poor regions.

The centrality of the argument is that regional development is considered as an outcome of geography for which economic aid is needed. Now it is clear in the age of economic determinism, geography is to be looked at like poor are looked at for ages as it is geography that creates the disparity and makes it burden for the economist. It is this centrality of understanding that needs to be done away with. The process of regional development is to be understood not as geographically determined but economically created and the regional development is not a dole but an economic requirement for creation of demand and also for the territorial security of the nation-state.

Space in the sense of real geometrical entity and also in the form of economic, cultural, urban or abstract space is production of human activity on the earth surface (Lefebvre 1991). The activities those are located over space are often decided on the basis of political and social consideration especially in terms of power equation of social and local elites and here Rostow (1960) could have been correct to argue that social and cultural factors are equally important as economic factors had he considered the existing power relation amongst different groups. This is not to suggest that the non-communist manifesto is more real than the Marxist understanding of the fact that it is the material condition of individual or group that decides the social, political and cultural being of the individual and society.

The restructuring of space based on changing economic activities needs to be linked with the process of regional development if regional development is conceived as more abstract terms as entitlement, capabilities and sustainability which are rather difficult to operationalize. The studies have established beyond doubt that capitalism (the prime mode of development today) can not stop to enhance growing concentration or to say can not ever be in the "later stage of development" (Foster 2003). Geographers in constructing their theories have drawn freely on the concepts and perspectives of different schools of economics; however, economists have often tended to accord little attention to the role of geography in economic progress. It has been the dominant trend in economics to consider international trade an exchange mechanism devoid of any geography, "a world where goods and services move between dimensionless points at zero or uniform transport cost" (Martin and Sunley 1996). This brings us to the margin of the scenario that what needs to be done to minimise the existing rather increasing relative deprivation of the region.

Where Do We Need to Focus Today?

The regional concept involves the idea that there are associations of phenomena to be observed on the earth that these associations and the patterns of their individual elements are significant of the interacting processes at work on the earth (James 1952). The ignorance or avoidance of spatial factors in the development process is one amongst many reasons for the existing lopsided development and economic progress in different parts of the world and in various regions of the world. Krugman (1991) wrote that space matters and he tried to bring geography back into economic analysis. Economic geography does consider that uneven regional development is a central part of the process by which national economic prosperity and trade are created and maintained. However, there has been very little study about finance and financial implications in different regions. Thrift (1990) argued in his writing when he wrote that "I simply can not understand why so little attention is paid to the matters of money and finance in human geography". The space as production of social process, as Lefebvre (1991) considers, does possess characteristics that decide the amount of resource allocation, the amount of resource exploited and also that for whom the resource are exploited and which resources are being exploited. The understanding of space in this sense does attract the study of power structure that decided the identification of regions at different scale and the differential nature of resource allocation for different purposes. Lefebvre (1991) noted that "each new form of state, each new of political power, introduces its own particular way of partitioning space, its own particular administrative classification of discourses about space and about things and people in space". Regional and urban planning policies are conceived as spatial tactics- techniques deployed by the state to regulate, produce and reproduce configurations of social space sufficient for continuation of accumulation of capital (Brenner 1997).

The national prosperity and intra-national trade forms a crucial link in understanding the nature of restructuring of economic space over territorial space. The study of development based on regional concept needs to recognise that focus in the spatial science should be on place rather than on isolated economic process that are growth-centric and are guided towards profit making and accumulation of wealth. The process of accumulation of wealth and profit making also needs to be analysed if a better understanding of place is to form the basis for regional development. The study of the process in a particular place explains with greater details that why such process is induced in a particular place only and not on other places. These decisions are not based on natural characteristics of space rather they are dependent on the type of people inhabiting the space and the social, political and economic characteristic of the place.

It is rightly argued that the emergence of new industrial geography and impact of economic integration does affect the process of regional development and need to be in the core of the theory building on which's basis new methodology needs to be developed so that the process of political and social dynamism can be integrated in the study of regional development. These kinds of writings have time and again argued for a closer relationship between economic theories and the process of regional development and also between regional economics and economic geography (Martin and Sunley 1996).

How Geographers Can Contribute to This?

The relationship between economic geography and economics has remained as asymmetric one. The problem of the regional development theories so far has been that most of the theories consider development and region separately and then try to forge a balance between the two trying to impose region over development or the opposite. But regional development is not either way. Many have tried to establish link between geographical space and economic space but economist are yet to consider geography as important factor in ameliorating persistent regional inequality. For regional development, region is to be understood in totality, region as a resource endowment region, region as potential for some aspect, as ready recipient of certain technology, as a hindrances to other advancement needs to be addressed. We need to understand trade and regional development in one framework as suggested by Krugman (1991). One needs to draw explanations from the existing pattern of industrial development that is necessarily uneven. Geographers as a spatial science owes to the society to put forward analysis of the differential nature of development and highlight the growing inequality and provide explanation beyond the existing paradigmatic norms. Even when accepting the existing paradigm of regional development where geography is considered as a reason for underdevelopment that economics tries to minimise, geographers can provide analysis on following count:

Spatial location

Spatial location of economic activity is decided on many counts other than only spatial characteristics that is normally said to be geographical. The discipline needs to highlight the untold stories for the selection of a particular location for any developmental activity. For example, when large dams are planned at a particular location, different kinds of argument for and against are made, spatial scientist need to provide an explanation not only looking at environment or social cost at macro level but also the cost of development in terms of location and the possibility of smaller dams in alternate locations, where larger benefits can be harvested. Analysis of spatial analysis in traditional sense does tell us about the amount of funds that needs to be sanctioned for a particular region, or identification of region that is depressed on the basis of selected characteristics.

Ecology

Geography as a spatial science is different from the cardinal spatial science as it studies the location of human in the environment, their interaction with the ecosystem, the expected results and also the forthcoming changes due to restructuring of space over time. As a spatial scientist, geographer will look into the dynamics of ecosystem sustainability so that ecology does not get disturbed beyond repair with the adoption of different kind of economic activity and will be able to suggest the suitable economic activity for a particular location.

Employment

The level of employment in the particular region is an indicator of development. Geographers not only identify the opportunities of employment but also identify the potential source of employment by identifying the resources and the technologies that need to be in place to harness the specific employment potential.

Nature of the project

Nature of project can be suggested on the basis of the requirement of the region and the available resources in the region. The nature of project should be in tune with the ecological limits in the ecosystem and geographers are the one who on the basis of their background will be able to formulate sustainable project(s) for the region.

Fixed asset investment

Once the local resource base is identified on the basis of the identified resources, the identification of the sector in which fixed asset investment is needed can be made.

Potential benefits for the region

In the conventional sense, the potential benefits of the project needs to be identified to attract the fund without the critic of the economist who care too much dependent on the cost-benefit analysis. The geographers' understanding of cost-benefit analysis needs to be on a long-term sustainable basis and improvement of resource entitlement for the people in the region rather than on factor-price or constant-price.

Industry segment

On the basis of above, the amount of industrialization can be identified. The nature of industry that can be suitable for the region without putting stress on local environment and identified resource and can be amalgamated with the available resource and available human resource potential can be suggested for the region.

Research content

The above identified steps can be adopted only if there is enough research to identify the priority area of intervention in various regions depending on the nature of ecology, nature of people's practices and processes that can not be altered and analyzing the expectation of the people from the developmental activity. The research should fix the priority area in which investment can be made so that we can expect desirable result in limited time.

Towards Conclusion

For all these to be a success, there is a need to move beyond the framework in which regional development has been looked at so far, where the aim has been understood only as identifying the depressed area and suggesting the amount of aid. The branch of study will have to develop itself as an integrated research field forging a balance between physical geography, resource geography, regional economics and human geography. The importance of resource identification for different population group is paramount. Geographers will now have to develop the theoretical model that explains the process of selection of resource and determination of technology. The study of geography of power needs to be explained and linked with the present distribution of developmental activity in Baconian sense of power. Various models and indices like "localized income inequality" (Wang and Arnold 2008), "staple-theory" (Gunton 2003), "theory of social Polarization" (Sassen 1994), and "theory of spatial mismatch" (Wilson 1987) and so on using methodology developed in different field of knowledge can be used for such explanations. There can be effort to develop a simulation model to capture the impact of spatial location on developmental process in different environmental conditions.

Need of the hour is that geographers should broaden their horizon to venture out of their traditional field and start seriously looking into the problems of disparities between different regions at various levels. The change to this effect needs to be brought in the curriculum itself so that the scholars trained in geography can understand and use the models and theories across the field of social and natural sciences. Space should not be considered as physical reality and human landscape separately; rather geography should study space that incorporates cultural features, when taken culture as a way of life rather than the art and architecture. The study of the interplay of economic forces, like resource, finance, market, technology, etc. on the geo-ecological space with humans as prime dynamic provides us the understanding of the existing economic scenario. By incorporating historical materialism as a method geographers will be better equipped to analyse the processes that are responsible for the current scenario that deprive many regions and more individuals due to their so-called locational disadvantages. This will help in understanding the explicit and implicit power relations of local and international dominant player that influence the developmental decisions. The study of the process of technological change, and the process of technology transfer breaks the myth of advantageous and disadvantageous locations and makes it clear that at any one point of time only certain stuff are resources and rest acts as hindrance. Geographers need to identify the possible and probable resources in deprived region and also expose the hindrances including dynamics of global capital and state policies. Some of these adaptations will make path for brighter geography that is more intervening in formulation of state policy rather than remaining a mere descriptive discipline that is more informative than analytical.

References

Agnew, J. 2000. From the political economy of regions to regional political economy. *Progress in Human Geography* 24 (1): 101-110.

Anonymous 1982. *Regional Planning Efforts in India: A Monograph*. Delhi: TCPO.

Barkin, D. and T. King. 1970. *Regional Economic Development: The River Basin Approach in Mexico*. Cambridge: Cambridge University Press.

Basu, K. 2000. *Analytical Development Economics: The Less Developed Economy Revisited.* Delhi: Oxford University Press.

Behrman, J., and T. N. Srinivas. 1998. *Handbook of Development Economics*, Vol. III. Amsterdam: Elsevier.

Berberoglu, B. 1992. *The Political Economy of Development.* Albany: State University of New York Press.

Bhat, L.S. 1985. Spatial perspectives in socio-economic development from national and regional angle. In *Geography and Planning*, ed. K.V. Sundaram, 309-321. New Delhi: Concept Publishing Co.

Chakravorty, S. 2005. The history and geography of regional development theory: A futile search for a paradigm. In *Development, Displacement and Disparity*, ed. N. Banerjee and S. Marjit, 29-52. New Delhi: Orient Longman.

Crush, J. 1995. *Power of Development.* New York: Routledge.

Das, K. 2004. Uneven development and regionalism: A critique of received theories. *Economic and Political Weekly* 39 (45): 4917-4925.

Datta, B. 1997. *Indian Planning at Crossroad.* Delhi: Oxford University Press.

Demurger, S. *et al.* 2004. *Geography, Economic Policy and Regional Development in China.* Discussion Paper No. 1950, Harvard Institute of Economic Research.

Dewar, D., A. Todes and V. Watson. 1986. *Regional Development and Settlement Policy: Premises and Prospects.* London: Allen and Unwin.

Escobar, A. 1994. *Encountering Development: The Making and Unmaking of the Third World.* New Jersey: Princeton University Press.

Friedmann, J., and W. Alonso. 1964. *Regional Development and Planning: A Reader.* Massachusetts: MIT Press.

Glikson, A. 1955. *Regional Planning and Development.* Leiden: Netherlands Universities Foundation for International Cooperation.

Gunton, T. 2003. Natural resource and regional development: An assessment of dependency and comparative advantage paradigm. *Economic Geography* 79 (1): 67-94.

Harrish, J. 2004. *Depoliticizing Development.* Delhi: Left World.

James, P.E. 1952. Towards a further understanding of the regional concept. *Annals of the American Geographers* 62 (3): 195-221.

Khan, M.B.A. 1988. *Regional Planning and Economic Development.* Delhi: Pointer Publishers.

Kuklinski, A. 1974. Regional development, regional policies and regional planning. In *Regional Disaggregation of National Policies and Plans*, ed. A. Kuklinski, 1-17. Paris: Mouton.

Larrain, J. 1989. *Theories of Development: Capitalism, Colonialism and Dependency*. Cambridge: Polity Press.

Lipietz, A. 1992. *Towards a New Economic Order: Postfordism, Ecology and Democracy*. New York: Oxford University Press.

Mackay, R.R. 2003. Twenty-five years of regional development. *Regional Studies* 37 (3): 303-317.

Martin, R., and P. Sunly 1996. Paul Kurgan's geographical economics and its implications for regional development theory: a critical assessment. *Economic Geography* 72 (3): 259-292.

Misra, R.P. 1982. *Regional Development and Planning*. New Delhi: Concept Publishing Co.

Morris, A. 1998. *Geography and Development*. London: UCL Press.

Munshi, S.K 1984. *India Resource Regions and Regional Disparity*. New Delhi: People's Publishing House.

Myrdal, G. 1957. *Economic Theory and Underdevelopment Regions*. London: Duckworth.

Nash, M. 1963. Approaches to the study of economic growth. *Journal of Social Issues* 19 (1): 1-15.

Pathak, C. R., and K.G. Bagchi. ed. 1974. *Regional Development and Planning*. Calcutta: Calcutta University Press.

Pathak, C.R. 2002. *Spatial structure and Process of Development in India*. Calcutta: RSAI.

Peet, R. 2005. *Theories of Development*. Jaipur/New Delhi: Rawat Publications.

Rajalakshmi, N. 2000. *Environment and Economic Development*. New Delhi: Manak Publications.

Ray Choudhary, J. 2001. *An Introduction to Development and Regional planning*. New Delhi: Orient Longman.

Rees, J. 1989. Regional development and policy. *Progress in Human Geography* 13 (4): 576–588.

—. 1992. Regional development and policy under turbulence. *Progress in Human Geography* 16 (2): 223-231.

Sassen, S. 1994. *Cities in a World Economy*. California: Pine Forge Press.

Saunders, P. 1998. *Capitalism: A social Audit*. Delhi: World View Press.

Sundaram, K.V. 1985. *Geography and Planning*. New Delhi: Concept Publishing Co.

Sen, A., and M. Nussbaum 2000. *Quality of Life*. Oxford: Clarenden Press.

Sharp, J. and Briggs, J. 2006. Postcolonialism and development: New dialogue? *The Geographical Journal* 172(1): 6-9.

Schumacher, E. F. 1973/1993. *Small is Beautiful: A Study of Economics as if People Mattered*. London: Vintage.

Thomas, D. 1981. United Kingdom. In *Regional Development in Western Europe*, ed. H.D. Clout, 241-263. London: John Wiley and Sons.

Thomas, M.D. 1964. Resource and regional development: some comments. *Regional Science* 13 (1): 201-205.

Thrift, N.J. 1990. Commentary: The perils of the international financial system. *Environment and Planning A*, 22: 1135-1137.

Tulpule, B. 1996. Redefining development and alternative paradigm. *Economic and Political Weekly* 31(45): 2991-2996.

Wang, F. and M.T. Arnold. 2008. Localised income inequality, concentrated disadvantage and homicide. *Applied Geography* 28: 259-270.

Wilson, W.J. 1987. *The Truly Disadvantaged: The Inner City, the Underclass and Public Policy*. Chicago: University of Chicago Press.

Zimmermann, E. 1951. *World Resources and Industries*. New York: Harper & Brothers.

PART V:

THE *SUMMUM BONUM*

Two roads diverged in a wood, and I –
I took the one less travelled by,
And that has made all the difference.
—Robert Frost (1874-1963) in "The Road Not Taken" (1916)

CHAPTER NINETEEN

CONCLUSION: DESIGNING A WORKING AGENDA FOR THE 21ST CENTURY

RAVI S. SINGH,
BANARAS HINDU UNIVERSITY

Introduction

Introspectory exercises are healthy for the growth of a discipline and they also indicate the vibrancy level in the concerned field of knowledge. In the main essays spread across four sections, contributing authors have tried to find out the lacking(s) in respective field, identify the respective need(s), and have in turn also suggested the ways of accomplishment. It is however not the first instance where such an exercise has been undertaken by the "Indianist" or "dweller" Indian geographers. Of course, the history of such endeavours is relatively new. The purpose of this last Chapter of the book is to intertwine the emerging thoughts from the included essays here. Accordingly, the remaining portion of this Chapter takes up four main issues: forgotten agendas, what to do, why to do, and how to do.

Forgotten Agendas

As mentioned in Chapter two, Spate (1956) stressed "demand-mapping" rather than overworked "problem-mapping". It is still a valid call to which we have not attended sufficiently and continue overworking the already "overworked" problem-identification exercise. The "highly desirable and indeed essential" fields of research identified by him; of course, need to be seen in the present context and the ensuing challenges to the Indian economy, culture and society. Sopher (1973/2009) was perhaps the first to make a clarion call, to "dweller" Indian geographers especially, to bring

"Indian-ness" in their studies and to seek what is already there in the landscape and folk tradition. His emphasis was also on the use of indigenous concepts and terms, neglected in the works of Indian geographers, which can explain the local reality in the best manner possible. Similarly, Schwartzberg (1983/2009) too indicated to a few weaknesses crippling healthy growth of Indian Geography and ultimately affecting the (international) status. In the same vein Misra (1983, 1-10) pounced over a variety of problems and recommended four issues needing attention of Indian geographers especially: quality of textbooks, reinforcing physical geography, areas of concentration, and ways of implementation. Since then, one did not hear any voice of that sort until 1991 when Mukerji (1992/2009), one of the senior most geographers, honestly affirmed "ailments" the Indian Geography was suffering from. The sectional titles reflected issues like senior geographers' regressive dominance, methodological problems, physical geography's position, research stereotyping, funding issues, over-application of choroplething method, weakening tradition of fieldwork, intellectual import from overseas, stress on relevance, etc. These points ably demonstrated his deep concerns and considered view on the health of Indian Geography. Some of these issues like deprivation of physical geography have been raised by others, like Gosal (1980) and Dayal in Kapur (2002, 318), too. In this book, Dey (Chapter 8) also talks about the same which is an indication of how unfortunately nothing has perceptibly changed and we tend to forget our fallacies as if they were no fallacy at all. It is pertinent to remember that many among us doubt and question our expertise in efficiently handling the problems of physical environment (Lahiri-Dutt 2005/2009). Such an observation should instead of being simply rubbished need to be considered in view of the contemporary scenario. The intension, of citing both types of views here, is to clarify the nature of difference in the perspective. It is high time that we do away with dualistic vision which mars the image of the discipline as much as it has multiple detrimental effects on the future course of development. I believe 21st century is an opportunity to return to our disciplinary roots where geography was not seen partitioned or divided and it had a "unified" character which not only defined it but also placed it uniquely among systematic disciplines. Under the influencing wave of dominant paradigm of systematic branching off the body of knowledge, we were trapped and the physical drifted away from human. Matthews and Herbert (2004) argue for the "unity in geography" and the other essays in their volume too remind us of our disciplinary heritage and impress with fresh arguments for this cause.

In the meantime, formation of a "national" association in the name of National Association of Geographers, India, popularly known as NAGI took place in 1978 in the line of American association of Geographers. NAGI may boast to be the largest body of professional Indian geographers with a membership of nearly 1500 life members which is larger than the aggregate number of life members of all other major geographical societies/bodies in India (cf. Kapur 2002, xii); but, the membership number does not provide answer to few points which question its national representative-ness, etc. Anyway, with it started an annual event called "The Indian Geography Congress" with a customary tradition of delivering Presidential Address. These addresses should/could be taken as "vision statement" of the incumbent presidents of NAGI. Till the dawn of the 21st century altogether 23 addresses were delivered each setting a future agenda for the Indian Geography; of course, a few of them had a very specific focused theme to elaborate upon (see, Kapur 2002). Nevertheless, they tried to bring a variety of issues ranging from the problem of immigration in Assam (H.P. Das) to the role of voluntary agencies in rural development (K.N. Singh) and school geography (B. Arunachalam). Aijazuddin Ahmad's address at Thiruvananthapuram (Kerala) on the other hand took stock of Indian Geography vis-à-vis international developments during the post-World War period of the 20th century. It is an indication of the fact that Indian Geography has been taking stock of the contemporary disciplinary positions and setting agendas for self. But, the agendas have been put on the back burner and thus never pursued in a committed manner. What did hurdle the committed pursuance is a problem worthy of a full analysis.

A few recent writings have tried to raise issues unheard in Indian Geography. Raju (2004) for example argued for openness in research and tried to exhibit her own experience in which practicing critical geography was scorned by colleagues for not being "real geography". Such attitudes' persistence, is indicated by Ravi Singh (Chapter 13) in this book too, has taken heavy toll. And, it is one of the reasons why Indian Geography is yet to make a foothold at par with the international intellectual developments in geography. Kapur's (2004) article can not be termed as addressing any of the issues in Indian Geography, rather she tried to argue for the social science tag for geography and in the course of doing so discussed about the neglect of geography by social scientists and also its "exclusion from stock-taking of social sciences in India". Joining the discussion, Lahiri-Dutt (2005/2009) seconded Kapur's views and added gender dimension to the professional Indian Geography again drawing basically from own experience. Ravi Singh (2008) have tried to bring vital issues to the fore which have either been overlooked or underestimated but definitely contribute to qualitative downfall of Indian Geography. Rana Singh (2009a & b) have looked at the critical issues confronting Indian Geography and has also drawn a big canvass for the revitalisation in the 21st century.

What to do?

The essays in this book identify a few tasks for the young geographers who would be in the helm of affairs in the 21st century. The starting point here emerges as making of Indian Geography. The debate on Indianness of Indian Geography, initiated long back (Sopher 1973), could not attract even lukewarm response from the Indian geographers for around two decades. Contending the idea of "Indian Geography", Ahmad, in his presidential address to the NAGI in 1997, did not find logical validity in it. However, the dialogue has got revived once again recently. While Ravi Singh (2009) argues and tries to make a clear case for Indian Geography; Rana Singh (2009a & b) critically analyses "Indianness" and also "deconstructs" fallacies that mar the true image of Indian Geography and the deeper meanings emerging. He has also advocated a dialogue between the East and West for a healthy growth of the discipline (Rana Singh 2009b).

Barring all sorts of quantitative progresses, nothing else appears satisfying a curious and concerned Indian geographer. In a way, I find this dissatisfaction a positive sign for the health of "maturing" Indian Geography. It will intensify the hunger for a better future, as attested by the way of it is unfolding which is bound to give a new shape in the present century. The "dissatisfied" young geographers yearn to seek a better future. Reclaiming Indian Geography however is not a simple task. As urged by Arunachalam in his presidential address in the Indian Geography Congress, the beginning has to be made from the school level, as "The real pillars of strength for university geography have to firmly rest on the plinth of school geography" (cf. Kapur 2002, 126). To "bring back geography alive at school level" a host of interlinked reforms in school geography are required, which are discussed by Alam (Chapter 5) and Ryngnga (Chapter 7). However, it would be sufficient to note here that one of the major changes needed is in the style of school text book writing and giving due place and sufficient space to non-dominant side of the issues dealt with. This issue is well elaborated by Sunny in her essay (Chapter 6) in this book and elsewhere (Sunny 2006/2009).

At the higher (tertiary education and research) level, we need to inculcate rigour in training, develop critical eyes, promote 'deep' researches and equally important is promotion and practice of "openness" in research. For some reason, we lag behind our sister-disciplines in these respects. Perhaps that is the reason why we do not get our due. We, being strongly field-based discipline, should not waste time in (re-)deployment of field based skills to observe, explore and monitor. For unknown reasons, field work is growing weaker in Indian Geography and is increasingly becoming synonymous with questionnaire based data collection—the basis for statistical analysis to make a study scientific. Lahiri-Dutt (2005/2009) has narrated the dilemma regarding "compulsory" employment of statistical tools in geographic research in India. To train students and the young researchers, development of team work involving younger students in the line of laboratory based sciences is a good idea however with some practical difficulties.

In the same vein, a considered attention towards ethical questions too is required. It includes a long wish list which many of us would not relish but it is a truth that we can not get way with it either. Self-plagiarism is one into which majority of us indulge without knowing that it is violation of professional ethics and morality even if we are not caught infringing upon the copyright acts. Production of quality research work is desirable to get published in international refereed journals with good impact factor. It is no more a secret that propensity of Indian geographers to publish internationally is very low (cf. Kapur 2004, 4193). Though, it requires a separate enquiry to analyse reasons behind it, a common sense reason appears to be apparent poor quality of our works (cf. Chapter 14) not withstanding the demand of international standards. Of course, there remains a consistent problem of language barrier as our general proficiency in English is not sufficient enough to express as we understand. Then an alternative is to raise the standard of Indian journals to earn reputation and international recognition, as suggested by Dey (Chapter 8). We all are aware that publishing is a well organised industry in the contemporary time and journals too are increasingly becoming a product of this system. Most of the active and reputed (geography) journals today are being brought out by publishing houses; of course, the academic input comes from members of the academic community. This process has already gripped India too. Reference to this process intends to indicate that publishing good journals is an affair which requires openness, transparency, professional commitment, etc. Most of these attributes are found missing in the Indian geography journals. Schwartzberg (1983, 235) lamented the problem from the root, "Geographic societies, each with its own journal, abound in south Asia.... While one might suppose this to indicate the intellectual vigour of the field, that is not the case. Rather, most societies may be viewed as the academic fiefs of their respective founding departments, dominated by small coteries of senior professors or their anointed successors. The journals are virtually unrefereed, poorly edited, ...serve largely as vehicles for the publication of writings by the resident faculty, post-graduate students, and alumni. Society membership and non-library journal circulation are overwhelmingly regional; and the topics addressed....also tend to reflect this parochialism". Things have barely changed during last quarter century as a cursory look at the content sufficiently reveals. Rather lack of academic leadership and shrinking vision has made the things worst. To my mind, it is one area needing immediate intervention by young geographers provided they are ready to "bell the cat", if I am allowed to phrase it in this way.

India has continuously evolved and in the contemporary experience, India witnesses entirely new circumstances and environment in the post-liberalisation phase in the era of globalisation. Hence it is worth a call to the young geographers to take up studies in cutting-edge research frontiers and deal the older issues in the new perspective. Mishra (Chapter 3) rightly concludes that young geographers should explore emerging relationship between resource demand and the nature of resource production, identify social orders and hierarchies and at the same time analyse emerging constraints to agricultural land use when the forces and processes of globalisation invade India. Satpati's call (Chapter 9) for giving primary focus on water in geography and to undertake future projection of land use patterns commensurate with water availability, development of non-spatial data base on water for planning purpose, and associated issues is valuable for the 21st century agenda. Disasters and hazards are inseparable from any civilisation. However, despite their ever increasing frequency, they have been not given due attention by geographers in India. Jana (Chapter 10) identifies hazard study as a major issue for the present century where geography can contribute to the study of regional variation in the frequency and magnitude of hazards, risk zone mapping, and beyond.

Globalisation has apparently affected people's movement which is not merely changing location by people. Since the people on move are causing the movement of human resource, it becomes crucial not only to study the forces, processes and patterns of migration but to interlink migration with gender and changing scenario of economy, environment and policy which is missing, migration and related issues need to be considered integrated, argues Roy (Chapter 15). Migration, gender, amenities, quality of life and question of sustainable development with special reference to cities, integrated study of small and medium towns in terms of their potential role as "migration-absorbers" and future development, etc. should be given due place in the future urban geography as discussed by Khairkar (Chapter 16) simultaneously with conventional issues concerning inner city and city-fringe. Arun Singh (Chapter 17) sees the future role of geographers. He stresses specialised academic input and training and capacity building. Keeping the future (employment) prospects in mind, he contends for training geographers as urban planners and managers.

Any discipline which aspires to grow sustainably can not afford to shirk away from social responsibility (cf. Ahmad 1997). We know about several debates on the question of "social relevance", in the preceding Chapters too, the authors keep on giving call for a socially "meaningful" Indian Geography. It is rightly felt that we need to relate our agenda with the people's agenda and (re-)orient towards human welfare and ensure social justice. By doing so we can directly contribute towards meeting the common man's aspirations from which we are miles away, in honest sense. Such an approach gives scope to employ the unified geography—convergence point of physical and cultural dimensions—for not only the human well-being but also to understand regional development nuances. Bansal (Chapter 11) rightly makes a case to give due recognition to women and her role in society, culture and economy, construct feminist perspectives on gender and geography, focus on the interactive and creative relation between gender and environment, focus on cultural representation of gender, etc. like issues in the future academic agenda of mainstream Indian geography. Ravi Singh's views (Chapter 13) could be seen in the similar vein where he stresses on bringing back "culture" to the centre stage and respond to the shifts from "race" to "racism" or in India, from "caste" to "casteism" and questions of inclusion/exclusion together with emerging new urbanism perceptible in metros which will have greater ramification for India which is poised to leap into an urban future. It is found that the regional dimension was forgotten in the national/regional developmental plans and in the 21st century it is highly needed to bring that back in the centre. The process of restructuring of space needs to be linked with regional development process and the region in this context is to be understood in totality in which trade and regional development are considered in a single frame. To undertake analysis of differential nature of development to explain regional inequality, study of wealth-accumulation and profit is crucial, focus be on "place" rather than isolated economic processes, notes Choudhary (Chapter 18).

Advent of computers led to the process of technology feeding technology. Naturally, the tempo of technological change is touching new heights with every passing moment with direct effect on the end-users. Not only the younger generation (cf. Kapur 1998) but also the seniors (Dayal in Kapur 2002) have equivocally advocated use of modern information technology (MIT). Various MIT based tools have been developed with immense utility for geographers and other geo-scientists as well. One has to admit, that the traditional surveying chain and tapes, plane table survey sets, drawing pens and stencils etc. are already replaced by remote sense data in the form of air photographs and satellite imageries and mapping softwares, computer aided designs, and geographic information system (GIS) with wider application. That convinces younger geographers to not only "build bridges between MIT and geography" but also to strengthen that by using them in different geographical studies where spatial patterns are to be deciphered. It is also indicated that we should focus on interactive mapping which is going to be largely employed in the near future itself. Friedman (2006), in *The World is Flat*, has elaborated the growth of computer based information technology and its impact in all walks of the world society. Geographers can not remain in isolation nor lose the opportunity knocking at their doorstep. Yet, they need to remember that the modern techniques shall ever remain aid only and can neither become geography nor its purpose *per se*.

Why to do?

One may ask why we need to do what had been spoken about earlier and identified by the authors in this book. No single justification forwarded can make a satisfactory answer forget a complete one. The starting point could be pursuance of the agenda identified for 21st century Indian geographers, to make Indian Geography intellectually prosperous. Because despite of "illustrious achievements" by the pioneers, we could neither make recognised international presence nor earn respectability in general. So, if we wish to become the academic leader at least in developing countries, as Mishra hopes (Chapter 3), a systematic approach, sincere and committed and concerted effort becomes obvious requirement. The changes are desired at many fronts and different levels. One can wish to begin with changing pedagogic perspectives in designing a text book, especially at all levels of education, to include ethos of mass education and meet the aspirations of education for equality which is argued for the schools particularly by Sunny (Chapter 6). Similarly, issues like water and hazards and disasters can not be overlooked. They need prioritised treatment in this century. In a country like India embarking to become urban very soon, anew perspective to solve them is required as Problems of urban areas can not be solved independent of rural/countryside. Similar gaps need to be identified carefully and addressed gaps within a sub-field of geography about which Roy is concerned with reference to migration (Chapter 15).

Contemporary paradigms have made it clear that the things are not as simple as they appear. Therefore to develop a deeper understanding of a problem, it is highly required to expose the power relations and raise ethical questions. Of course, Indian geography has been refraining from it, but the call of "critical approach" makes it imperative. Bringing the geography of women to mainstream Indian Geography can provide opportunity for engaging in critical geography, to uncover and eradicate gender based biases and prejudices against females, as per the belief of Bansal (Chapter 11). To make a niche in the employment opportunities, geographers have the potential of playing a diversified role about which Arun Singh discusses with special reference to urban management and governance (Chapter 17). For that development of professional skill with applied face is required. Some recent examples of increasing geographers' employability through professional training in GIS and remote sensing like skills is one indicator. Thus, it would be possible to directly contribute or intervene in policy making for which geographers have been yearning (cf. Kapur 1998). We can make path for a "brighter geography that is more intervening in formulation of state policy rather than remaining a mere descriptive discipline", as believed by Choudhary (Chapter 18). "Brighter geography" is possible if we commit to "unified geography".

How to do?

A discipline is sculptured primarily by two tools—teaching and research. The young geographers dealing themes other than education in this book show dissatisfaction in this regard and urge for urgent changes. Even a senior geographer like L.R. Singh (1996/2009), while identifying the tasks ahead gave first priority to evolving a national policy for geographic teaching and research, called for establishing a national institute of geography. Though not impossible, this issue of a national institute prerequisites a few things basic among them being a "national" body of geographers to lobby with the government and its bodies. We have several bodies with the word "national" or "Indian" prefixed with their name; yet, the million dollar question is do we have a unanimous representative body? As far as the workable ideas are concerned, "improve the quality of the product" may be taken as initial remark on the ways to do what have been spoken about and identified for the 21st century Indian Geography. Starting should be from school education reforms: reform in curriculum, relevant syllabus, training of the school teachers and their trainers, new technologies for teaching spatial concepts, need for quality school books, re-focusing on the regional geography, teaching tools and technologies, reinvent and reintroduce field teaching, restructuring practical geography, involvement of school teachers, geographers for geographical education, as suggested by Alam (Chapter 5). Some of these are already indicated by Misra (1983). Reinvention of curriculum suggested by most of the authors is worthwhile and one of the urgent tasks to be taken up on priority. No issue or problem remains constant over time and space, so should be the response of academic disciplines including geography to solve them. Era of globalisation for example have altogether changed the context and has overwhelmingly influenced our lives, may that be employment, production-distribution-consumption, migration, education, cultural behaviour or some thing else. Hence, the conventional courses too require incorporation of relevant topics accordingly. Since, that can not address the whole set desirable changes, offering new courses, which we call "papers" in the Indian parlance, on "hydrogeography", hazards and disasters, migration, urban governance, etc. Therefore becomes imminent. It would create opportunity to study and research the issues in holistic way. Moving beyond traditional framework, use of cross disciplinary methodology to broaden horizon moving beyond own traditional field, develop theoretical models, explain geography of power with reference to present development activity, engagement with critical issues and methodology which deconstruct visible reality through questioning. The practice of questioning however is very poor in Indian geographic tradition (Ravi Singh 2009, 7-9). Somehow we have missed rapport with common people compared to our sister-disciplines despite immense potentiality for the same. That is yet another way to improve our outreach. "Professional commitment" is the catchword for all of our actions—curriculum framing and implementation, conducting research, raising journals quality, etc.—which we miss for several known reasons.

Some Conclusions

The recent "indicative list of a web of human problems that thirst for a geographical inquiry" proposed Kapur (2004, 4195) is interesting enough and reinforces the need of shifting research focus with changing times. And, therefore, one may say what is new about the agenda that authors identify here. All issues on the agenda outlined may not be exactly new; however, the details which they cover are fresh and give a better insight into the problems they have dealt with. In one sense, the respective Chapters are complete as they are first raising issues, line of required action and then discussing the implementation. I do not think them to be "too ambitious" against which Misra (1983, 10) had warned while talking about implementation of the suggested changes by him. Moreover, the line of action proposed is not by outside rather it is by those who are going to become academic leaders in respective fields and centres in the near future. A recent analysis with respect to the postcolonial British academic geography (Noxolo, Raghuram and Madge 2008) metaphorically communicates many relevant messages for us too. The 21st century is fertile with many post-isms—an opportunity for building and exerting identity free from external biases—which the Indian Geography can benefit from (Rana Singh 2009b, 5). The ideas presented here should not be taken as "pontificating" or "inculpation". At the end, nevertheless, we can not help asking ourselves: Will the dreams come true!

References

Ahmad, Aijazuddin. 1997. Fifty years of geography in India and abroad. Presidential address delivered at the 20th Indian Geography Congress, December 28-31, in Thiruvananthapuram (Kerala).

Friedman, Thomas L. 2006. *The World is Flat: A Brief History of the Twenty-First Century*. London: Penguin Books.

Gosal, G.S. 1980. Frontiers of geographic research and teaching in India. Presidential address delivered at the 1st Indian Geography Congress, January, in Chandigarh.

Kapur, Anu. 1998. *Indian Geography: A Future with a Difference*. New Delhi: Allied Publishers Limited.

—. 2004. Geography in India: A languishing social science. *Economic and Political Weekly* 39(37): 187-4195.

Kapur, Anu. ed. 2002. *Indian Geography: Voice of Concern*. New Delhi: Concept Publishing Company.

Lahiri-Dutt, Kuntala. 2005/2009. Geography as a marginal science. *Economic and Political Weekly* 40 (7): 689-691. Reprinted in *Indian Geography: Perspectives, Concerns and Issues*, ed. Ravi S. Singh, 242-250. Jaipur/New Delhi: Rawat Publications.

Matthews, John A. and David T. Herbert. 2004. Unity in geography: Prospects for the discipline. In *Unifying Geography: Common*

Heritage, Shared Future, ed. John A. Matthews and David T. Herbert, 369-393. London: Routledge.

Misra, R.P. 1983. The status and prospects of Indian Geography. In *Contributions to Indian Geography: Concepts and Approaches*, ed. R.P. Misra, 1-10. New Delhi: Heritage Publishers.

Mukerji, A.B. 1992/2009. What ails Indian Geography? Some statements. *National Geographical Journal of India*, 38 (1-4): 205-218. Reprinted in *Indian Geography: Perspectives, Concerns and Issues*, ed. Ravi S. Singh, 223-241. Jaipur/New Delhi: Rawat Publications.

Noxolo, P., P. Raghuram and C. Madge. 2008. 'Geography is pregnant' and 'Geography's milk is flowing': Metaphors for a postcolonial discipline? *Environment and Planning D: Society and Space* 26(1): 146-168.

Schwartzberg, Joseph E. 1983/2009. The state of South Asian geography. *Progress in Human Geography* 7 (2): 232-253. Reprinted in *Indian Geography: Perspectives, Concerns and Issues*, ed. Ravi S. Singh, 305-330. Jaipur/New Delhi: Rawat Publications.

Singh, L.R. 1996/2009. A hundred years of Indian Geography: Development and change. In *New Frontiers in Indian Geography*, ed. L.R. Singh, 1-57. Allahabad: R.N. Dubey Foundation. Reprinted in *Indian Geography: Perspectives, Concerns and Issues*, ed. Ravi S. Singh, 109-171. Jaipur/New Delhi: Rawat Publications.

Singh, Rana P.B. 2009a. Identity of "India" and "Indianness", and geography in India: Sensitive appraisal from the past and move to the 21st century. In *Indian Geography: Perspectives, Concerns and Issues*, ed. Ravi S. Singh, 25-64. Jaipur/New Delhi: Rawat Publications.

—. 2009b. *Uprooting Geographic Thoughts in India: Toward Ecology and Culture in 21st Century*. Newcastle upon Tyne: Cambridge Scholars Publishing.

Singh, Ravi S. 2008. Challenges to Indian Geography: A contextual analysis. *Geographical Review of India* (Platinum Jubilee Volume) 70 (1): 57-70

—. 2009. Introduction: Concerns and debates in Indian Geography. In *Indian Geography: Perspectives, Concerns and Issues*, ed. Ravi S. Singh, 1-22. Jaipur/New Delhi: Rawat Publications.

Sopher, David E. 1973/2009 Towards a rediscovery of India: Thoughts on some neglected geography, in: Mikesell, Marvin W. (Ed) *Geographers Abroad: Essays on the Problems and Prospects of Research in Foreign Areas,* pp. 110-133 (Chicago: Department of Geography, Research Paper No. 152, University of Chicago). Reprinted in *Indian*

Geography: Perspectives, Concerns and Issues, ed. Ravi S. Singh, 305-330. Jaipur/New Delhi: Rawat Publications.
Spate, O.H.K. 1956. Modern geography and its meaning for India. *National Geographical Journal of India* 2(4): 175-185.

APPENDIX 1

A CHRONOLOGY OF SELECT EVENTS AND PUBLICATIONS IN INDIAN GEOGRAPHY: 1761-2009[1]

1761. Systematic geography survey by Plastid on the coasts of Chittagaong.
1767. James Rennell appointed first surveyor General of Bengal. Survey of India comes into existence.
1788. First map of Hindoostan published.
1793. 1st use of plane table.
1799. William Lambton laid foundation of the great trigonometric survey of India.
1802. Great Trigonometrical Survey commenced near Madras.
1823. 1st government lithograph press established at Calcutta.
1836. Bombay Geographical Society Transactions established.
1840. Geomagnetic observatory at Calcutta.
1851. Geological Survey of India,
1857. The first three modern universities at the presidency headquarters – Bombay, Calcutta and Madras established.
1861. Archaeological Survey of India.
1861. Census of India established.
1871. (Sir) Alexander Cunningham, *The Ancient Geography of India.*
1872. 1st Census conducted in India.
1875. Indian Meteorological Department established at Pune.
1877. Allahabad University comes into existence.
1881. W. H. Hunter, *Imperial Gazetteer of India* (First edition, 9 vol. set)
1889. W.W. Hunter, *Imperial Gazetteer of India* (Second edition, 14 vol. set).
1890. Botanical Survey of India constituted.
1894. R.D. Oldham, "The making of the Indian Geography" (*Geographical Journal*, 3).
1894. W.W. Hunter, *Atlas of India.*
1901. Shimla Education Conference.

1906. Cameron Morrison, *A New Geography of the Indian Empire and Ceylon.*
1906. Forest Research Institute comes up at Dehradun.
1907. Hunter, W.W. *Imperial Gazetteer of India* (Second edition, 26 vol. set).
1914. Indian Science Congress Association inaugurated.
1916. Banaras Hindu University established.
1916. Zoological Survey of India formed.
1920. Geography makes entry in higher education via Lahore College, Punjab University, Lahore (now in Pakistan).
1924. Geography established in Aligarh Muslim University under teachers training programme.
1925. Curzon Geographical Society, Aligarh.
1925. The Inter-University Board (later known as the Association of Indian Universities) established.
1926. *Geographer* (Aligarh) launched.
1926. *Indian Geographical Journal* (Madras/Chennai) launched.
1926. Intermediate courses in geography at Madras University started.
1926. J. Rennel, *Regional Geography of Bengal.*
1927. Geography established at Patna
1928. Arthur Geddes, *Au pays de Tagore*: *la civilisation rurale de Bengale occidental et ses facteurs géographiques.*
1929. Patna College Geographical Society, Patna.
1931. Indian Statistical Institute, Calcutta founded by P.C. Mahalanobis.
1931. Post-graduate department of geography at Aligarh Muslim University.
1932. B.C. Law, *Geography of Early Buddhism.*
1935. R.N. Dubey awarded D.Litt at Paris.
1936. *Geographical Review of India*, Calcutta launched.
1936. S.P. Chatterjee awarded D.Litt at Paris.
1937. B.C. Law, *Geographical Essays*, Vol. I.
1937. Geography established at Agra.
1937. R.N. Dubey joins Allahabad and undergraduate course in geography offered there.
1938. A. Geddes, "Dynamic problems of Geographical Reconnaissance in India" (*The Indian Geographical Journal* 13).
1938. Radhakamal Mukerjee, *The Regional Balance of Man.*
1939. Calcutta Geographical Society established.
1939. Geography established at Calcutta.
1940. M.H. Rahman awarded D.Litt at Paris.

1940. Radhakamal Mukerjee, *Man and His Habitation: A Study in Social Ecology*.
1942. Indian Science Congress Association merged Geography and Geology section at Baroda.
1944. Geography introduced in Banaras Hindu University at intermediate level.
1945. Anthropological Survey of India formed.
1945. N. Ahmed, "The Task before the Indian Geographers" (*Geographical Review of India* 7, 4).
1946. Geography department at Banaras Hindu University.
1946. Research bulletin series of National Geographical Society of India (Varanasi) started.
1947. India attains freedom.
1947. Post-graduate department established at Allahabad.
1948. M. Shafi joins Aligarh Muslim University.
1948. Post-graduate course started at Madras/Chennai under the headship of George Kurien.
1948. *The Geographer*, Aligarh launched
1949. First university department of geography in the north-east India at Gauhati University (Guwahati, Assam) under the leadership of H.P. Das.
1950. M.P. Rajagopala, "The Role of Geographer in Free India" (*The Indian Geographical Journal* 25, 3-4).
1950. Moonis Raza joins lectureship at Aligarh Muslim University.
1951. A.B. Mukerji joins a college at Modinagar.
1951. C.B. Fawcett, "Challenges to Indian Geographers" (*The Indian Geographical Journal* 26).
1951: P. Dayal joins Patna University.
1952. Geography at M.S. University, Baroda.
1952. Masters in Regional Planning introduced at the Indian Institute of Technology, Kharagpur.
1953. Bombay Geographical Magazine launched.
1953. Geography department established at Sagar.
1953. Inauguration of University Grants Commission (New Delhi).
1953. P. Dayal, *Bihar in Maps* (Patna,).
1954. B.C. Law, *Historical Geography of Ancient India*.
1954. Oskar H. K. Spate, *India and Pakistan: A General and Regional Geography*.
1955. Demise of H.L. Chhibber, R.L. Singh takes over the headship of the geography department at Banaras Hindu University.
1955. *Geographical Outlook*, Ranchi launched.

1955. *National Geographical Journal of India*, Varanasi, launched.
1955. R.L.Singh, *Banaras: An Urban Geography.*
1955. Town and Country Planning Organisation (TCPO) established at Delhi.
1956. 1st international seminar on geography, Aligarh.
1956. All India Soil Survey Organisation comes into existence.
1956. International Institute of Population Sciences (IIPS) established at Bomaby/Mumbai (known as Demographic Training & Research Centre prior to 1970).
1956. National Atlas and Thematic Mapping Organisation (NATMO) established.
1956. O.H.K. Spate, "Modern geography and its meaning for India" (*National Geographical Journal of India* 2, 4).
1956. University Grants Commission (New Delhi) formally established.
1956. V.L.S. Prakasa Rao appointed the first director of the regional planning unit at the Indian Statistical Institute, Calcutta.
1957. A.B. Mukerji leaves for the US to work under Fred B. Kniffen.
1957. Kashi N. Singh joins Banaras Hindu University.
1957. O.H.K. Spate, *India and Pakistan: A General and Regional Geography* (Second Edition).
1957. S.P. Chatterjee (ed.), *National Atlas of India.*
1958. B. Subbarao, *The Personality of India: Pre-and Proto-Historic Foundation of India and Pakistan* (Second revised edition).
1958. G.B. Singh, *Agricultural Planning Atlas of Punjab.*
1958. K.V. Sundaram the first geographer appointed at Town and Country Planning Organisation (TCPO).
1958. *National Geographer* (Allahabad) started.
1959. G.S. Gosal joins Punjab University.
1959. Prof. George Kurien joins Delhi School of Economics as the founder of the department of (human) geography.
1960. A.B. Mukerji joins Osmania University.
1960. Department of Geography comes into being at Punjab University (Chandigarh).
1960. Geography established at Ranchi University, Enayat Ahmed assumes headship.
1960. S. P. Chatterjee becomes International Geographical Union (IGU) Vice-President.
1960. S. S. Harrison, *India, the Most Dangerous Decades.*
1961. Calcutta Metropolitan Planning Organisation established.
1961. Kusum Nair, *Blossoms in the Dust: the Human Factor in Indian Development.*

1962. 1st summer school of Geography at Shimla.
1962. M.R. Chaudhuri, *Indian Industries: Development and Location—An Economic and Geographical Appraisal.*
1962. Roy Turner (ed.), *India's Urban Future.*
1962. *The Deccan Geographer* launched.
1963. A.B. Mukerji moves to Punjab University (Chandigarh).
1963. S. P. Chatterjee, *Fifty Years of Science in India: Progress of Geography.*
1963. S.M. Alam, *Hyderabad-Secunderabad: A Study in Urban Geography.*
1964. A.B. Mukerji, "Cultural landscape of a Telangana Village: A morphogenetic study (*Geografia* 3).
1964. C.R. Pathak appointed first faculty member in regional planning at the Indian Institute of Technology (Kharagpur).
1964. P.D. Mahadev joins University of Mysore.
1964. S.P. Chatterjee becomes the first Asian President of the IGU (1964-1968).
1965. R.P. Misra takes over headship of the geography department at Mysore University.
1965. *Uttar Bharat Bhugol Patrika* (Gorakhpur) launched.
1966. 1st Demographic Atlas.
1966. Autumn School of Geography, Aligarh
1966. D.N. Wadia, *Geology of India.*
1966. Indian Institute of Remote Sensing (IIRS), formerly Indian Photo Interpretation Institute (IPI), set up at Dehradun.
1966. Marvin Harris, "The Cultural Ecology of India's Sacred Cattle" (*Current Anthropology* 7).
1966. P.P. Karan, "Recent contributions to the geography of south Asia" (*Cahiers de Geographie de Ouebec* 9).
1966. S.M. Ali, *The Geography of Puranas.*
1966. Summer School of Applied Geography, Varanasi.
1967. B. Dube, *Geographical Concepts in Ancient India.*
1967. Bombay Metropolitan Regional Planning Board established.
1967. Regional Science Association founded; *Indian Journal of Regional Science* started.
1967. Robert I. Crane (ed.), *Regions and Regionalism in South Asian Studies.*
1967. *Transactions*, Bhubaneshwar, launched.
1967. U. Singh takes the charge at Gorakhpur University.
1968. David E. Sopher, "Pilgrim circulation in Gujarat" (*Geographical Review* 58).

1968. Frederick J. Simoons, *A Ceremonial Ox of India: The Mithan in Nature, Culture and History, with notes on the Domestication of Common Cattle.*
1968. IGU congress held first time in Asia at Delhi.
1968. Irawathy K. Karve, *Kinship Organization in India* (Third edition).
1968. Kashi N. Singh, "The territorial basis of towns and village settlement in eastern Uttar Pradesh, India" (*Annals of the Association of American Geographers* 58, 3).
1968. Milton Singer and Bernard S. Cohn (eds.), *Structure and Change in Indian Society.*
1968. Phulrani Sengupta and Galena Sdasyuk, *Economic Regionalization of India: Problems and Approaches.*
1968. R.L. Singh (ed.), *Applied Geography.*
1968. R.P. Misra, *Diffusion of Agricultural Innovations.*
1968. S. P. Chatterjee, *Progress of Geography in India (1964-1968).*
1968. S. P. Chatterjee, *Progress of Geography in India* (IGU Congress, New Delhi).
1968. S.C. Malik, *Indian Civilization: The Formative Period.*
1969. Brian J.L. Berry and Philip H. Rees, "The Factorial Ecology of Calcutta" (*The American Journal of Sociology* 74).
1969. Geography established at Bombay University, Mumbai.
1969. Indian Council of Social Science Research (ICSSR) was established.
1969. Jawaharlal Nehru University (New Delhi) established.
1969. R. Ramachandran, *Spatial Diffusion of Innovations in Rural India: A Case Study of Spread of Irrigation Pumps in the Coimbatore Plateau.*
1969. R.P. Singh, *Geomorphological Evolution of Chotanagpur Highlands.*
1969: Joseph E. Schwartzberg, *Occupational Structure and Level of Economic Development in India: A Regional Analysis.*
1970. B. Arunachalam joins Bombay University.
1970. David G. Mandelbaum, *Society in India.*
1970. Geography department comes into existence at Utkal University, Bhubaneswar.
1970. Geography started at newly established Centre for Regional Development (later renamed as Centre for the Study of Regional Development, CSRD), Jawaharlal Nehru University.
1970. Richard G. Fox (ed.), *Urban India: Society, Space and Image.*
1970. S. Wanmali, *Regional Planning for Social Facilities: A Study of Eastern Maharashtra.*

1971. Brian J.L. Berry and Howard Spodek, "Comparative Ecology of Large Indian Cities" (*Economic Geography* 47 supplement).

1971. Geography established at Jamia Millia Islamia, New Delhi.

1971. L.K. Sen *et al.* (ed.), *Planning for Growth Centres for integrated Area Development.*

1971. R.D. Dikshit returns from the Australian National University (Canberra) after completing doctoral research.

1971. R.L. Singh (ed.), *India: A Regional Geography.*

1971. R.P. Misra becomes founder Director of the Institute of Development Studies, Mysore University.

1971. Taneja, K.L.: *Morphology of Indian Cities.*

1971: Richard G. Fox, *Kin, Clan, Raja and Rule: State-hinterland Relations in Pre-industrial India.*

1972. B.N. Sinha, *Industrial Geography of India.*

1972. Enayat Ahmad, *Coastal Geomorphology of India.*

1972. K.V. Sundaram, *Towards A National Urban Policy in India.*

1972. L.K. Sen *et al.* (ed.), *Readings in Micro-level Planning and Rural Growth Centres*

1972. L.S. Bhat, *Regional Planning in India.*

1972. N.B.K. Reddy joins Sri Venkateswara University (Tirupati) as founder chairman.

1972. R.B. Singh, *Regional Clan Settlements in Varanasi District.*

1972. R.P. Misra, *Regional Planning and National Development.*

1973. Ashish Bose (ed.), *Studies in India's Urbanization (1901*-1971).

1973. Ashok K. Dutt *et al.* (eds.), *India: Resources, Potentials and Planning.*

1973. David E. Sopher, "Towards a rediscovery of India: Thoughts on some neglected geography".

1973. *Indian Geographical Studies*, Patna

1973. *Indian Geography*, Raipur started.

1973. North-Eastern Hill University, Shillong established.

1973. R. L. Singh (ed.), *Rural Settlements in Monsoon Asia.*

1973. R. Vaidyanadhan switches over from geology to geography department at Andhra University (Visakhapatanam).

1973. S. P. Chatterjee, *A Decade of Science in India* (1963-72).

1973. Surinder Mohan Bhardwaj, *Hindu Places of Pilgrimage in India: A Study in Cultural Geography.*

1974. A.B. Mukerji, "Morphogenetic analysis of rural settlements of the Chandigarh Siwalik Hills (*Punjab University Research Bulletin*, Arts, 5).

1974. Asish Bose *et al.* (ed.), *Population in India's Development, 1947-2000.*
1974. B.H. Farmer, *Agricultural Colonization of India since Independence.*
1974. Jasbir Singh, *Agricultural Atlas of India: A Geographical Analysis.*
1974. R.D. Dikshit becomes founder head at Punjabi University, Patiala.
1974. R.P. Misra, K.V. Sundaram and V.L.S. Prakasa Rao, *Regional Development Planning in India: A New Strategy.*
1975. K. Bagchi, "Geography at the Cross Roads" (Geographical Review of India 37).
1975. K.L. Rao, *India's Water Wealth.*
1975. Piers M. Blakie, *Family Planning in India: Diffusion and Policy.*
1975. R. D. Dikshit, *The Political Geography of Federation: An Inquiry into Origins and Stability.*
1976. A.B. Mukerji "Choe terraces of the Chandigarh Siwalik Hills, India: A morphogenetic analysis" (*Revue de Geomorphologie Dynamique* 25).
1976. K.R. Dikshit, "Drainage basin of Konkan: Forms and characteristics" (*National Geographical Journal of India* 22).
1976. K.R. Dikshit, "Geomorphic features of the west coast of India between Bombay and Goa" (*Geographical Review of India* 38).
1976. L.S. Bhat *et al.*, *Micro-Level Planning: A Case Study of Karnal Area, Haryana.*
1976. R.L. Singh, K.N Singh, and Rana P.B. Singh (ed.), *Geographic Dimensions of Rural Settlements.*
1976. S.M. Alam, *Planning Atlas of Andhra Pradesh.*
1977. Ashok K. Dutt and A.G. Noble, *Indian Urbanization and Planning.*
1977. B. Ghose, S. Singh, A. Kar *Geomorphological Investigations of Rajasthan Desert.*
1977. C.P. Singh, "Indian Geographers and Foreign Area Studies" (*Geographical Review of India* 39).
1977. K.V. Sundaram, *Urban and Regional Planning in India.*
1977. Rana P.B. Singh, *Clan Settlements in the Saran Plain (Middle Ganga Valley): A Study in Cultural Geography.*
1977. Robert C. Eidt, K.N Singh, and Rana P.B. Singh (ed.), *Man, Culture and Settlement.*
1977-78. C. D. Deshpande, "Discovering (Indian) Geography" (*Geographical Outlook* 13).
1978. 1st Indo-French seminar in Geography.
1978. C.D. Deshpande, "Recent Trends in Geographical Thought and the Indian Context" (*The Indian Geographical Journal* 53).

1978. *Indian Journal of Landscape Studies* launched.
1978. National Association of Geographers, India (NAGI) comes into existence at Delhi.
1978. R.P. Misra (ed.), *Million Cities of India.*
1979. G.S. Bhalla, *Performance of Indian Agriculture: A Districtwise Analysis.*
1979. Indian National Cartographic Association founded at Hyderabad.
1979. J.E. Schwartzberg, *Historical Atlas Of South Asia* (First edition)
1979. Moonis Raza, (ed.) *A Survey of Research in Geography 1969-1972.*
1979. *Population Geography* (Chandigarh) launched.
1979. R.L. Singh *et al.* (eds.), *Place of Small Towns in India.*
1979. *Transactions of the Institute of Indian Geographers* launched.
1980. David E. Sopher (ed.), *An Exploration of India: Geographical Perspectives on Society and Culture.*
1980. *Hill Geographer* launched, Shillong
1980. R.P. Misra and K.V. Surdaram, *Multi-level Planning and Integrated Rural Development in India.*
1981. *Annals of the NAGI* launched.
1981: Barbara D. Miller, *The Endangered Sex: Neglect of Female Children in Rural North India.*
1981: Deryck O. Lodrick, *Sacred Cows, Sacred Places: Origins and Survivals of Animal Homes in India.*
1982. Allen G. Noble and Ashok K. Dutt (eds.), *India: Cultural Patterns and Processes.* Boulder, Co Westview Press.
1982. S.M. Alam founded Centre for Area Studies at Osmania University (Hyderabad).
1983. A.B. Mukerji, "Dispersal and resetting of refugees in Independent India: A cultural ecological model" (*Indian Geographical Journal* 58).
1983. B. H. Farmer, *An Introduction to South Asia.*
1983. Joe Schwartzberg, "The state of South Asian geography" (*Progress in Human Geography* 7).
1984. G.B. Singh, *Agricultural Atlas of Punjab.*
1984. M. Shafi became second Indian Vice-President of the IGU (1984-1992).
1984. S. Wannmali and Tim Bayliss-Smith Eds.), *Understanding Green Revolutions: Agrarian Change and Development Planning in South Asia.*
1985. A.B. Mukerji, and A. Ahmad, *India: Culture, Society and Economy.*
1985. H. Ishihara, *Rural Markets in South India*
1985. S.P. Chatterjee awarded Padma Bhushan.

1986. B.N. Sinha (ed.), *Trends in Geographical Research in India.*
1986. F. J. Costa *et al.* (eds.), *Urbanization in Asia.*
1986. Savindra Singh and R.S. Pandey, "Morphological analysis and development of slope profiles over Bhander scarps" (*International Geomorphology*, Part II).
1987. Indian Institute of Geomorphologists formed at Allahabad.
1987. Savindra Singh and S.P. Agnihotri, "Rilland gully erosion in the subhumid tropical riverine environment of Teonthar Tahsil, M.P., India" (*Geograsika Annaler* 694).
1988. W. Tietze (ed.), *Asian Urbanization: Problems and Processes.*
1989. Death of S.P. Chatterjee
1989. Geography at Himachal Pradesh University, Shimla.
1990. Moonis Raza and Aijazuddin Ahmad, *An Atlas of Tribal India.*
1991. A.B. Mukerji, "What ails Indian geography?"
1991. H.S. Sharma (ed.), *Indian Geomorphology.*
1991. J. Diddee (ed.), *Indian Geography.*
1991. K.R. Dikshit, *Environment, Forest Ecology and Man in the Western Ghats.*
1992. C.D. Deshpande, *India: A Regional Interpretation.*
1992. R.L. Singh and Rana P.B. Singh (ed.), *The Roots of Indian Geography.*
1992. Savindra Singh *et al.* "Morphogenesis of cavernous topography in sub-humid tropical environment—A case study of Guptadham Cave, India" (*Z. Geomorphology*, NF Suppl. Bd. 85).
1993. Aijazuddin Ahmad, *Muslims in India.*
1993: National Institute of Ocean Technology (NIOT) established.
1994. K. R. Dikshit, V. S. Kale, and M. N. Kaul (eds.), *India: Geomorphological Diversity.*
1994. Moonies Raza expired.
1995. R. D. Dikshit, *Geography of Elections: The Indian Context.*
1995. S.R. Jog (ed.), *Indian Geomorphology: Landforms and Processes*, 2 vols.
1996. S. Chaturvedi, *The Polar Regions: A Political Geography.*
1998. Prasad P. Gogate; B. Arunachalam, "Area maps in Maratha cartography: A study in native maps of Western India" (*Imago Mundi The International Journal for the History of Cartography*, 50).
1999. Aijazuddin Ahmad, *Social Geography.*
1999. Gurudeo Singh Gosal (ed.), *Fourth Survey of Research in Geography.*

1999. KR Dikshit, "Paradigm shifts in Geomorphology: trends and implications" (*Transactions Institute of Indian Geographers*, 21).
1999. S. Raju *et al.*, *Atlas of Women and Men in India.*
2000. Vishwas S. Kale, *et al.* "Sedimentary records and luminescence chronology of late Holocene palaeofloods in the Luni River, Thar Desert, northwest India" (*Catena*, 40).
2001. Death of R.L. Singh.
2001. M. Shafi awarded Padma Shri.
2001. Ramesh Dutt Dikshit, "Indian Geography: An encounter with reality" (*Transactions, Institute of Indian Geographers* 23).
2002. Anu Kapur (ed.), *Voice of Concern: Indian Geography.*
2002. Arunachalam, B. *Heritage of Indian Sea Navigation.*
2002. Robert Bradnock and Glyn Williams, *South Asia in a Globalising World.*
2002. Singh, J.P. and Mumtaz Khan *Mythical Space, Cosmology and Landscape: Towards Cultural Geography of India.*
2002. Vishwas S. Kale "Fluvial geomorphology of Indian rivers: An overview" (*Progress in Physical Geography*, 26).
2004. Savindra Singh *et al.* (ed.), *Geomorphology and Environment.*
2004. Anu Kapur, "Geography in India: A languishing social science" (*Economic and Political Weekly*, 39).
2004. Arunachalam, B. *Mumbai by the Sea.*
2004. Neelam Grover and Kashi Nath Singh (eds.), *Cultural Geography: Forms and Patterns.*
2004. Saraswati Raju, "Contextualizing critical geography in India: Emerging research and praxis" (*Geoforum*, 35).
2004. Singh, Rana P.B. *Cultural Landscape and the Lifeworld: Literary Images of Banaras.*
2004. Swapna Banerjee-Guha (ed.), *Space, Society and Geography.*
2004. Vishwas S. Kale and P. S. Hire, "Effectiveness of monsoon floods on the Tapi River, India: Role of channel geometry and hydrologic regime" (*Geomorphology*, 57).
2005. K.R. Dikshit, "The roots of Indian Geography" (*The Transactions, Institute of Indian Geographers* 27).
2006. 1st International Indian Geography Congress, Hyderabad.
2006. Aijazuddin Ahmad expired.
2006. S. Raju, M. Satish Kumar and S. Corbridge (eds.), *Colonial and Post-colonial Geographies of India*, Sage, Delhi.
2007. Death of M. Shafi
2007. Vishwas S. Kale, "Geomorphic effectiveness of extraordinary floods on three large rivers of the Indian Peninsula" (*Geomorphology*, 85).

2008. Arunachalam, B. "Technology of Indian Sea Navigation (c. 1200–c. 1800)" (*The Medieval History Journal,* 11).
2008. 2nd International Indian Geography Congress, Udaipur.
2008. Indian parliament passes the Central Universities Act to establish 12 central universities and upgrade 4 state universities into central universities.
2008. Ravi S. Singh, "Challenges to Indian Geography: A Contextual Analysis" (*Geographical Review of India* 70, Platinum Jubilee issue).
2009. Rana P.B. Singh, *Geographical Thoughts in India: Snapshots and Vision for the 21st Century.*
2009. Rana P.B. Singh, *Uprooting Geographic Thoughts in India: Toward Ecology and Culture in 21st Century.*
2009. Ravi S. Singh (ed.), *Indian Geography: Perspectives, Concerns and Issues.*

Note:

1. It is the preliminary outcome of an ongoing work; hence, need not be treated conclusive and final (© Ravi S. Singh).

(*Sources*: Compiled by from various print and web sources)

CONTRIBUTORS

Sarfaraz Alam, PhD, is Assistant Professor in the Department of Geography, Banaras Hindu University (Varanasi, UP). His research interests include political and social geography, geography education and geographical thought. He is also actively involved in the advocacy campaign for popularization and promotion of Geography in schools of India.
Email: sarfaraz05@yahoo.co.uk, sarfarazalam05@gmail.com

Taruna Bansal, Assistant Professor, Department of Geography, Jamia Millia Islamia, New Delhi. She has several authored/co-authored papers to her credit presented in academic seminars/conferences and published. Her basic areas of interest are regional development, urban and gender studies.
Email: btaruna@gmail.com

Aparajita Chattopadhyay, PhD, is Assistant Professor, Department of Development Studies, International Institute for Population Sciences, Mumbai. She has published widely in reputed international and national journals, and also contributed popular articles. Her areas of interest are women issues, urban issues, education, health and environment. She has visited United States, United Kingdom, Brazil and Pakistan for academic presentation.
Email: apachat@rediffmail.com

Bikramaditya K. Choudhary, PhD, is Assistant Professor in Geography, MMV, Banaras Hindu University, Varanasi, Uttar Pradesh. He specializes in areas of political ecology, disease and health and regional development and planning. He has publications on theme of waste-picking and disease ecology including a book entitled *Tuberculosis in India: A Political Ecology Approach* (2008).
Email: bikramadityac@gmail.com, bkc_bhu@rediffmail.com

Sudip Dey, PhD, is an Assistant Professor of Geography and Disaster Management, Tripura University. He won the prestigious Prof. N.P Aiyyer Young Geographer Award (2004) and has published widely in peer reviewed national and international journals and scientific articles in edited books/volumes. His fields of research interest are coastal

morphodynamics, spectral expressions of surface geophysical signatures, tertiary-quaternary microstructure and landform evolution.
E-mail: sudip_geo@rediffmail.com

Sohel Firdos, PhD, trained as a Geographer, is Fellow in Development Studies, Centre for Studies in Social Sciences, Kolkata, West Bengal. He was a visiting faculty to Massey University, New Zealand in 2005 and 2007. His initial research focused on the relationship between environmental resources and tribal economy. Currently, he works on issues like urban governance and socio-spatial exclusion in cities.
E-mail: sohel@cssscal.org, sfirdos@hotmail.com

N. C. Jana, PhD, is Reader in the Department of Geography, University of Burdwan, Burdwan, West Bengal. He has published in various reputed journals and edited volumes; and, authored/edited the books *The Land: Multifaceted Appraisal and Management*, *Transformation of Land: Physical Properties and Development Initiatives*, *Tsunami in India: Impact Assessment and Mitigation Strategies*, and *Disaster Management and Sustainable Development: Emerging Issues and Concerns.*
Email: jana.narayan@gmail.com

Vijaya P. Khairkar, PhD, Reader, Department of Geography, University of Pune, Pune, Maharashtra. She has published widely in her interest areas and handled many research projects. Her areas of interest are population problems, urban planning and rural development, use of GIS/RS in social issues.
Email: vpkhairkar@unipune.ernet.in

Ramkrishna Maiti, PhD, Senior Lecturer, Department of Geography & Environment Management, Vidyasagar University, Paschim Medinipur, West Bengal. He has completed two research projects and working one major project. His contributions have appeared in many journals and books. He has also published five books including school textbooks. His main area of interest lies in Advanced Geomorphology.
Email: ramkrishnamaiti@yahoo.co.in

A. P. Mishra, PhD, is Associate Professor at the Department of Geography, Banaras Hindu University (Varanasi, UP). He has published several research papers in journals of repute, three books, and a novel in Hindi. He has also completed three projects, research monographs and

reports which have attracted planners and policy makers. His major fields of interest are Marxist Geography, rural poverty and regional development.
Email: adeepayan@yahoo.com, adeepayan@gmail.com

Archana K. Roy, PhD, is working as Assistant Professor in the Department of Geography at Banaras Hindu University (Varanasi, UP). She has contributed to NCERT textbook writing and has research publications on health issues of left behind wives of migrants in books and journals of national and international repute. She has also visited France, Thailand, and Singapore for academic presentation.
Email: royarchana@yahoo.com

P. K. Ryngnga, PhD, is Assistant Professor, Department of Geography, North-Eastern Hill University, Shillong (India). She has several published papers to her credit and a book *Fundamentals of Astronomy for School Children (Ka Kot Jingtip Shaphang ka Astronomi: Na ka bynta ki khynnah skul*, 2009, in Khasi). Her areas of interest are regional development, urban planning, remote sensing and GIS.
Email: kryngnga@rediffmail.com

L.N. Satpati, PhD, is Lecturer, Department of Geography, University of Calcutta, Kolkata. With an over a decade long a teaching experience he has a number of research papers to his credit He is also associated with a number of professional organizations of geographers in different capacities. At present he is the Honorary Assistant Secretary (Publication) of the Geographical Society of India, Kolkata.
E-mail: lsatpati@yahoo.co.in, satpati.ln@hotmail.com

Arun K. Singh, PhD, is Reader in Geography, MMV, Banaras Hindu University (Varanasi, UP). He has published several papers and completed projects sponsored by government departments and prestigious agencies like UNDP and HUDCO. He had been editor of *Governance*—a bi-annual journal. Areas of his specialisation include urban geography, population geography, area planning strategies and participatory rural appraisal.
Email: skarun66@yahoo.co.in, aksmmvbhu@gmail.com

Ravi S. Singh, PhD, is Associate Professor, Department of Geography, Banaras Hindu University (Varanasi, UP). He has several national and international publications to his credit including *Paths of Development in Arunachal Pradesh* (NBC, New Delhi; 2005); *Indian Geography: Perspectives, Concerns and Issues* (ed., Rawat Publications, Jaipur/New

Delhi; 2009), *Indian Geography in the 21st Century: The Young Geographers Agenda* (ed., CSP, UK, 2009), and *Environment-Development Interface* (ed., APH Publishing Corporation, New Delhi, *forthcoming*).
Email: ravisingh.geog@gmail.com, ravi_geog_bhu@yahoo.com

Yemuna Sunny, MPhil, is a Researcher engaged with issues of geography education since last several years. This involves research and material development in Eklavya (Madhya Pradesh) and teaching of social science at post graduation in MAEE, TISS (Mumbai). Her recent publications include a co-edited book *Proselytisation in India*: *The Process of Hinduisation in Tribal Societies* (2009, Aakar Books, Delhi).
Email: yemunas@gmail.com, yemunas@yahoo.com

INDEX

C

D

E

H

I

J

K

L

O

P

S

V

W

About the Editor

DR. RAVI S. SINGH (b. 1971) is an Associate Professor in the Department of Geography, Faculty of Science at Banaras Hindu University, BHU (Varanasi, India). Trained in geography mainly at BHU and Jawaharlal Nehru University, JNU (New Delhi), he is author/editor of four books *Paths of Development in Arunachal Pradesh* (2005, NBC: New Delhi) and *Indian Geography: Perspectives, Concerns and Issues* (ed. 2009, Rawat Publications, Jaipur/New Delhi), and *Indian Geography in the 21st Century: The Young Geographers Agenda* (ed. 2009, Cambridge Scholars Publishing, UK), and a forthcoming volume entitled *The Environment-Development Interface* (ed. 2010, APH Publishing Corporation, New Delhi). Over fifty academic works of Dr. Singh, many of them cross-disciplinary, are published internationally as well as nationally in reputed journals and research series like *South Asian Affairs* (CSAS, Gifu Women's University, Japan), *The Himalayan Review* (Kathmandu), also in Göteborg University Development Studies Series (Sweden), Heidelberg's Ethno-Indology Series (Germany), *Geographical Review of India*, *Annals of NAGI*, *National Geographical Journal of India*, *North-Eastern Geographer*, *Journal of Geography*, *Gandhian Perspectives*, *Oriental Anthropologist*, etc. His latest publications have appeared in *Journal of Geography in Higher Education* (Routledge, Taylor & Francis Ltd.) and a special volume of the reputed *Man in India* (2009).

He is Life Member of several prestigious national academic bodies, a Full Member of IRF-SAARC (Tokyo), and a Senior Member of the IACSIT (Singapore). Dr. Singh has been associated with international academic programmes like New York State Independent College Consortium for Study in India, NYSICCSI (USA) and the consultant to two major projects of the Department of Tourism & Culture (Government of India) on the history, culture and oral tradition tribes of Arunachal Pradesh. He has participated in several international and national seminars on cross-disciplinary themes in various capacities, given invited lectures; and, presented over three dozen conference and seminar papers.

The main areas of his academic interest are geographical thought, social and cultural geography, and development geography. He has taught/teaching geographical thought, population geography, settlement geography, regional and rural planning & development, economic geography, and an interdisciplinary course on population & development.

Email: ravisingh.geog@gmail.com